Atlas
of Plants and Animals
in
Baltic Amber

Die Deutsche Bibliothek - CIP-Einheitsaufnahme

Atlas of plants and animals in Baltic amber /
Wolfgang Weitschat ; Wilfried Wichard.
[Transl. from the German By Maryann Onofrietto]. -
München : Pfeil, 2002
Dt. Ausg. u.d.T.: Atlas der Pflanzen und Tiere im baltischen Bernstein
ISBN 3-931516-94-6

Flyleaves:

Sections of "Der Bernsteinwald" (The Amber Forest) – oil painting by Otto Frello, 1998
The original painting is housed in the "Ravmuseet" (The Amber Museum), Vestergade 25, DK-6840 Oksbøl.
Administration: Museet for Varde By og Omegn, Lundvej 4, DK-6800 Varde, Danmark.

We are grateful to Mr. Otto Frello and the museum for the permission
to reproduce part of the painting in this book.

Translated from the German by Ms. Maryann Onofrietto (Basking Ridge, USA)

Dr. Friedrich Pfeil, Wolfratshauser Straße 27, D-81379 München

Druckvorstufe: Verlag Dr. Friedrich Pfeil, München
CTP-Druck: grafik + druck GmbH Peter Pöllinger, München
Buchbinder: Thomas, Augsburg

Printed in Germany

– printed on permanent and durable acid-free paper –

ISBN 3-931516-94-6

Atlas of Plants and Animals in Baltic Amber

Wolfgang Weitschat
Wilfried Wichard

Verlag Dr. Friedrich Pfeil · München

Preface

The authors and the publisher have responded to countless requests from readers interested in amber by publishing this second edition of the "Atlas der Pflanzen und Tiere im Baltischen Bernstein" in the English language. We hope to appeal to international readers who, after discovering the beauty of amber or studying its bizarre inclusions, have experienced the subtle fascination it inspires. We would like to provide readers with a well-researched standard reference that gives an informative and interesting overview of the plant and animal groups in Baltic amber on the basis of illustrations, photographs and accompanying texts. English-language books on amber have been published by Grimaldi 1996, Poinar 1992, 1999 and Ross 1998, who review amber from all over the world, particularly Dominican amber and its inclusions. We take up our place in this line of authors, exclusively presenting Baltic amber in this English edition.

Georg Carl Berendt published "Die im Bernstein befindlichen organischen Reste der Vorwelt" (The organic remains of the prehistoric world in amber) in two volumes between 1845 and 1856. With its illustrations and descriptions of the plants and animals in Baltic amber, this pioneering work served as the authoritative reference for quite some time. As many as one hundred years passed before a comparatively brief and readily comprehensible monograph was published by Bachofen-Echt (1949). Just thirty years later, the first comprehensive review of Baltic amber to appear in English was presented by Larsson (1978), who had compiled a work of great scientific precision and detail. Since then, the periods between publications have become increasingly short. Our Atlas of 1998 and 2001 makes use of modern techniques of color and close-up photography, takes decades of progress in taxonomy into account and draws conclusions about the lifestyles of the fossil species enclosed in Baltic amber based on the biology of living plants and animals. This Atlas will also be replaced one day, because our knowledge of Baltic amber and its inclusions will advance, and because new, exciting discoveries will be made concerning the taphocoenoses in amber and the biodiversity, biogeography and paleoclimate of the Eocene Epoch.

This Atlas is thus an interim report, which depended on the contributions of all those to whom we expressed our gratitude in the first edition. We would vicariously like to thank Mr. Bruno Endrusseit (Stralsund) for his constructive suggestions for the second, English edition. We are particularly indebted to Ms. Maryann F. Onofrietto (Basking Ridge, USA, currently residing in Bonn) for the competent translation of the revised edition, to Dr. Michael S. Engel (Lawrence, USA) for being kind enough to edit and proofread the text, and to Ms. Agnes Gras (Cologne) for revising the literature references and the index of scientific names.

January 2002

Wolfgang Weitschat, Wilfried Wichard

Contents

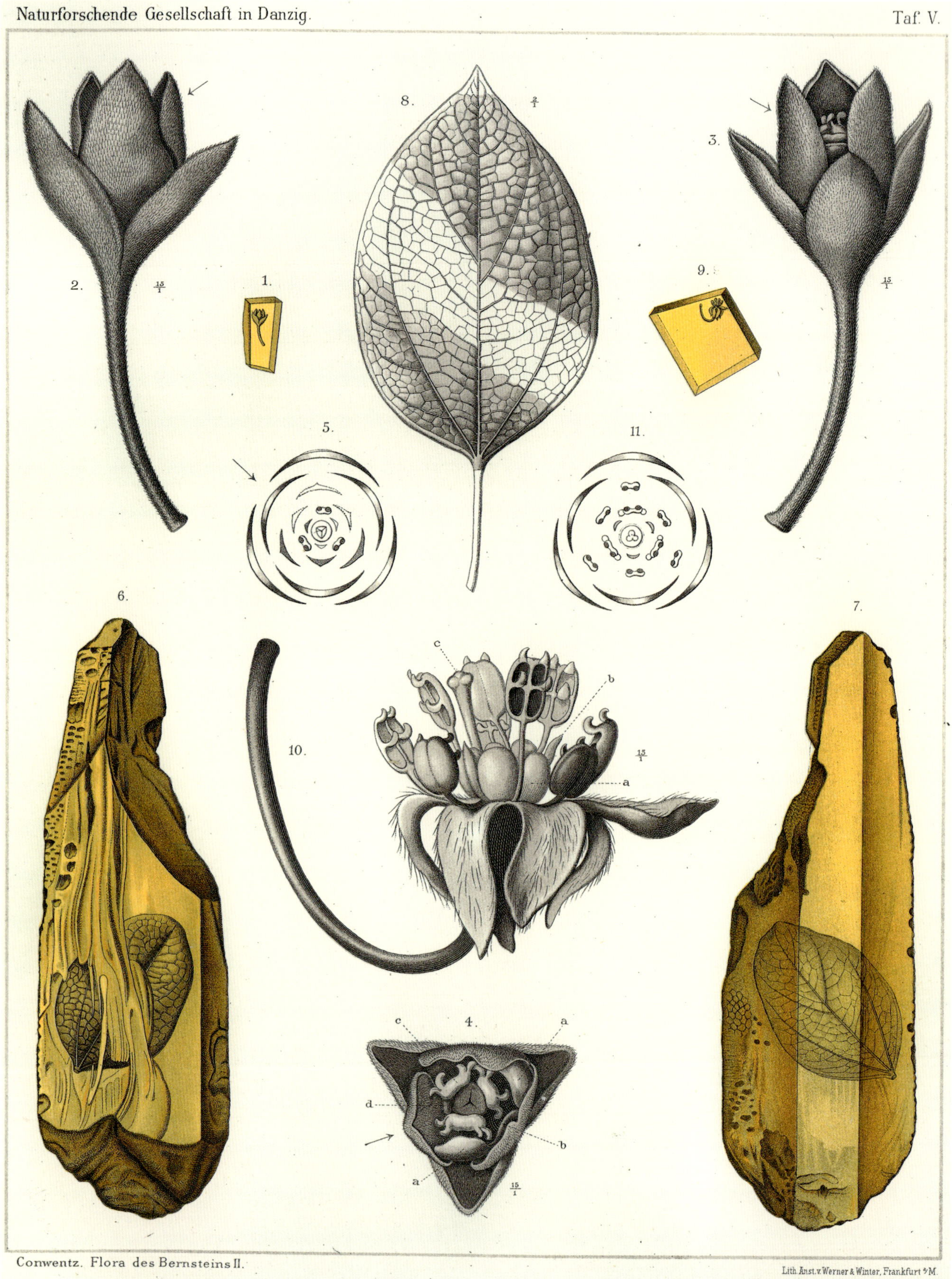

Plate V from CONWENTZ, H.: Die Flora des Bernsteins. – 2. Die Angiospermen des Bernsteins. – Danzig 1886.

1 Introduction

1.1 Basic information on amber

1.1.1 Resin – Copal – Amber

The term "amber" is a general designation for fossil resins or "petrified" plant sap. The time, conditions and sequences of chemical reactions required for the process of turning resin into amber have yet to be explained conclusively. Only the initial phases of amber formation can be understood on the basis of the resin secretions of contemporary plants. Many conifers and a number of deciduous trees secrete resins spontaneously or as a result of injury. In chemical terms, resins are mixtures of various organic compounds belonging to the family of terpene derivatives. They are insoluble in water, have a more or less aromatic odor and are initially adhesive. When exposed to air, the viscous resin mass can harden relatively quickly due to the loss of volatile components. Polymerization then follows with the formation of carbon chains. The resin continues to transform and harden for millions of years, during which time the filamentary molecules of the resin mass become progressively "knotted" and the solvents and volatile constituents gradually disappear. The formation of amber could be referred to as a process of maturation, comparable to the carbonization of plants into brown coal and then hard coal.

The process also includes intermediate forms of different age between resin and amber, which are generally referred to as copals. Copals are often transparent, hard and usually odorless, but "smear" when ground. Typical copals from tropical deciduous or coniferous trees are known from East Africa, Madagascar, New Zealand, Colombia and the Dominican Republic. Africa and New Zealand in particular used to export copal by the ton for paint production. Studies based on carbon-14 dating have shown that the majority of commercial copals have their origins during the historical age.

1.1.2 Amber – The "burning stone"

The German word for amber, "Bernstein" or "burning stone", is derived from its generally known property of flammability. "Börnen" is the Low German word for "burning". Based on its lithic character, amber is classified as a caustobiolite, a group that also includes coal and oil shale. This group encompasses rocks that possess the property of flammability (from the Greek "kaio" = to burn) and are of biogenetic origin (from the Greek "bios" = to live).

Even today, a certain degree of confusion is caused by the fact that attempts were made in the past to restrict the term "amber" to Baltic amber. The reasoning was based on the content of succinic acid, which distinguished Baltic amber (succinite) from all other fossilized resins (retinites). This debate later led to designations, such as "true amber" or "amber in a broad or narrow sense". These ambiguous definitions stemmed from the fact that many of the worldwide deposits of fossilized resins had not yet been discovered. However, as these deposits also yield resins containing succinic acid, this characteristic can no longer be reserved for Baltic amber alone (Helm 1885, Beck 1996). In addition, more recent results from the field of resin chemistry have shown that knowledge of the geological history of every fossilized resin can be significant for its botanical classification, the reason being that resin secretions from identical "mother plants", which are subject to different geological conditions, can result in ambers with a distinctly different chemical structure.

We agree with the view of Schlee & Glöckner (1978) and Schlee (1980, 1984, 1990), who used the term amber for all natural fossil resins that are "several million years old". The location name should be used for more precise reference, e.g. Siberian amber, Borneo amber, New Jersey amber, Dominican amber and Baltic amber.

1.1.3 Amber deposits

At present, there are nearly 100 known amber deposits throughout the world, and new ones are discovered every year. Today, amber can be found on all continents except Antarctica. However, this broad distribution is no indication of the quality and quantity of the individual deposits. Only a few of the deposits discovered recently have achieved economic importance (e.g. Dominican amber, Bitterfeld amber and Ukrainian amber). The number of known, fossiliferous amber deposits has risen dramatically in the last several decades.

The geological age of the amber deposits varies greatly, as does the nature of the resin-producing plants. It was assumed for quite some time that the plants of the Paleozoic Era did not produce resin at all (SCHLEE & GLÖCKNER 1978). This assumption has since been refuted by discoveries of resin in hard coal of the Upper Carboniferous (320 to 285 million years ago). These finds concerned very minute particles that probably formed inside the plants and thus could not be considered as resin traps for insects. The same applies to several amber deposits from the Early Mesozoic Era (Triassic/Jurassic).

The oldest known fossiliferous amber comes from Lower Cretaceous deposits in the Lebanon Mountains and is estimated to have an absolute age of 120 to 130 million years. Although these deposits are not of economic significance, the fossils they contain make them very valuable to science, because the origin of these resins falls in an extraordinarily interesting epoch from a paleontological standpoint. While dinosaurs, ammonites and belemnites still existed, mammals, birds and flowering plants had also already appeared. Scientific analysis of the rich fauna of Lebanese amber has led to some surprising results concerning the evolution of various insect groups (SCHLEE 1970).

A number of additional amber deposits from the Late Mesozoic Era have just been discovered in the last few years. Those containing fossiliferous amber include Jordanian amber (approx. 120 million years old), Taimyr amber (approx. 80 million years old), New Jersey amber (approx. 80 million years old), Cedar Lake amber (approx. 75 to 80 million years old), French amber (approx. 70 million years old) and the amber recently discovered in the Pyrenees (approx. 100 million years old).

The real "amber age", however, is the Cenozoic Era, from which the large majority of fossil resin deposits originated. These deposits are linked to the development of the modern plant world in the Late Mesozoic and Early Cenozoic, when the angiosperms and gymnosperms (specifically the conifers) diverged.

1.1.4 Geographical origins of amber

Many questions that remain largely unanswered relate to the precise age and geographical classification of the various types of amber. This is due to the fact that nearly all amber deposits known today are not located at the site where the resins were originally produced and deposited in the forest floor. Contrary to what one might assume, an amber deposit generally does not represent a fossil forest floor. In most instances, sediments containing large amounts of amber usually indicate a more marine environment, i.e. an ocean deposit.

Amber has very low resistance to the weathering effects of oxygen, meaning that this land-to-sea journey protected and preserved amber from destruction. If the resin had been deposited in the forest floor, it would have had only a very limited life expectancy and been destroyed by oxidation and drying processes after just a few thousand years. According to a model developed by SAVKEVICH (1969), one essential prerequisite for the formation of fossil resins and their transformation into amber is, therefore, rapid embedding without exposure to air. On the basis of this theory, amber deposits can only form, if the resins are flushed out of the forest floor by rivers during or shortly after their formation and transported into oceans or large lakes, where they are protected from weathering processes after deposition.

If the preservation of amber depends on its being transported by running waters, the location of present-day sites can hardly provide precise information on the areas of origin. Since amber easily floats in moving water because of its low specific weight, it is protected during transport against comminution and wear and usually retains its original shape. Of course, this also makes it impossible to draw any conclusions about the possible transport distances, which probably varied greatly.

Once deposited in this so-called secondary deposit, amber can be transported yet again by geological processes and redeposited elsewhere. As a result, resins from a single area of origin can be found in very different amber deposits that are also of different ages. Thus, the location of an amber deposit usually does not correspond to the habitat of the resin-producing trees.

1.1.5 The age of amber

What are the methods for determining the age of amber? Many existing problems could be resolved, if it were possible to date individual amber specimens directly. At present, this cannot be achieved using either physical or chemical methods. Using carbon-14 dating, the only radioactive element in amber, is only adequate for up to 60,000 years.

In addition to the absolute measurement of time, the science of geology also has another kind of "calendar" based on the developmental history of the organisms embedded as fossils in the rock sequences. With the help of so-called index fossils, which are characteristic of a preferably short period of geological history, generations of geologists throughout the world have developed a universal stratigraphic scale, or time scale of sorts, that continues to be expanded and refined. Since only relative ages can be determined with this method, it is also referred to as a relative time measurement. The floras and terrestrial faunas encased in the various types of amber cannot be used as index fossils, thus eliminating another possibility for directly determining the age of a piece of amber containing inclusions.

The only remaining possibility is the geological dating of the amber-bearing stratum. The relative age can only be defined if this stratum itself, or the underlying or overlying ones, bears index fossils. As mentioned, the geological age of a respective amber deposit determined in this manner does not correspond to the age of the location of origin of the resin or that of the habitat of the resin-producing trees. Most current data on the age of amber deposits refer only to the approximate time and generally only indicate the minimum age of their generation.

1.1.6 The source of amber

No less problematic than determining the geological age of amber is linking the various fossil resins to the original plants that produced them. Initial information is provided by the range of plant inclusions in the respective amber deposits. The resin-producing plant should be among them. However, direct observations that provide further detail are only possible on rare occasions. Such observations require amber preserved "in situ" with wood fragments that are suitable for cell and tissue structure analysis and exhibit highly species-specific characteristics (SCHUBERT 1939, 1953). Specimens of this kind are extremely rare in the secondary amber deposits we know of today. The wood was probably already separated from the amber during the process of redeposition.

Various methods are used to obtain information on the original plants that produced the amber. Infrared spectroscopy has proven to be particularly effective right up to the present time (BECK 1996, KOLLER et al. 1997). In this method, the origin of the resin is determined on the basis of its physicochemical properties. The method is primarily derived from studies by HUMMEL (1958), MOENKE (1962), BECK et al. (1964, 1965) and LANGENHEIM & BECK (1965). The infrared spectroscopy of a few milligrams of amber permits reliable conclusions to be drawn about the identity of amber and present-day resins, when their spectra, i.e. diagrams of their infrared spectra, are compared. Similarities between the curves of recent resins of known botanical origin and the curve of an amber specimen provide information on the resin-producer. For example, the infrared spectroscopic diagram of the present-day resin from a tropical deciduous tree, the leguminous *Hymenaea courbaril*, matches that of both Mexican and Dominican amber almost to the last detail. Although this method can be applied successfully, its limitations also become apparent. The main source of difficulty lies in the fact that the chemical structure of a fossil resin can change as a result of its geological history. SAVKEVICH (1969, 1996) pointed out that the chemical composition, and thus also the characteristics of resins from the same plant species, change under the influence of pressure and temperature in particular. SAVKEVICH found intermediate forms of amber and experimentally demonstrated the transformation of *succinite* into *rumanite*. On the other hand, resins of different botanical origin that have been subject to similar pressure and temperature conditions can take on the same chemical composition, which is reflected by both a number of properties and the infrared spectra. Based on these observations, various ambers usually reflect their specific geological histories and not always their botanical origin.

1.2 Genesis of Baltic amber

1.2.1 Overview

Baltic amber has been the subject of intense research in a wide variety of scientific disciplines for nearly 150 years. One might assume that only a few general questions would be left to clarify after all that time. In fact, the great number of articles published in recent years, in which we have experienced a kind of "amber Renaissance" that persists to the present, indicates just the opposite. Fundamental issues, such as the question of the resin-producer, the exact age of Baltic amber and the associated location and life span of the legendary amber forest, continue to be the subject of controversial discussion. Indeed, it almost appears as if the intense research conducted over the last decade has led to an increase in the number of unresolved problems!

Although this atlas focuses on the fossil floras and faunas in Baltic amber, the following sections cover several of the unsolved mysteries relating to the deposits, genesis and age of Baltic amber, particularly in cases where the composition of the fossil floras and faunas can contribute to resolving these problems. Of the numerous subjects relating to Baltic amber, only those aspects directly or indirectly related to its inclusions are presented and discussed here.

1.2.2 Natural shapes of amber

If one more piece of evidence of the resinous nature of amber were required, it could be provided by the natural shapes of individual amber pieces as they occur in the deposits. The most remarkable of these are the natural resin flows or drops, such as amber "stalactites" which we can still observe today on many resin-producing forest trees (Figs. 2, 3). In principle, however, every natural piece of amber has a specific, non-random shape reflecting its formation on or in the tree. It is even possible in many cases to elaborate precise details on the history of its development based on the shape, morphology and surface characteristics (Fig. 1). This information can be used to determine natural shapes of fossil resins (KATINAS 1971). In this context, a rough distinction is made between "external" resin accumulations on the outside of the trees and "internal" accumulations that formed inside the trunks between the heartwood and the bark. Internal shapes, which make up by far the greater percentage, are usually opaque and do not contain any inclusions. The larger pieces of amber are also found among this group. The largest known specimen to be found to date weighs 9,750 grams (BARFOD et al. 1989).

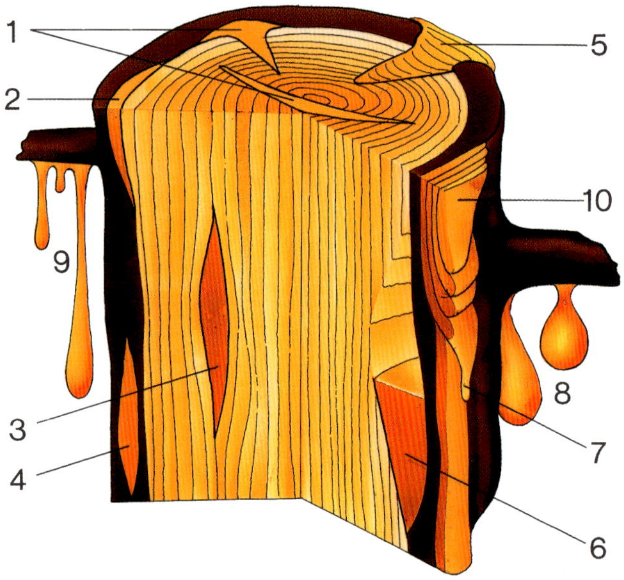

Fig. 1: Natural shapes of amber (after KATINAS 1971 and GANZELEWSKI 1996): 1. Resin-filled fissures, 2. Shapes under the outer bark, 3. Resin pocket, 4. Shapes in the outer bark, 5. Filled-in wounds, 6. Filled-in hollow space with flat top (resin level), 7. Drops and stalactite-like shapes, 8. Drops, 9. Stalactites, 10. *Schlauben*.

Fig. 2: Pear-shaped "amber drop" with shrinkage cracks and protrusions on the neck. Drops of this kind are indicative of a highly viscous resin.

Naturally, it is the external types that must be expected to supply fossils in Baltic amber. There is one specific type of specimen that almost exclusively contains the inclusions of animals and plants that are so valuable to science. It is referred to as a "*Schlaube*" in German, or also as fossil-bearing amber due to its numerous inclusions. A *Schlaube* is a typical, layered amber consisting of successive resin flows. These specimens are usually not very large, only rarely reaching the size of a fist, and their structure is one of transparent laminations of amber (Fig. 4 a, b). ANDRÉE (1937) suggested that the key characteristic of transparency was attributable to intense exposure to the sun, which warmed the resin and ultimately caused it to clarify.

In contrast to the large, oval amber drops (Fig. 2), which are indicative of a more viscous consistency of the secreted resin, everything about the *Schlauben* indicates the material was of an extremely low viscosity. This is particularly evident in the long, slender, so-called amber stalactites, with drops hanging from thin threads of resin (Fig. 3).

Schlauben develop when batches of resin from a single source flow down the tree bark at intervals. The thin layers of resin do not mix in the process, but rather form consecutive layers. Each resin flow follows the previous one in such a way that the contact surface is concave, but the outer surface convex. We can only speculate about the rhythm of the individual resin flows. The interval between flows certainly often amounted to only a few hours. However, some examples of mold formation in enclosed insects indicate that individual resin flows must have sometimes occurred at fairly long intervals, perhaps several days. The thickness of the layers generally measures no more than several millimeters, and their distinct boundaries appear as more strongly refractive, reddish or yellow lines (Fig. 4 b). If the resin flows occurred at relatively long intervals, the surface of the older layer may already have hardened slightly. An intermediate space filled with air is often found between *Schlauben* elements formed in this manner. A *Schlaube* easily splits apart in the direction of flow at this point when subjected to mechanical stress.

These sticky resin surfaces on the tree bark must have literally acted as flycatchers. The majority of inclusions are, of course, concentrated at the individual boundary layers that were formerly resin surfaces. However, a considerable number are also located inside the individual layers, thus giving the impression that the animals that were captured in the resin had sunk quickly and "drowned". This can be considered yet another indication of the extremely low viscosity of the secreted resin.

Stalactite-like amber shapes (Fig. 3), whose genesis does not essentially differ from that of *Schlauben*, can also contain inclusions. In rare instances, inclusions also occur in non-laminated, extremely clear, massive pieces that must have been the result of a single, uniform, rapid flow of resin.

Fig. 3: "Stalactite" with thickened end; formed by several successive resin flows. Indication of the extremely low viscosity of the resin material.

1.2.3 Color and transparency of amber

Baltic amber is primarily yellow, ranging in shade from very light yellow, to orange and dark yellow, all the way to brown. However, green, blue, gray, black, white and red ambers also exist, where even a single specimen can often display a variety of color variations (Fig. 5 b). Transparent amber shows that the basic substance itself is yellowish. Opaque amber has a "foamy" internal structure caused by microscopic air bubbles. In this case, the various color tones are produced by optical effects, such as interference, light scattering, reflection and absorption, and thus depend largely on the number, size and arrangement of these bubbles. Minerals, such as pyrite, enclosed in the amber are another, albeit secondary, factor that can affect the color. Much greater is the influence of secondary weathering processes, which turn yellow amber to an intense red on the outer crust and along holes and hollow spaces.

1.2.4 The resin-producer

For a long time, evidence that the mother plant of Baltic amber could only be a single type of pine tree was considered to be conclusive. As early as the middle of the nineteenth century, the botanist H. R. GOEPPERT (1836) believed on the basis of microscopic studies of wood fragments enclosed in amber that a fossil pine tree was probably the resin-producer of Baltic amber. However, as he recognized certain differences compared to the wood of recent *Pinus* species, he established a new genus, *Pinites succinifera*, for the fossil form. CONWENTZ (1890), a student of GOEPPERT, described the resin-producer as *Pinus succinifera* (GOEPPERT & BEHRENDT 1845) in his revision of the wood contained in amber, even though he noted "that the

Fig. 4a: Amber *Schlaube* with visible boundary layers of the individual resin flows. *Schlauben* are also indicative of a low-viscosity resin.

Fig. 4b: Cross-section of an amber *Schlaube*. The red lines indicate the boundaries of the individual resin flows.

structure of the amber wood displays characteristics of both pines and spruces, although primarily tending towards the former". More recent paleohistological studies confirm the hypothesis that the resin-producer of Baltic amber was apparently a pine tree (Schubert 1961, Turkin 1997, Pielinska 1997) (cf. Chap. 2.04).

Initial doubts about the nature of the resin-producer originated in a completely different scientific discipline, namely archaeometry. In the field of archaeology, the presence of succinic acid was long considered to be a reliable property for determining the origin of Baltic amber (Helm 1877). Based on this characteristic component, Baltic amber was classified as *succinite*, distinguishable from all other amber-like fossil resins *(retinites)*. However, once succinic acid was also discovered in Romanian amber (Helm 1885) and later in several other fossil resins, Helm's method for determining the origin of Baltic amber lost its value in archaeology.

This led to a search for another method, infrared spectroscopy (Hummel 1958, Beck et al.1964, 1965). According to studies by Beck (1986) involving 2,500 infrared spectra of the known amber deposits in Europe, all spectra of Baltic amber (succinite) display the same characteristic curve. Succinite is distinguished by a single absorption band between 1,250 and 1,175 cm^{-1}, which is preceded by a broad absorption shoulder, the so-called "Baltic Shoulder" (Fig. 6). According to Beck (1996), this shoulder is nearly horizontal in well-preserved amber and increasingly slopes downward with increasing weathering. No other European amber deposits display similar spectra. Beck emphasized, however, that the infrared spectra should only be viewed as "fingerprints" for determining the origin of individual amber deposits and generally do not support any conclusions about the botanical classification of the resin-producer.

A comparison between the infrared spectra of Baltic amber and resins of recent conifers surprisingly showed that they display only minimal similarity to pine resin and are much more similar to the resins of araucarian

Fig. 5a: Pieces of raw amber from the "blue earth", Primorskoje strip mine, Yantarnyi, (Kaliningrad district).

Fig. 5b: Unusual color variations of Baltic amber.

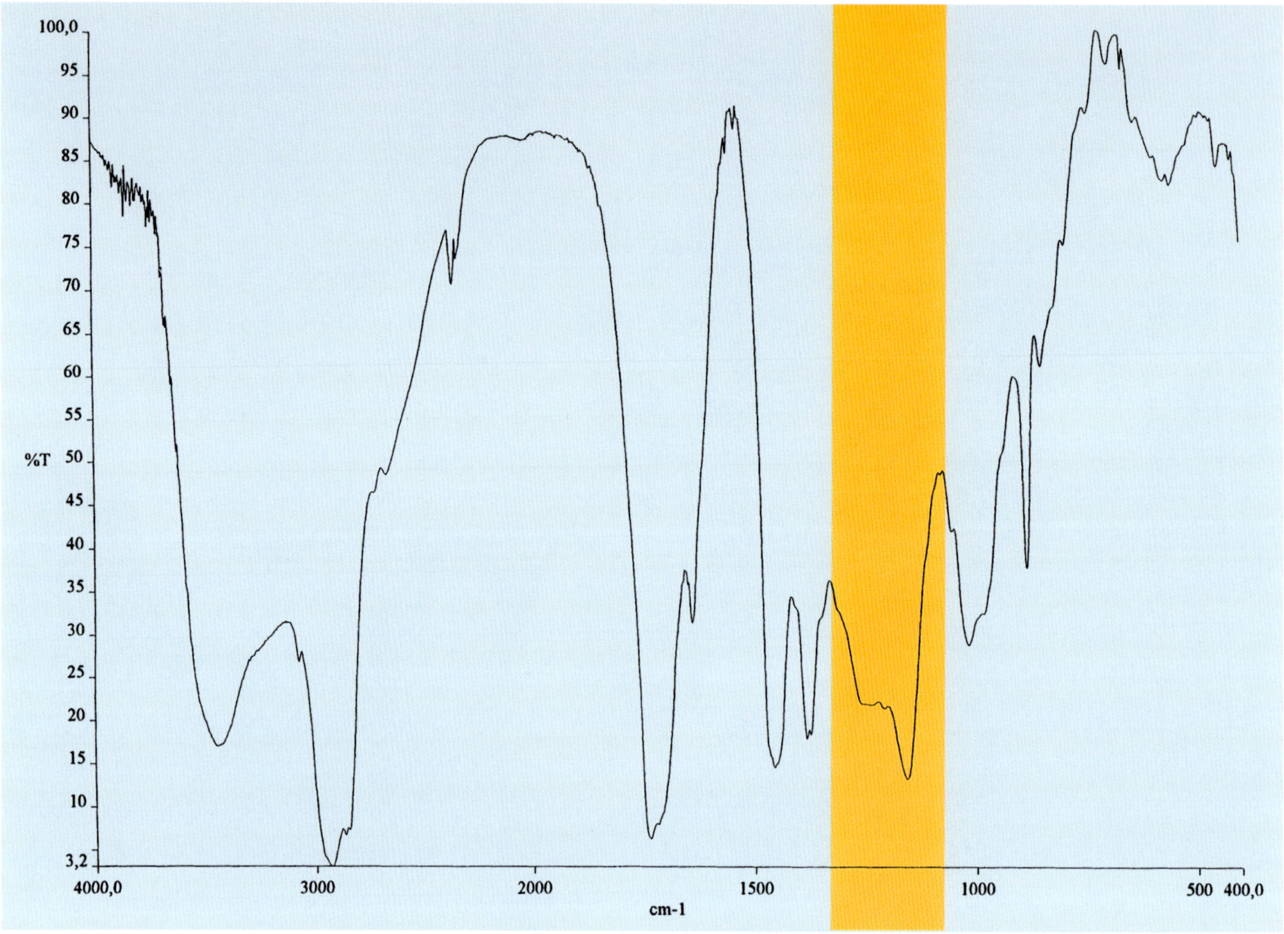

Fig. 6: Infrared spectrum of succinite, with the typical "Baltic Shoulder" (yellow region) (after KOLLER et al. 1997).

trees *(Agathis australis)* growing today in New Zealand (LANGENHEIM 1969). THOMAS (1969) also pointed out the great similarity between the infrared spectra of Baltic amber and copal from *Agathis australis*. In addition, New Zealand araucarian trees produce large quantities of resin that hardens rapidly on the ground. The spectroscopic similarity and the high resin production have given several authors (KUCHARSKA & KWIATKOWSKI 1978, POINAR 1992) reason to revive speculation about the resin-producer of Baltic amber and to suggest the probability of an araucaria being the resin source. The resins of a recent cedar tree *(Cedrus atlantica)* from the Atlas Mountains of North Africa also display remarkable similarity to succinite, so that a cedar has also recently been discussed as the potential resin supplier (KATINAS 1988).

A controversial situation has thus arisen, where all previous morphological and paleohistolgical studies clearly point to a pine tree as the plant of origin, while analytical methods indicate araucarians and cedars as potential resin suppliers. However, the latter assumption is strongly contradicted by the present range of species in the macroflora and microflora known from Baltic amber inclusions today. While needles and male cones from pine trees have been known to exist for quite some time, there are no records of araucarians or cedars. The microflora presents a similar picture. Studies of pollen in Baltic amber (ARNOLD 1998) show a distinct dominance of vesiculate pine pollen. The pollen studies, which have often been neglected in the past, provide remarkable evidence that the potential resin supplier is a pine relative.

1.3 Distribution of Baltic amber

As is the case with the majority of all fossil resin deposits, Baltic amber is today no longer found at the site where the resins were originally produced and deposited in the forest floor. Its relocation already began during or shortly after its formation in the amber forest. In this first, most vital step for the survival of the resins, large, possibly seasonal rivers flushed the resin masses out of the forest floor and transported them to a depositional environment, where they were protected against the effects of weathering. The process probably only affected a very small portion of the total amount of resin, produced in the entire amber forest. The much larger amounts that remained in the forest floor were probably destroyed after a relatively short time by processes of oxidation and drying.

Although the resins preserved by transport and redeposition are already in their second (secondary) deposit in the new environment, only these beds permit statements and conclusions to be made about the minimum age and place of origin of Baltic amber.

The oldest deposits known today in eastern Scania and in the "wild earth" of the Samland Peninsula indicate that the amber forest must have already existed in the Lower Eocene (about 50 million years ago). Paleogeographical maps prepared by geologists give us greater insight into the geographical conditions that prevailed in Northern and Central Europe at that time. Our concept of the position and changes of the individual continents throughout the history of the earth was radically changed by modern theory of plate tectonics (i.e. continental drift).

Baltic amber could have originated anywhere in an enormous region (Fig. 8 a). Its western border could be drawn roughly along the north-south axis South Scania – Rügen – Berlin (eastern edge of the Lower Eocene ocean). To the east, it may have extended as far as the Ural Mountains. The northern border was certainly determined by climate, while the southern border was probably formed by the central German brown coal forests.

This European continent was undoubtedly covered with forests wherever climatic conditions permitted, and there is no evidence to contradict the assumption that resin-producing trees were very widely distributed. Consequently, amber deposits could theoretically be expected to exist all over this vast region. However, this was prevented by numerous geological processes and certainly also by the low preservation potential of fossil resins discussed above.

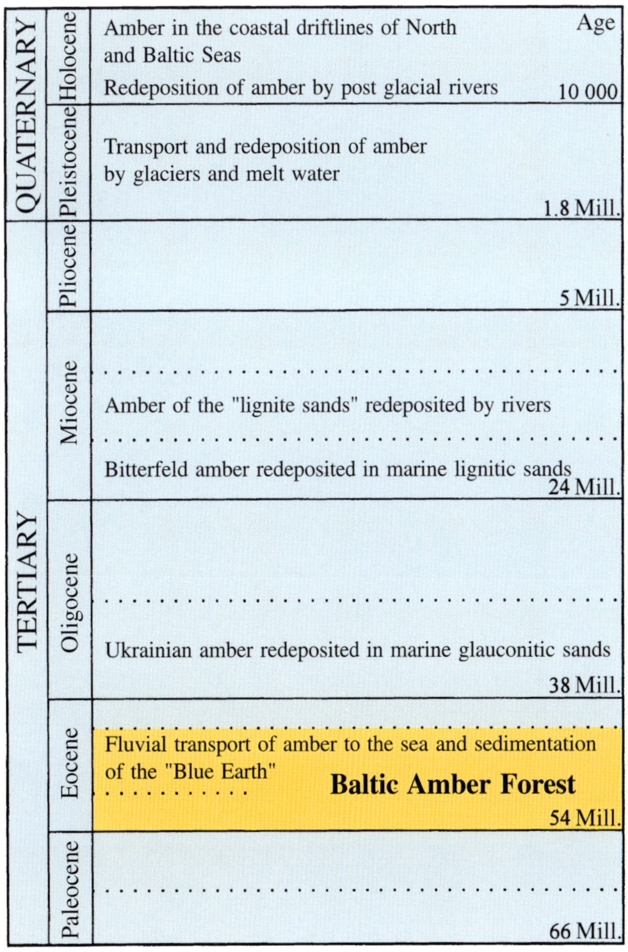

Fig. 7: Geological division of the Cenozoic and the history of Baltic amber.

The ocean already began to advance eastward during the Eocene, thereby flooding and destroying substantial areas of the amber forest (Fig. 8 b). On the other hand, large rivers made it possible for the resins to be transported from the North directly into the Paleogene shallow sea, where they settled in enormous delta deposits and were protected against weathering processes. The largest known amber deposit, the "blue earth" of the Samland Peninsula, is also assumed to have formed in this manner (Kosmowska-Ceranowicz 1996a). Experimental drilling confirmed that these delta deposits continue far to the west, all the way to the region around Danzig (Leba district, Chlapowo) (Kosmowska-Ceranowicz 1987). However, the amber-bearing strata there are so deep that mining operations would not be worth the expense.

The paleogeographical situation changed utterly once again in the Neogene. The ocean had receded

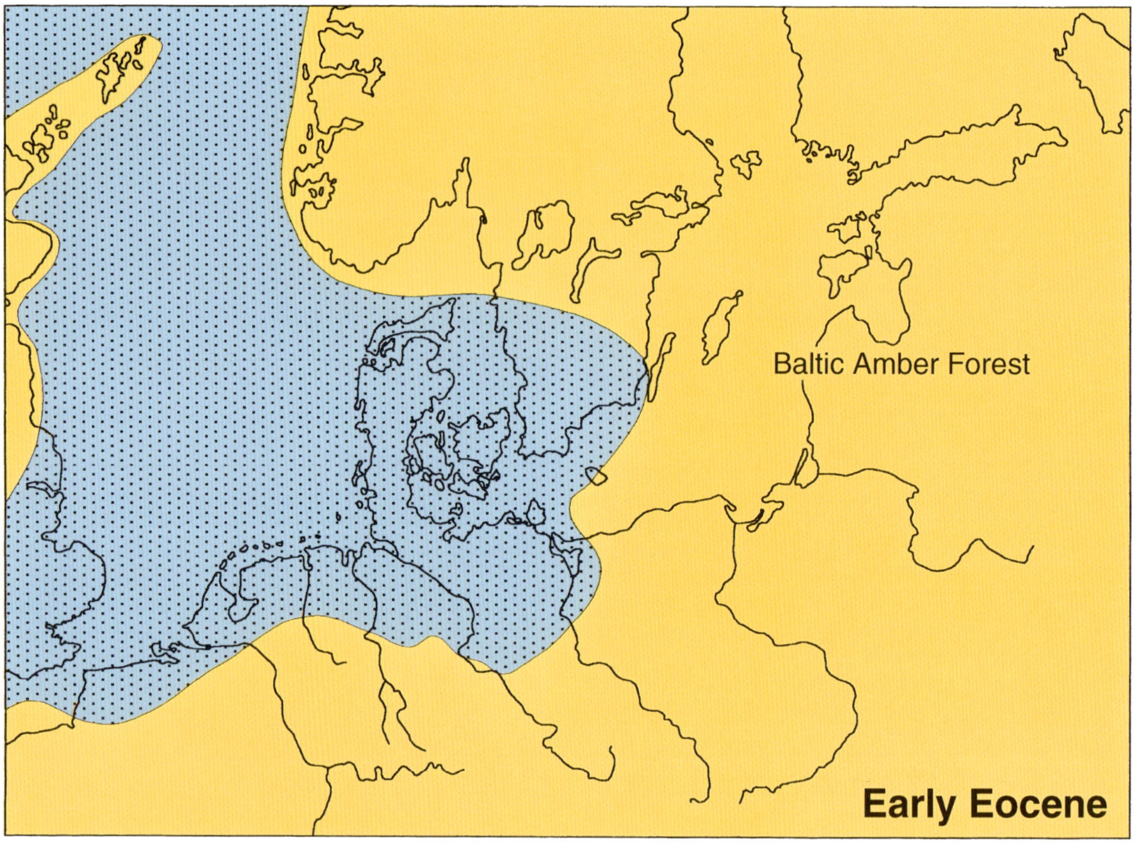

Fig. 8a: Distribution of land and sea in the Early Eocene (about 54 million years ago) and presumed position of the "amber forests" (after VINKEN 1987).

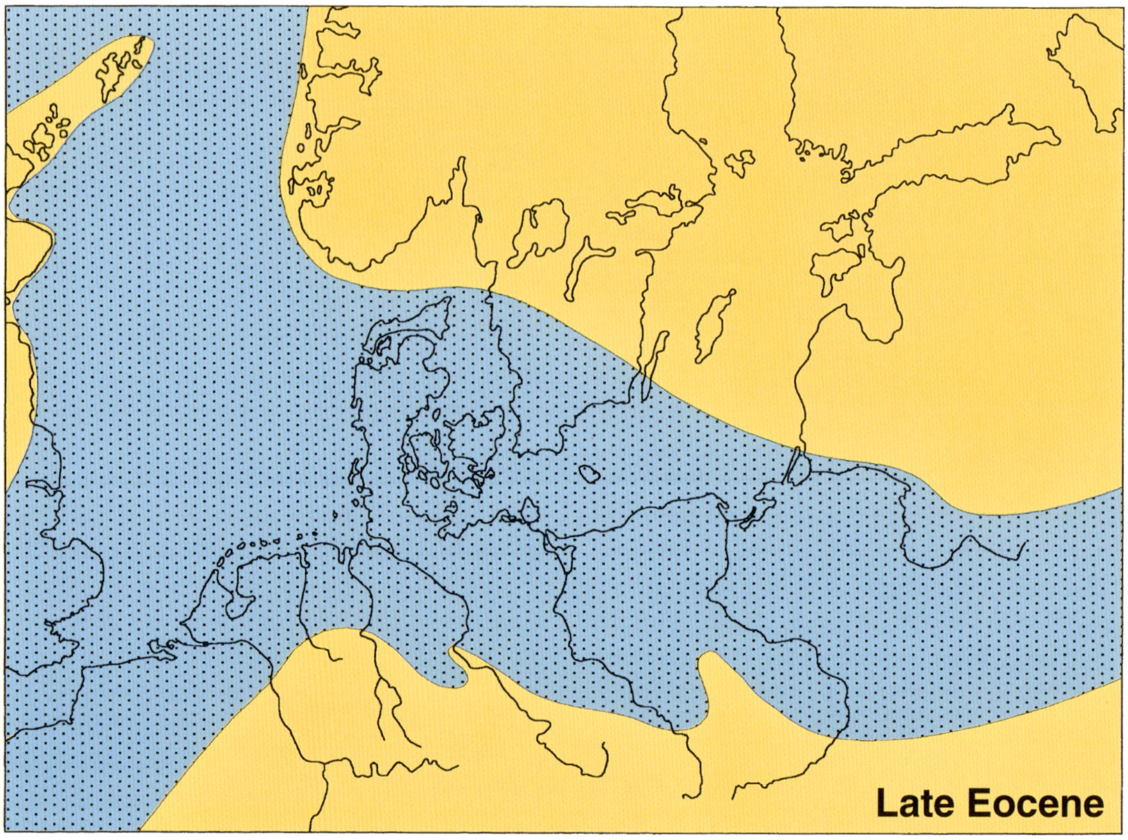

Fig. 8b: Distribution of land and sea in the Late Eocene (roughly 39 million years ago) (after VINKEN 1987). In the Middle and Late Eocene, the ocean advanced far to the east and flooded substantial areas of the "amber forest".

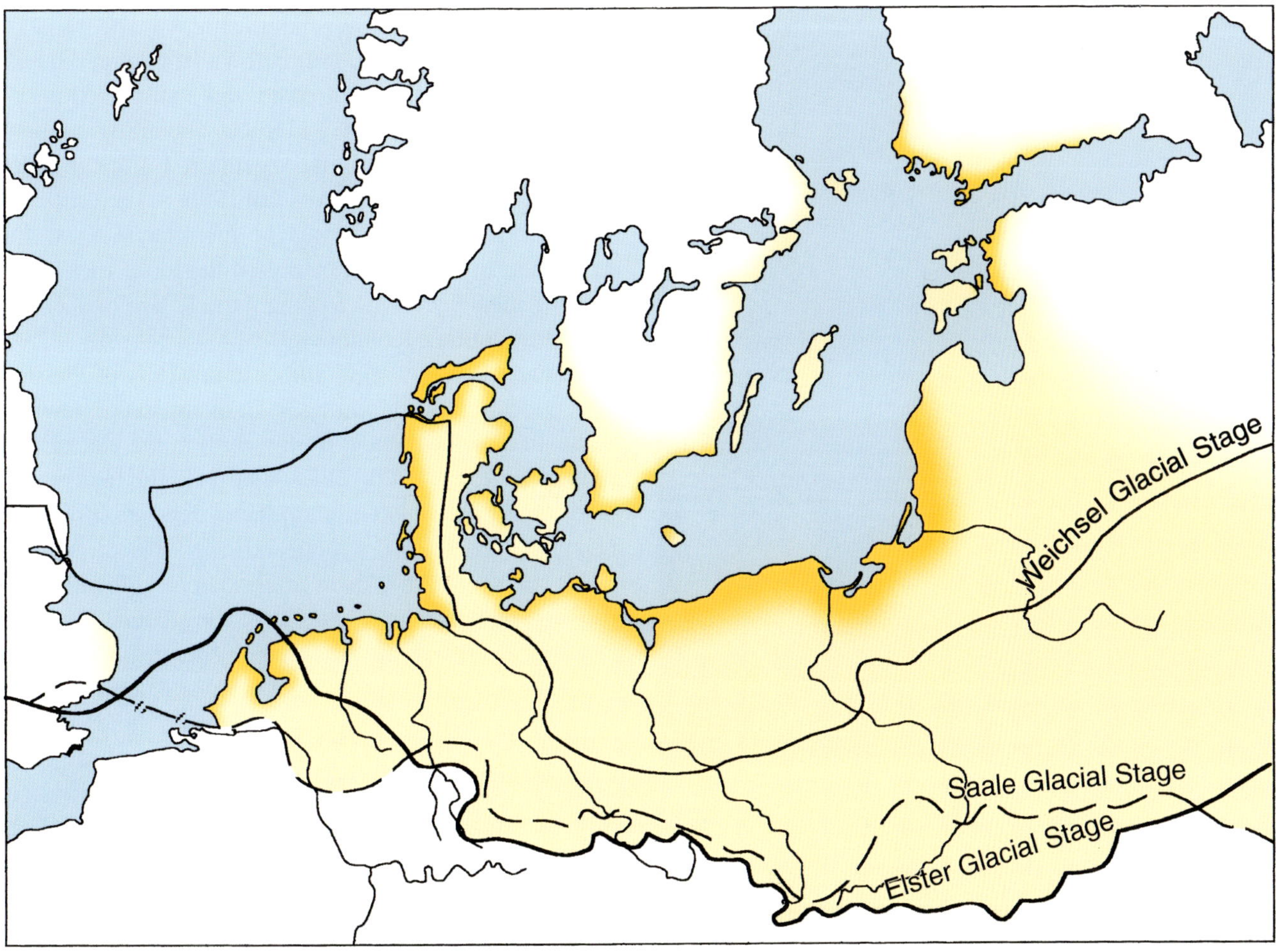

Fig. 9: The current distribution of Baltic amber (orange) and the extent of the different glacial stages during the Pleistocene (after BISMARCK 1987).

almost entirely from the northeastern European basin, and the Baltic Sea region had again become part of the mainland. Since the beginning of the Miocene, large river systems of the Scandinavian shield and the Baltic platform had been transporting clastic sediments to the south into the northwestern European and eastern German-Polish basins, depositing brown coal sand containing considerable amounts of amber in the process. The concentrated amber deposits nested there led to the first systematic mining operations (RUNGE 1868). The brown coal sand undoubtedly drew its amber from redeposited, Paleogene strata, meaning that the amber thus lies in at least its third deposit (ANDRÉE 1951, WEITSCHAT 1987, KOSMOWSKA-CERANOWICZ 1991/92).

These considerable Mid Miocene amber deposits led WETZEL (1939) to suggest the existence of a Neogene amber forest roughly in the region of the present-day western Baltic Sea. To date, no evidence has been uncovered to support this assumption. ANDRÉE (1943, 1951) was opposed to this position. WETZEL's theories of a Miocene amber forest have been revived in recent years due to major amber discoveries in the central German brown coal region near Bitterfeld. Referring to the genesis and age of the deposits, BARTHEL & HETZER (1982) and SCHUMANN & WENDT (1989) emphasized the independence of this deposit, which reportedly formed in the Lower Miocene and indicates an absolute age of roughly 22 million years.

WEITSCHAT (1987, 1997) strongly opposed the assumption that independent amber forests existed during the Miocene. The primary argument he presented was the identity of the flora and fauna with that of amber from the Samland deposits, in conjunction with paleoclimatic findings. If this argument is correct, the Bitterfeld deposits show that Baltic amber was redeposited to a previously unknown extent during the Neogene.

The most massive redepositions of Baltic amber certainly took place during the Pleistocene, meaning within roughly the last two million years. Closer examination of the current range of amber, which extends far eastward into Russia, westward to the Netherlands and the English coast, and southward to the German hill region, reveals that these borders are virtually identical

Fig. 10: Amber with "scratched" surface – evidence of transport by glaciers; Hude gravel pit near Bremen.

to those of the cover of glacial till (Fig. 9). The massive continental glaciations coming down from Scandinavia carved out the Baltic Sea basin and transported enormous masses of debris and rock, including amber-bearing sediments, into the northern European flatlands. When the ice masses came to a standstill, or retreated, as they did during the interglacial ages, the melt-water eroded and transported the loose rocks, depositing them according to their particle size or specific weight at another location.

Consequently, we find amber in a variety of Pleistocene deposits of different ages: in one instance "swimming" in an entire block of blue earth in the glacial till; another time washed together in nests in the melt-water sands, and finally also in interglacial deposits (Figs. 10, 11).

At the same time, there are apparently distinct differences between Late and Early Pleistocene deposits in terms of how much amber they contain – an observation made initially by Meyn (1876) and confirmed by the work of Neubauer (1994). However, this is understandable, as it must be assumed that the oldest glaciations of the Elsterian Glacial Stage, which traveled over a Tertiary relief, took up more Tertiary material,

Fig. 11: "Till amber" with cracked, yellowish-brown weathered crust. The thick, flaky crust was formed by embedding in melt-water sands above the groundwater level (weight: 160 g), Hoisdorf gravel pit near Hamburg.

Fig. 12: Amber in the drift line of the Baltic Sea coast near Nidden (Kurische Nehrung).

Fig. 13: Sea amber ("Seestein") with balanid growth. Sea amber is characterized by the lack of a weathering crust.

and thus also amber, than the later Saale and Weichsel Glacial Stages.

Not least, there is also the amber which, from the glacial periods up to the present, has been washed out of one of the numerous, geologically older deposits by running waters on the mainland (chiefly the major Pleistocene watercourses) or by the ocean and redeposited at another location. These processes in the Holocene are primarily responsible for the amber we find in the drift line on the shores of the North and the Baltic Seas (Figs. 12, 13). As long as man does not interfere, amber continues to be washed into "deposits" there, even today.

1.4 Baltic amber deposits

Despite its enormous range, the number of "real" (i.e. minable) Baltic amber deposits is relatively small. This is primarily a result of the low preservation potential of fossil resins, as well as economic factors in individual cases. For example, a rich amber deposit discovered during prospecting work in the Leba district west of Gdansk was not profitable because of the very thick overburden.

The Baltic amber deposits are presented below, giving particular emphasis to their genesis and age.

1.4.1 Blue earth of the Samland Peninsula

The most significant deposits of Baltic amber, also the largest known amber deposits in the world, are located in the northern and northwestern areas of the Samland Peninsula in the former territory of East Prussia, today the Kaliningrad district of the Russian Republic. The amber located here in the "blue earth" is so concentrated that it has been extracted since the end of the 19th century from large underground and strip mines, most importantly from the pits near Palmnicken (currently Yantarnyi). The thickness of the "blue earth" layer in Samland varies between two and ten meters with an average amber content of roughly 2.5 kg per cubic meter. At the edge of the Baltic Sea, the blue earth is located almost at sea level, while the inland layers are covered by 30 to 40 meters of thick Tertiary and Pleistocene sediments. The "AG Russian Amber" company currently operates two large strip mines, from which several hundred metric tons of amber are extracted every year (Fig. 14). The amber reserves in this area are estimated to be about 640,000 metric tons, or many times the amount already extracted (Kosmowska-Ceranowicz 1996a).

Blue earth, the most important bearer of Baltic amber, is part of the Paleogene strata series of Samland, which were previously referred to collectively as the amber formation or "glauconite formation". The divisions and designations of the individual layers originate from age-old amber diggings and were introduced in the literature by Zaddach (1868). They are based less on lithological differences than on the profitability of the amber deposit from the standpoint of mining. Although these designations are still used today, they have since been incorporated into a modern lithostratigraphic concept (Katinas 1971, Kaplan et al. 1977).

Blue earth received its name from its greenish-blue color, which stems from the significant content of glauconite, an iron-aluminum silicate. In lithological terms, it is a slightly clayish, micaceous fine sand. It contains an irregular distribution of amber in pieces of all shapes and sizes (Figs. 15 a, b). However, blue earth is not the oldest amber-bearing horizon in the Paleogene Samland strata series. There are also two underlying horizons containing amber, albeit not in minable quantities.

The age dating of the Samland amber formation goes back to studies by Noetling (1883, 1888) and is based on a diverse marine fauna consisting primarily of mollusks, echinoderms, crustaceans and bryozoans. The entire strata series was classified in the Lower Oligocene (approx. 38 million years ago) by comparing fauna of similar age from the northern German Tertiary (von Koenen 1865). This biostratigraphic classification, which was principally dependent on mollusk faunas, was considered to be indisputable for quite some time. However, new methods of scientific study have led to justified doubts about its validity. It became evident that mollusk faunas are not suitable for the accurate breakdown of the Paleogene period in question. Modern zoning is chiefly based on various microfossil groups (calcareous nannoplankton, dinoflagellate cysts) that are linked to absolute (radiometric) ages (Odin & Lutterbacher 1992). Application of this method to the Paleogene Samland strata series has led to results that deviate considerably in some instances from previous beliefs.

Using new, absolute dating analyses of glauconites, Ritzkowski (1997) concluded that the formation period of blue earth falls in the Mid Eocene (Lutetian) and is thus significantly older than previously assumed. The

Fig. 14: Yantarnyi, western view of the lower side of the Primorskoje strip mine, Yantarnyi (Kaliningrad district).

two additional amber-bearing horizons in the underlying beds of the blue earth indicate that amber had already been transported to secondary deposits during the Lower Eocene (Ypresian) nearly 50 million years ago. If we accept these theories, then the minimum geological age of Baltic amber has been pushed back nearly ten million years by the new method.

1.4.2 Ukrainian amber

As we assume that a single, continuous, expansive amber forest existed in Northern and Central Europe during the Paleogene Period, the amber deposits in the Ukraine should also be included in order to get a complete picture.

The amber deposits in the northern and western regions of the Ukraine were discovered long ago and described in detail by TUTKOVSKI (1911). Comprehensive geological amber explorations were carried out between 1978 and 1985 in northern Ukraine and southern Byelorussia. Over 50 amber-bearing deposits were found in Paleogene sediments, thus making it possible to narrow down the occurrence of amber in the Pripyat basin (VASSILISHIN & PANTSCHENKO 1996). The main area of distribution is located on the northwestern slope of the Ukrainian shield and consists of Paleogene sediments in the peripheral zone of the crystalline rock. Amber from the Pripyat basin is currently mined from the Klesow deposit, where it is concentrated in strata series of up to six meters in thickness consisting of slightly glauconitic, fine to medium-grained quartz sand with humus-rich, clayish intermediate layers. The sands are interpreted as deposits of lagoonal-deltaic facies, thus clearly indicating that this amber is also in at least a secondary deposit. The average amber content amounts to roughly 50 grams per cubic meter sediment, although it can increase to over 400 grams in certain depressions ("traps"). The amber-bearing strata are biostratigraphically classified in the Mid and Upper Oligocene.

Like the amber from the Samland blue earth, Ukrainian amber is a succinite (KOSMOWSKA-CERANOWICZ 1991/92). It is primarily distinguished from blue earth amber by the weathering crust. Most raw pieces are covered with a dark brown to black oxidized crust several millimeters thick. The crusts often have a polygonal crack pattern extending into the intact amber (Fig. 16 a). This crust, which sometimes easily breaks away from the core,

Fig. 15a: "Blue earth" with amber; Primorskoje strip mine, Yantarnyi, (Kaliningrad district). The clayish fine sand is greenish-blue when wet and turns light gray when dry.

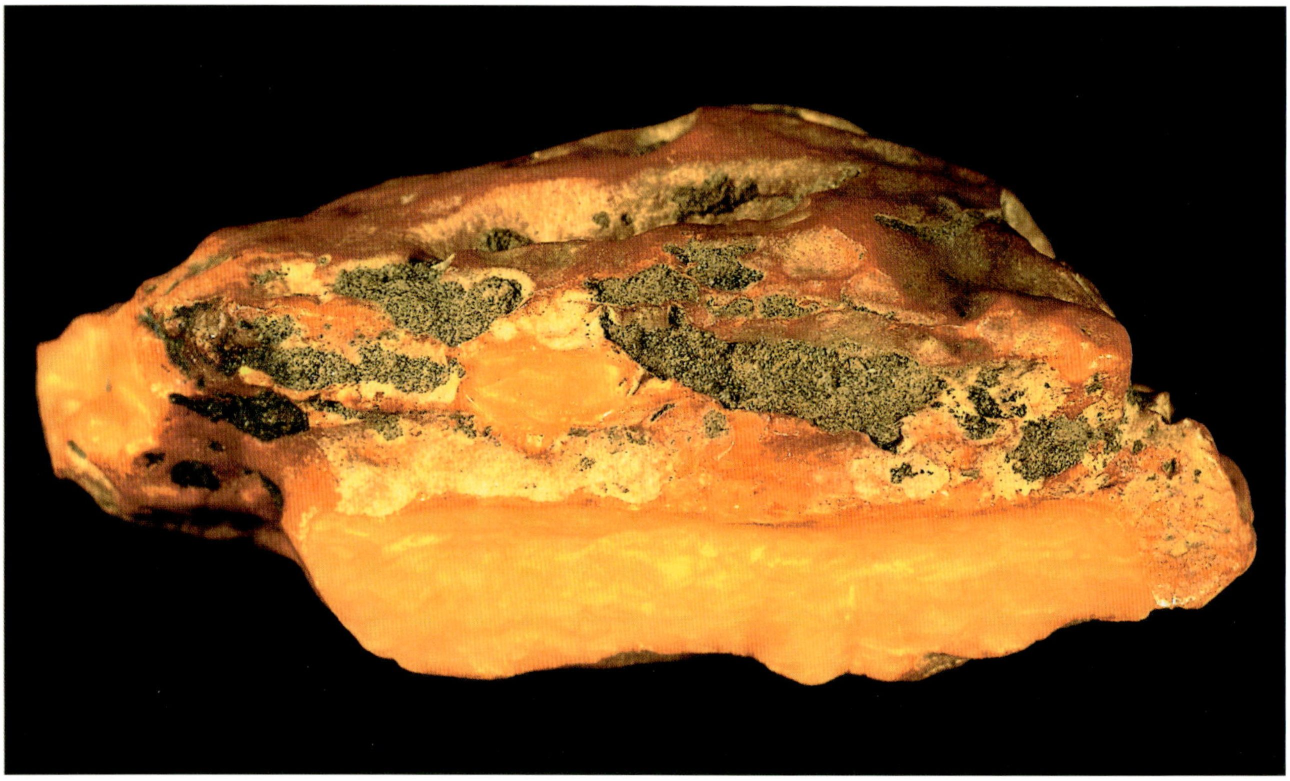

Fig. 15b: Raw amber with residual "blue earth" in small depressions (weight: 1,050 g). Primorskoje strip mine, Yantarnyi (Kaliningrad district).

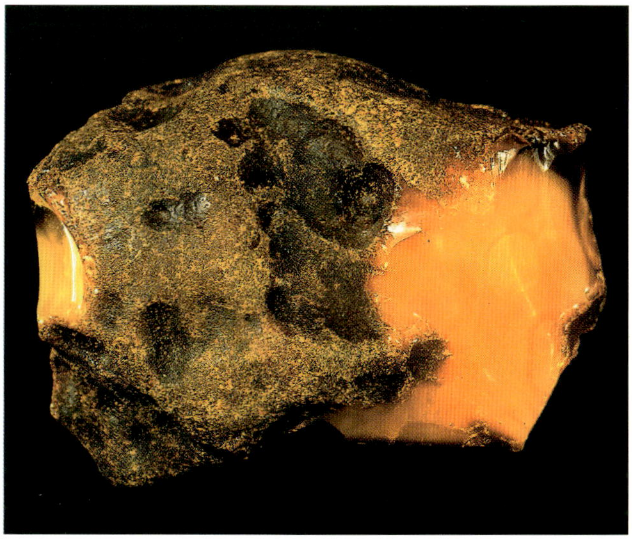

Fig. 16a: Ukrainian amber (weight: 205 g); strip mine near Klesow, Ukraine. In contrast to amber from the "blue earth", Ukrainian amber has a typical brownish-black oxidation crust several millimeters thick. The formation of these often polygonal crusts can be explained by the substantial effects of oxygen on amber in the Ukrainian deposits.

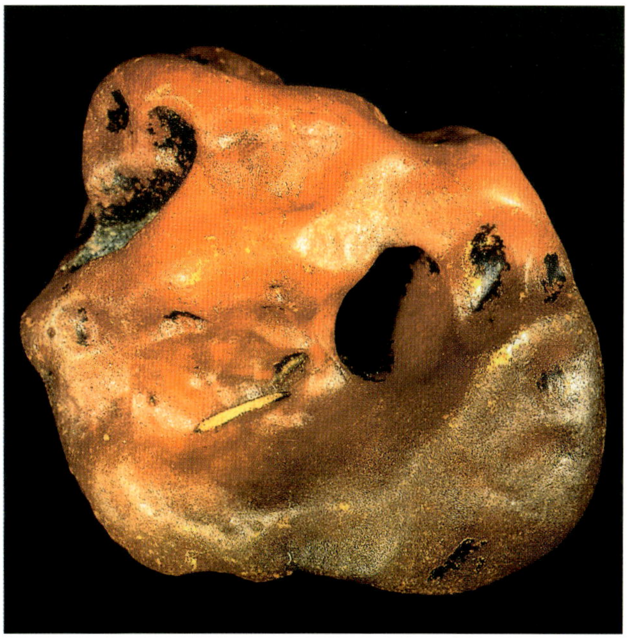

Fig. 16b: Ukrainian amber without an oxidation crust and with the characteristic range of dark red colors; Klesow strip mine, Ukraine.

is followed by another dark red zone that lends Ukrainian amber its characteristic color spectrum (Fig. 16b). The formation of this oxidized crust is probably attributable to the significantly greater effect of oxygen in these amber-bearing deposits compared to the blue earth.

Plant and animal inclusions appear to be considerably more rare in Ukrainian amber than in the Samland and Bitterfeld deposits. Personal observations indicate an unusually low proportion of *Schlauben* material. For the most part, inclusions found to date in Ukrainian amber are limited to fairly large, clear pieces consisting of a single resin flow without any identifiable lamination. It is conceivable that the relatively fragile *Schlauben* were largely destroyed during redeposition.

Previous studies of the animal and plant inclusions indicate a high degree of similarity to those from the Samland and Bitterfeld deposits (Maidanovitsch & Makarenko 1988, Pantschenko & Kvasnica 1982, Serebrickij 1979).

1.4.3 Bitterfeld deposits

In 1955, amber-bearing layers were discovered in the seat rock of the Bitterfeld main seam at the "Goitsche" brown coal strip mine near Bitterfeld (greater Halle, Leipzig region). The geological and mining exploration of this bed in the following years revealed that it was minable and indeed a major amber deposit. By the time the mine was closed in 1993 for reasons of environmental protection, roughly 50 metric tons of raw amber had been extracted from it each year.

Bitterfeld amber is concentrated in silty, sandy, highly lignitic, micaceous sediments four to six meters thick. The microfauna and glauconite content provide evidence of an entirely marine environment and thus indicate that the amber is at least in a secondary deposit. Because of their spore and pollen floras, these layers are biostratigraphically dated as Lower Miocene, which corresponds to an absolute age of roughly 22 million years. However, it would appear to be more than questionable to conclude that Bitterfeld amber is 12 million years younger than Baltic amber. Neither of the two amber deposits is in its original location, but rather in marine sediments of at least secondary deposits. Only these deposits can be dated and directly compared. If the entire range of amber specimens occurring in the Bitterfeld deposits is taken into consideration, then the obvious variety of sizes and non-uniform character of the weathered crusts of individual pieces indicate that the material was more likely to have been redeposited several times.

Barthel & Hetzer (1982) and Schumann & Wendt (1989) published the first detailed scientific studies on the Bitterfeld deposits, which also formed the basis of most subsequent works. With regard to its genesis and age, the authors emphasized the independence of this deposit and introduced the term "Bitterfeld amber". However, they did not discuss the obvious question of whether these deposits could possibly consist of multiply redeposited Baltic amber. They specified the Lower Miocene as the period of formation, which corresponds to an absolute age of roughly 22 million years.

Only scientists who have studied the fossil content of Bitterfeld amber have contradicted these conclusions up to now (WUNDERLICH 1983, WEITSCHAT 1987, LOURENCO & WEITSCHAT 1996, WICHARD & WEITSCHAT 1996). Comparative studies of Bitterfeld amber inclusions with those from the blue earth of Samland demonstrated the identity of their flora and fauna on the basis of numerous taxonomic examples, thus raising doubts about the independence of Bitterfeld amber (WEITSCHAT 1997).

1.4.4 Bitterfeld amber inclusions

The question now arises of the extent to which Bitterfeld amber inclusions can be drawn into the discussion of its age.

In reference to similar issues involving autochthonic Miocene amber deposits in the western Baltic region (WETZEL 1939), Karl ANDRÉE (1943), probably the most recognized authority on amber of his time, claimed that "only the stratigraphic identification of distinctly Miocene amber inclusions would be proof of a different age". It is generally known that arthropods are used only very rarely to resolve biostratigraphical issues, let alone as index fossils. However, what ANDREÉ may have been trying to say is that evidence is required of unmistakable differences between the flora and fauna of two such inclusion groups that are so far apart in terms of time period and location.

If we expand the discussion to include the climatic history of the Tertiary, which was not known in such great detail in ANDRÉE's time, then the question he raised in 1943 becomes an absolute necessity today. The flora and fauna of a subtropical-tropical Eocene amber forest inevitably must have differed from those of a Lower Miocene amber forest. In particular, it is unlikely that the insect fauna could have survived unchanged in the period of drastic cooling between the Eocene and Oligocene, which led to a well-recognized episode of mass extinction worldwide.

However, if the detailed flora and fauna inventories of Bitterfeld amber now available (SCHUMANN & WENDT 1989, KRUMBIEGEL & KRUMBIEGEL 1994, 1996b) are compared to those of the blue earth of Samland, it becomes evident that the two deposits are nearly identical from a paleontological standpoint (WICHARD & WEITSCHAT 1996, WEITSCHAT 1997). This applies not only to their systematic diversity, but also, and above all, to the breakdown in percentages of both floral and faunal groups, particularly within the individual taxa. Naturally, this comparison greatly depends on the extent to which typical subtropical-tropical faunal elements of Baltic amber are represented in Bitterfeld amber.

In reference to the spider fauna, WUNDERLICH (1996) determined that "the majority of the spider families from Baltic and Bitterfeld amber are today restricted to tropical-subtropical regions". WEITSCHAT (1997) presented a number of similar examples based on representatives of various insect orders, which are not discussed here in any further detail. All paleontological evidence suggests that Bitterfeld amber can also be linked to the Eocene Epoch with regard to its period of origin. Consequently, it is at least already in its third deposit in the Lower Miocene.

1.5 The climate at the time of Baltic amber

The climatic history of the Tertiary Period should be examined more closely here, as we believe the climate not only plays a key role in determining the age of Bitterfeld amber, but also points the way to answering largely unresolved questions about the age and life span of the Baltic amber forest.

In the past few decades, the analysis of American deep-sea drilling activities has rapidly advanced our knowledge of the paleoclimate, i.e. the development of climatic conditions over the course of the earth's history. Oxygen isotope studies play a key role in this context and now make it possible to draw reliable conclusions about paleotemperatures (Fig. 17). The Tertiary climate curve (Fig. 18) has since been verified so thoroughly that the curve itself is no longer the subject of discussion, but rather its causes (Buchardt 1978, Zachos et al. 1994).

The end of the Mesozoic Era was characterized by very warm, but arid climate conditions. While temperatures changed only slightly at the beginning of the ▷

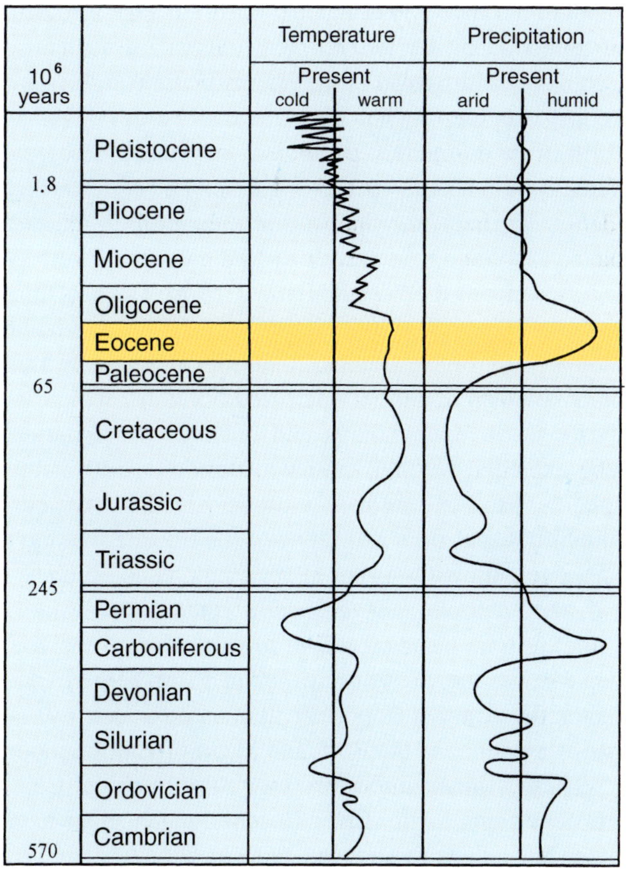

Fig. 17: Global climate development (temperature and precipitation) over the course of the earth's history (after Frakes 1979, Crowley & North 1991).

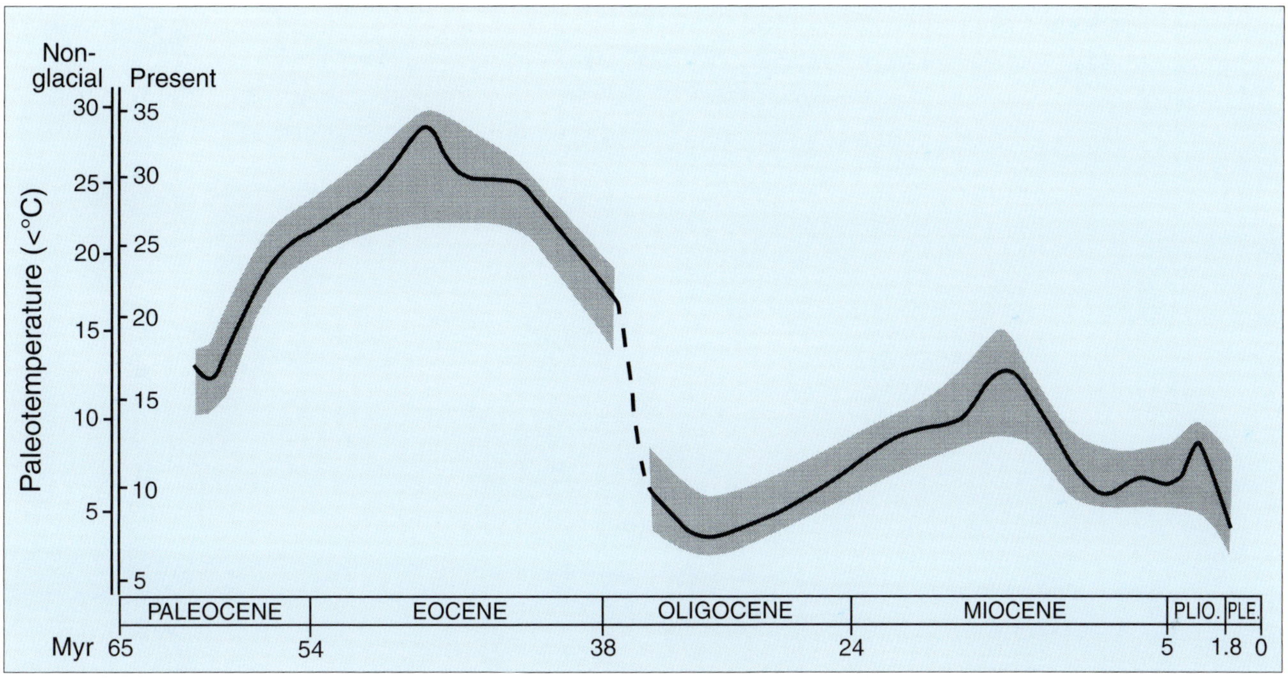

Fig. 18: Temperature curve of the Tertiary based on oxygen isotope measurements; shaded region: fluctuation range of Tertiary ocean temperatures. The different scales are due to the shifting oxygen isotope composition of the oceans during the glacial periods; the right-hand scale labeled "Present" applies to the periods in which the poles were covered with icecaps (after Buchardt 1978).

Tertiary, the overall level of precipitation increased considerably. Following a relatively brief cooling period in the Paleocene, temperatures rose significantly again in the Late Paleocene and Early Eocene. The Mid Eocene was by far the warmest interval in the entire Cenozoic, with global temperatures being roughly two to four degrees higher than they are today. This is verified by numerous geological and paleontological climate indicators.

Sea level and ocean bottom temperatures reached their peak during the Lower to Mid Eocene. The tropical flora from the "London clay", as well as the flora and fauna from the Geisel Valley and the Messel Mine indicate subtropical-tropical conditions at relatively high paleolatitudes. Lateritic weathered horizons, which indicate warm temperatures and seasonal precipitation, also extended beyond 45° paleolatitude in both hemispheres. Further emphasis must be given to the thermophilic fossil flora and fauna of the Eocene that have long been known to occur at very high paleolatitudes – e.g. the mammal and crocodile finds on Ellesmere Island (Arctic Canada) – and, in particular, to the coal deposits in the present-day Arctic (Spitzbergen) and Antarctic. Another important item to be added to this list of evidence is the flora and fauna of Baltic amber.

These climate indicators suggest subtropical-tropical conditions in the Early Eocene, which must have extended polewards beyond today's climate zones by more than 15 degrees latitude.

This warm period was followed by an interval of much lower temperatures. At the end of the Mid Eocene, tectonic events (separation of Australia from Antarctica) triggered drastic global cooling, considered to be the most significant event of the entire Cenozoic Era. This cooling was associated with a substantial decrease in annual precipitation. As a result, the earth's climatic zones shifted towards the equator, and the differences between them became more distinct. Towards the end of the Eocene, plants and animals attempted to retreat, and subtropical-tropical organisms disappeared from higher paleolatitudes. This led to the dramatic mass extinction that is known to have occurred about 40 million years ago during the Cenozoic Era and roughly coincided with the transition from the Eocene to the Oligocene.

A number of conclusions concerning the time and conditions of Baltic amber formation can be derived from these paleontological, paleogeographical and paleoclimatic data. The formation of amber in northern and central Europe probably already began in the Early Eocene and may have ceased towards the end of the Mid Eocene. Together with the Polish and Ukrainian deposits, the blue earth amber of Samland and Bitterfeld amber originated from a single "amber forest" that had a stable subtropical-tropical climate. It existed for a period of roughly ten million years and covered an enormous area of prehistoric northern Europe.

1.6 Fossils in Baltic amber

1.6.1 Inclusions

The term "inclusions" has been established as a special designation for fossils enclosed in amber. The embedding of plants and animals in resin is justifiably viewed as a unique phenomenon of nature. Inclusions differ from other fossils in many respects. Who would ever have dreamed of referring to an insect enclosed in amber as petrifaction? Innumerable attempts have been made in the past to separate such wonderfully preserved insects from the resin. How great the disappointment must have been to realize that this led to the complete disintegration of the inclusion. The reason is that amber inclusions are basically just thinly lined hollow spaces. Only the decay-resistant, chitinous integument is preserved. Apart from a few exceptions, inclusions are hollow on the inside. Despite this apparently unfavorable circumstance, the preservation and visibility of microstructures conserved in amber is unique in the field of paleontology. The nature of fossilization makes it possible to examine structures right up to the resolution limit of the light microscope, which relies most importantly on the transparency of the amber.

Another major difference in relation to other fossils is that the organisms were usually captured alive in the resin trap. The observer of an amber inclusion, particularly an animal one, of course, will immediately realize that not only has the resin entombed a life form, but also that it is a rare situation. Inclusions often appear to be a snapshot of life, even though only the last few seconds have been captured.

1.6.2 The fossilization process

After an animal became embedded in the resin, two processes took place simultaneously: the solidification of the resin and the decomposition of the tissue of the enclosed animal. Above all, the formation and long-term preservation of an inclusion required it to remain dimensionally stable until the resin solidified. Furthermore, it was important for the resin not to shrink during hardening. In order for the resin to turn into true amber, a number of additional geochemical and geological processes must have taken place. The death of the enclosed organism is followed by the microbial decay and decomposition of the tissue. The resultant gases and fluids are primarily released and diffused from the mouth, anus and other openings in the body (wounds), and through the body walls. These secretions can form an emulsion with the still liquid resin and thus hide the inclusion behind a cloudy, milky white deposit that makes detailed morphological study of the fossil impossible. Analyses with a scanning electron microscope have revealed no more than that the deposit is a foamy structure consisting of minute bubbles. Mierzejewski (1978) suggested that the cloudiness is caused by the escape of decomposition gases.

Fossils covered with a white coating formed as a product of decay (German: "Verlumung") are a phenomenon which has long been known to exist from the numerous observations of inclusions in Baltic amber and appears to be at least partially related to the size and volume of the enclosed body tissue. While it occurs only rarely with Diptera and mites, it is common and very pronounced with caterpillars, myriapods and isopods, for example. Based on personal observations, however, it is apparent that plant inclusions can also display a cloudiness of this kind, thus indicating that animal body secretions are not exclusively responsible for this occurrence, but that a certain degree of moisture can also cause a milky effect.

Attempts to explain this phenomenon are complicated by the fact that emulsion formations occur in all conceivable intensities, in all transitional stages or frequently not at all. Even large inclusions with plenty of body fluid can be preserved without any cloudiness at all.

Schlüter & Kühne (1975) were the first to point out that the cloudiness very frequently appears on only one side of an inclusion (Fig. 19 a, b). They provided conclusive evidence that the cloudy side is usually the one facing the concave interior of the *Schlaube*, i.e. facing the tree bark, and thus corresponds to the side of the inclusion facing away from the incident light. This is illustrated particularly well in specimens with several inclusions embedded in different positions. The cloudy zones are always in the shadow cast by an inclusion, corresponding to a light source radiating from a specific direction. The authors came to the conclusion that the intensity and duration of sun exposure must have made a decisive contribution to the clarification processes. The side of the inclusion facing the sun was warmed to such an extent that constituents escaped more easily by means of diffusion, and the resin became clear in that area. These observations are supported by the known process of artificial clarification, in which cloudy amber can be clarified at high temperatures and pressures. In this process, the minute bubbles combine to form larger ones that move to the surface of the resin and escape.

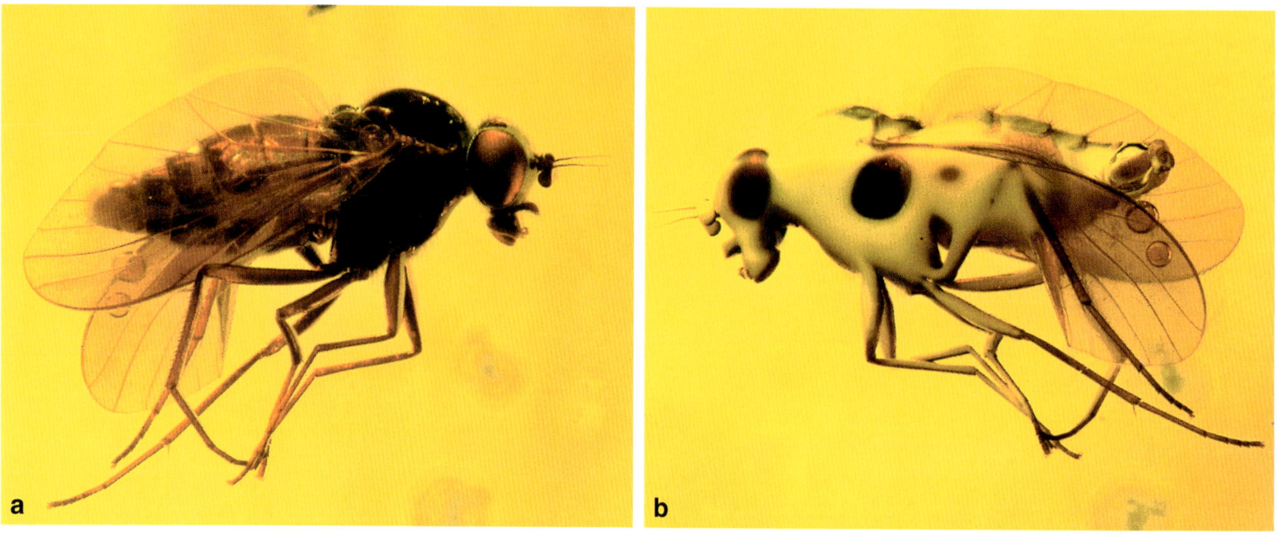

Fig. 19 a, b: Snipe fly (Rhagionidae), **a:** Ventral, **b:** Dorsal. Example of the phenomenon of one-sided turbidity ("Verlumung") in Baltic amber inclusions.

The working hypothesis of the clarification of amber by exposure to the sun would appear to be quite plausible, as it also offers potential solutions for a number of other unresolved issues. For example, it could conclusively answer the frequently discussed question of why only *Schlauben* of Baltic amber are routinely clear and transparent. It could also explain the noticeable differences in the intensity of the turbidity of the inclusions. The differences could be the result of seasonal effects, for example, or simply the diverse growing sites of the resin-producing trees in the amber forest.

1.6.3 Stability and aging

Although it is millions of years old, natural amber is not a stable substance. Under the natural effects of oxygen and light, it begins to change after just a few years as the result of polymerization, oxidation and drying. This process is indicated by a gradual darkening, with the color changing from red-yellow, to ruby red, brownish red and eventually brown. The surface of the amber becomes virtually opaque and dull. As the color changes, a network of fine, hairline cracks also begins to form on the surface and penetrates deeper into the piece as time progresses, making it brittle and ultimately causing it to disintegrate.

Archaeological finds from the Roman period confirm that this process even occurred in the span of historical time. Prehistorical amber generally only survived if it was protected against weathering processes in its deposit environment, such as in moors or water.

All natural pieces of raw amber also display a relatively thick, weathered crust and, depending on the type of embedding, all stages of this external transformation process can be found. The origin and amber-bearing layer of a specimen can often be determined simply by the appearance and thickness of the weathered crust. Amber that has been deposited below the groundwater level and thus protected against the effects of oxygen in the air, forms only a thin weathered crust, which protects the internal structures and can enable the amber to survive for millions of years. For example, amber in the blue earth has a relatively thin, uniform, firmly attached crust, which has a white, dusty appearance when dry. Amber from glacial deposits located above the groundwater level in sandy or gravelly beds has a thick, crumbly crust extending deep into the intact amber in a funnel-shaped fashion (Fig. 11). In contrast, the sea amber ("Seestein") found on northern European coastlines has only a very thin weathered crust or none at all, as it is worn off by the surf.

The worst fate that can befall a piece of amber is to be exposed to the dry air of a museum in a non-airtight showcase. Under these conditions, weathering can progress rather quickly. Inclusion collections most valuable to science, particularly the irreplaceable type material, face the greatest risk. Included are numerous specimens that have been cut into thin slices, some just a few millimeters thick, for the purpose of detailed scientific study. These specimens are thus highly susceptible to weathering and threatened by decay. If the internal weathering process reaches the inclusion, it turns dark red and finally brown (Fig. 20 a), usually making detailed study of the enclosed organism impossible. These thin pieces are also extremely susceptible to fracture due to the formation of deep cracks (Fig. 20 b).

Much of the type material of Baltic amber fauna and

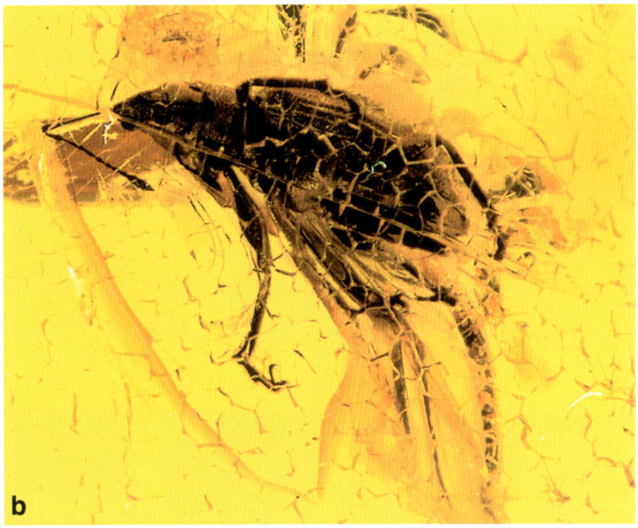

Fig. 20 a: Exuvia of a stonefly (Plecoptera, Perlodidae). The advanced oxidation process of this roughly 70 year-old specimen is clearly visible. The red discoloration has already reached the insect.
Fig. 20 b: Holotype of a weevil (*Paonaupactus sitonitoides* Voss, 1953). Decades of improper storage have caused a network of hairline cracks to form on the surface of the amber, making study of the morphological details very difficult.

flora stored in various European museums, some from the 19th century, is in an advanced stage of weathering and decay. Coordinated conservation programs are urgently required to save at least part of these collections.

The insidious decay of amber, which can take place within a relatively short period of time, is a well-know problem that has led to extensive scientific study. Even today, however, we still know relatively little about the physiochemical processes behind it. One thing that appears to be certain is that the drying and cracking leading to decay are caused by the evaporation of volatile substances in the amber and that these processes are promoted by the effects of light and heat (Koller et al. 1997).

Numerous attempts have been made to save amber from this fate by proper storage. The simplest method initially appeared to be storage in water, which would come closest to the natural conditions of deposit. However, it soon became evident that the antiseptic agents required to purify the water attacked the amber. Attempts at storage in alcohol and mineral oil resulted in the complete destruction of valuable collections.

No satisfactory solutions to the problem were found until modern synthetic resins were used. Today, there are two methods of conservation: embedding the amber in a two-component polyester casting resin and coating it with a one-component polyurethane resin. In addition to conservation, embedding in casting resin has the advantage that precise, plane-parallel surfaces can be ground for detailed morphological analysis. Coating is much less complicated and achieves the same result in conservation. Of course, there is still a lack of long-term experience with both methods concerning destructive interactions between amber and synthetic resins. However, no negative effects have been observed in specimens of amber inclusions from the Hamburg collection, which have been embedded in synthetic resin for more than 60 years.

1.6.4 Tissue preservation

Even though amber inclusions are basically just thinly-lined, hollow spaces, much effort has gone into investigating these spaces for preserved organ remains. In general, there are two procedures for finding organ remains in inclusions: to dissolve the amber along with the inclusion and analyze the remains in the solvent residue, or to open the inclusion mechanically and examine the two halves of the cavity under a microscope. Striated muscles in dipteran and neuropteran legs from amber inclusions were already described by Kornilowitsch (1903) in a Russian language publication that received little recognition. Other authors, particularly Tornquist (1910), vigorously opposed the suggestion, referring to the impossibility of the fossil preservation of tissue structures. They believed that copal inclusions had probably been mistaken for amber inclusions.

It was not until Eocene fossils in the brown coal from Geisel Valley were surprisingly found to contain tissue, skin remains, pigments and epithelia (Voigt 1935), that various scientists were induced to reconsider the conditions of preservation of animals in amber

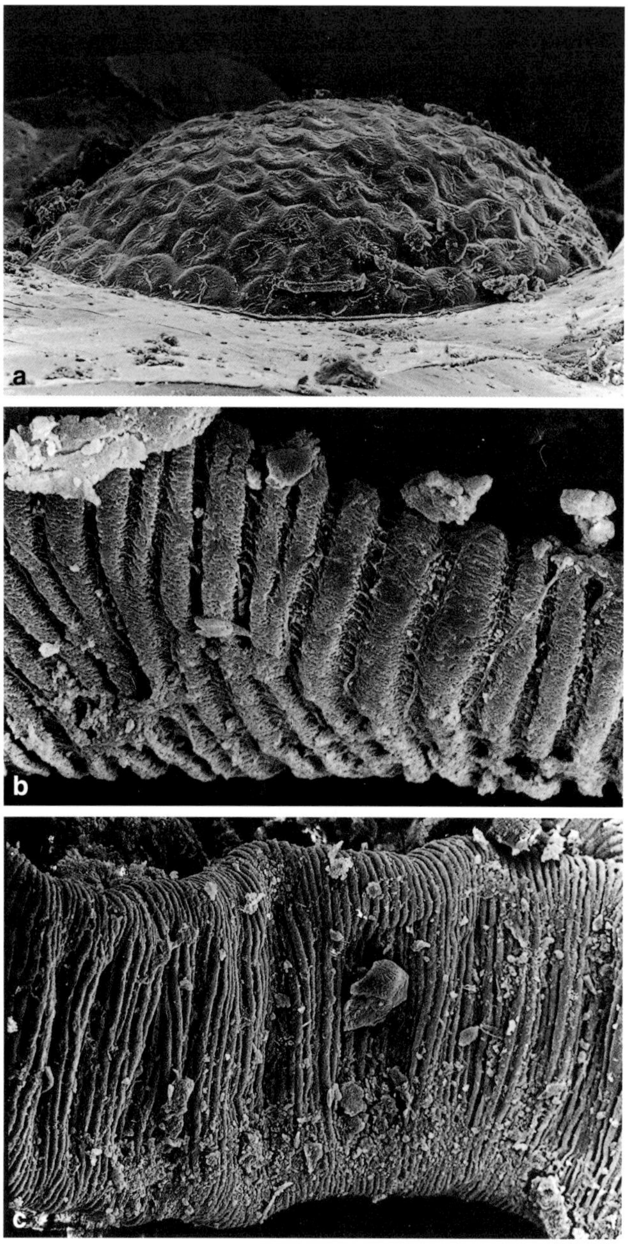

Fig. 21: Preserved tissue from a fossil termite *(Termopsis bremii)* from Baltic amber. **a:** Top view of a compound eye, **b-c:** Preserved tissue ("muscles") in the abdomen (after Kohring 1998).

inclusions. Dissolving the inclusions out of the amber led to impressive results that were reported in a lecture before the scientific society known as the "Physikalisch-ökonomische Gesellschaft zu Königsberg" (Andrée & Keilbach 1936) and thus dispelled all doubts about Kornilowitsch's results. Preserved tracheae, intestinal tissue and oviducts were found. The remains of striated muscle were identified most frequently and not just in large insects, but rather in varying quantities in virtually every specimen. The muscles were either lying loose in the cavities or attached to the chitinous exoskeleton.

Definitive proof that organ parts could be preserved in amber inclusions was finally provided by Voigt (1937, 1938) using the lacquer film method. He cut an inclusion down the middle using a fine-toothed saw, made small lacquer replicas of the two parts and was thus able to examine the transparent preparations at high magnification. Using this method, Voigt was able to identify the preserved remains of muscles and tracheae in a number of different insect groups.

The advent of electron microscopy led to promising new methods for researching tissue preservation in amber inclusions. The inclusion needed only to be opened and the cavity scanned for preserved tissues using a scanning electron microscope at the appropriate magnification (Fig. 21 a-c). Mierzejewski (1976 a, b) used this method on spider inclusions in Baltic amber to successfully identify lung remains and silk glands. Henwood (1992 a, b) described the flight muscles of various Diptera and the preservation of tissue in beetles of Dominican amber. Kohring & Weitschat (1994) and Kohring (1998) documented the complete preservation of organ tissue of various insects from Baltic amber (Fig. 21 a-c), particularly muscles, tracheae and eyes.

Poinar & Hess (1982) described muscle fibers and even cell structures of a fungus gnat from Baltic amber, after having managed to visualize the ultrastructure of a roughly 50 million year old animal cell with a transmission electron microscope for the very first time.

Motivated by such spectacular results, scientists began to consider the possibility of isolating DNA, the carrier of all genetic information, from amber inclusions. The studies were initially complicated by the excessively small amounts of tissue preserved in the amber insects. It was not until 1992 that two independent research groups in the USA succeeded nearly simultaneously in identifying DNA of a termite and a bee from Dominican amber (Cano et al. 1992, De Salle et al. 1992). Even though only very small DNA fragments measuring between 200 and 300 nucleotides were isolated, the identification of 25 million year old DNA was still a scientific sensation. Just one year later, the same research group (Cano et al. 1993) reported the successful isolation of DNA sequences from a Lebanese amber weevil, which is 120 to 130 million years old, placing it in the age of the dinosaurs. The publication of these scientific results curiously coincided with production of the movie "Jurassic Park", unexpectedly intensifying speculation about the possibility of cloning extinct creatures.

However, there are still doubts about the evidence of fossil DNA of such great age. British research groups were unable to confirm the results of the American scientists.

1.7 Flora and fauna of Baltic amber

1.7.1 Significance of resin preservation

Despite the incredible abundance and diversity of the flora and fauna passed down to us from the inclusions in Baltic amber, they should not be assumed to provide complete documentation of the various biotopes that existed in the Eocene Baltic amber forests. On the contrary, this record represents a fairly limited segment of the flora and fauna, namely those animals and plants that could come into contact with sticky resin, become embedded in it and then be preserved as fossils. As a result, specific biocoenoses of the amber forests are very well-represented, while others virtually do not appear at all. Consequently, the material preserved in amber inclusions reflects the fossilization potential of the different groups of organisms more than their actual numerical proportions. As fossil preservation can be strongly influenced by a number of selective factors, such as the size, habitat or behavior of the organisms, we ultimately obtain a distorted picture allowing only limited conclusions to be drawn about the various biotopes of the amber forest and their biological associations (Vavra 1982, Krzeminska & Krzeminski 1992). Any type of preservation of floral and faunal communities in resin is a prime example of the fact that the composition is chiefly determined by the different fossilization potentials of the individual groups of organisms.

An impressive example has been documented from the plant life of Baltic amber. The most common organic inclusions are minute, radially branched trichomes that probably originate from oak trees (Fig. 22). These stellate hairs, which are hardly visible to the naked eye, are found in virtually all *Schlauben* pieces and therefore considered to be characteristic of Baltic amber. If direct conclusions about the biotope were to be derived from this circumstance, then the amber forest must have been principally an oak forest and the likely resin-producer consequently an oak. However, as there are numerous arguments against this conclusion, the frequency of the oak stellate hairs can only be explained by their partic-

Fig. 22: "Stellate hairs" of the oak (size: approx. 0.5 to 1 mm). Stellate hairs are regularly found in *Schlauben* and are considered to be characteristic of Baltic amber.

Fig. 23: Male oak flower with stellate hairs preserved "in situ".

ularly favorable preservation potential. The chance of these minute, lightweight hairs being blown onto the tacky resin surfaces by the wind and remaining stuck there is naturally very high. On the other hand, as an essential prerequisite for this frequency, it would have to be assumed that oaks grew together with the resin-producer in the same plant community and that the buds formed during the period of greatest resin production. In present-day trees, natural sap secretion mainly occurs in spring and early summer (LARSSON 1978); the same can be assumed for the resin-producing trees of the Baltic amber forests.

1.7.2 Plant inclusions

The frequently discussed proportional imbalance between plant and animal inclusions in Baltic amber may be attributable to the different preservation potentials of the two groups.

Apart from the oak stellate hairs mentioned above, plant inclusions, especially larger plant parts, are extremely rare in Baltic amber. On average, only one out of a thousand inclusions is of botanical origin.

Nonetheless, the mining of enormous quantities of amber has given us insight into the flora of the amber forests and its impressive range of species, most of which were already monographically studied during the 19th century (GOEPPERT & BEHRENDT 1845, CONWENTZ 1886, 1890, CASPARY & KLEBS 1907). It must be kept in mind, however, that despite the host of inclusions, a significant number of the described taxa are based on just a few voucher specimens, or even on a single specimen in some instances.

No satisfactory explanation has yet been found for the extreme rareness of plant inclusions. The chance of plants or plant parts being preserved is certainly restricted to a great extent by the fact that they could only come into contact with the tacky resin surfaces in a passive manner, such as by wind. This particularly applies to plants that did not grow in the immediate vicinity of the resin-producing trees. Although this situation is understandable, it becomes absolutely incomprehensible in reference to the resin-producer itself. According to prevailing opinion, pine trees are considered to be the resin-producer of Baltic amber. But where are the needles? They must have been present in the billions in the amber forest and among the most common inclusions. However, the case is quite the opposite. Only an insignificant number of needles have been found up to now, thus repeatedly leading to doubts about the identity of the resin-producer. Oddly enough, inclusions of needles from the presumed resin-producer are rare finds in Baltic amber.

According to SCHLEE (1990), the length of the pine needles may be a possible, albeit unsatisfactory explanation for this occurrence. Due to their length, the needles may have been only partially enclosed by a resin flow and left protruding from the side of the *Schlaube*, thus providing a starting point for subsequent weathering processes. These would finally have led to total destruction "from the inside out", leaving just hollow spaces in the shape of the needles behind. However, this hypothesis is contradicted by the fact that needle impressions in Baltic amber are even rarer than the needles themselves. Consequently, there is still no plausible explanation for the scarcity not only of inclusions of the amber tree, but also of plant inclusions in general.

1.7.3 Animal inclusions

The fauna preserved for us in Baltic amber is almost exclusively restricted to the arthropods, with insects and arachnids making up a majority at nearly 99 %. Inclusions of representatives of other phyla (worms, mollusks, vertebrates) are extremely rare and only verified on the basis of a few, isolated finds in some cases. The numerical distribution clearly illustrates the different preservation potential of individual animal groups, which is influenced by multiple factors.

One significant selection factor is size. Most of the animals enclosed in Baltic amber are small, with a body length of between one and five millimeters. Strong, large animals were usually able to free themselves from the resin and are therefore found only rarely, if at all. Exceptions to this rule are exuviae and animals that got caught in the resin after they were already dead. A few inclusions of fairly large specimens show that the animals stuck to the resin became prey to others before being completely entombed by the resin. This can frequently be observed in caddisflies and termites, whose abdomen is completely eaten away and filled with the resin of the subsequent flow. In some instances, only traces of relatively large animals have been left behind in the form of individual wings and extremities, or molting remains. The preservation potential of the individual animal groups is primarily determined by their habitats in the amber forest (Krzeminska & Krzeminski 1992). Animals living on the resin-producing trees themselves or in the immediate vicinity were by far the most frequent victims of the resin traps. The numerous animal communities in the different ecological niches of the "amber tree", from the root zone all the way up to the treetops, have been preserved in great diversity and detail. In addition to real tree-dwellers, we also know of a considerable number of "visitors", who visited the amber tree for various reasons (e.g. hunting, reproduction, rest) and ended up getting captured in the resin traps. One important example in this context is the great number of small, winged insects (Diptera, Hymenoptera) that lived in the undergrowth of the resin-producing trees. They comprise nearly two-thirds of all Baltic amber inclusions. These animals are assumed to have actively flown into the resin traps. It is conceivable that the bright yellow resin surfaces acted as light or color traps and, together with the scent of the fresh resin, had a "fly-catcher" effect.

In special circumstances, the chance of coming into contact with resin can also be greatly influenced by an organism's specific habits (Krzeminska & Krzeminski 1992), such as the swarming flight of certain insect groups (Diptera, Isoptera). For example, evidence of the various termite castes (workers, soldiers, reproductives) is based almost exclusively on the winged reproductives, which landed in the resin during swarming flights or soon after in the search of a potential nesting site. The workers and soldiers living inside wood or in underground tunnels had no chance of becoming caught. Naturally, the preservation potential was also much higher for organisms moving about on the trunks of the resin-producing trees (ants, spiders) or inhabiting the crevices and cracks in the bark (certain Microlepidoptera) (Skalski 1973). In contrast, the danger of nocturnal arthropods getting entrapped by the resin may have been much less, as the resin surfaces hardened faster at lower nighttime temperatures or the resin flow stopped completely. This could explain the curious rareness of isopods (woodlice) and myriapods (millipedes) in Baltic amber. Finally, seasonal factors may also have had an influence on the composition of animal inclusions. Current observation reveals that resin secretion is also highest in spring and summer in subtropical climates, which could significantly limit the possibility of preserving typical fall or winter faunas. The extreme rareness of winter crane flies (Trichoceridae) in Baltic amber is offered as evidence (Krzeminska & Krzeminski 1992).

1.7.4 Scenes of life in amber

Fossils that provide information on their life history or behaviors are exceptional objects in the science of paleontology. They come under the field of paleoethology or ichnology, which reconstructs the behavioral patterns of fossil organisms. Moments of life "frozen" by fossilization require very specific embedding and preservation conditions. Resin preservation is especially suitable for this purpose and has provided us with a number of impressive examples. In addition to the developmental stages of numerous insect groups, amber has also passed down evidence of mating, egg-laying, brood care, feeding and many types of social behavior.

Copulating Diptera were frequently caught in the resin trap and offer the unique opportunity of observing reproductive behavior from a distant age. A number of Diptera and Trichoptera from Baltic amber expelled egg-strings after getting caught in the resin trap as a reflex or an act of desperation. The eggs of fungus gnats have an impressively delicate honeycomb structure on the shell.

The imagoes as well as the larval and pupal stages of various insect groups have been preserved in amber. A superb example is provided by wood gnats (Anisopodidae), of which not only the larvae and wood-dwelling pupae, but also the hatching process of the various life-stages have been documented in fossil form.

Ants living on the amber tree left behind a remarkable number of "snapshots". We have evidence not only of the various developmental stages (larvae, cocoon, imago) and castes (soldiers, workers, reproductives), but also of numerous scenes of everyday ant life (Bachofen-Echt 1935).

There are a number of examples of brood care among the arachnids of Baltic amber. For example, spider egg sacs have been preserved that served as protective shelter for the spiderlings after having previously been filled with eggs or a newly hatched brood (Plate 20). One inclusion even documents a sac of this kind being transported by a female daddy-long-leg spider. Several specimens in Baltic amber document typical hunting methods of the spiders, such as the use of nets or orbwebs with the remains of prey still preserved in them.

Our knowledge of the various biological associations (e.g. parasitism, phoresy) of Baltic amber was greatly expanded by mites. Parasitic mite larvae are found attached to different types of flies, midges, caddisflies, ants, beetles and cicadas.

Another example of paleoparasitism is the eggs of lice (nits) found on mammal hairs in Baltic amber (Voigt 1952). This discovery provided the first fossil evidence of animal lice (phthiriapterans).

The most famous example of parasitism in Baltic amber is undoubtedly the "amber flea" of the genus *Palaeopsylla* (Dampf 1911, Hennig 1939, Peus 1968). Two different species are represented in Baltic amber, both of which appear to be closely related to extant species of this genus (Peus 1968). This suggests that the primary characteristics of the fleas had already evolved 50 million years ago – a fact that may very well apply to the majority of arthropods. Present-day representatives of the genus *Palaeopsylla* live on insectivores, principally shrews (Soricidae) and moles (Talpidae). It is conceivable that their fossil ancestors also parasitized early insectivores.

Roundworms of the family Mermithidae are a good example of the evidence of endoparasitism in Baltic amber (Chap. 2.07). These small worms mainly parasitize diptera, particularly true midges (Chironomidae). Amber fossils indicate that this specific form of parasitism and dispersal in the aquatic worm already existed in the Eocene.

Another interesting type of behavior in the biocoenoses of present-day arthropods is the transport of wingless animals by a winged, transport host (i.e. phoresy). This "hitchhiking", which usually serves the dispersal of a species, is documented by a number of examples in Baltic amber. The passengers are often certain mites, whose "last ride" attached to beetles, Diptera or Hymenoptera ended in the resin trap.

Phoresy has also been recorded among pseudoscorpions in Baltic amber (Chap. 2.11). These small animals of just a few millimeters in size still expand their range of distribution today by using their pincers to cling to midges or the legs of harvestmen and letting themselves be carried off. There are also a few reports from Dominican amber of wood-boring beetles (Platypodinae) and crane flies acting as hosts (Schlee & Glöckner 1978, Schlee 1990, Poinar 1992).

1.8 Aquatic insects in Baltic amber

1.8.1 Aquatic fauna

At first glance the preservation of aquatic organisms in fossil tree resins appears to be rather improbable. However, if it is considered that the genesis of amber required the resins to be rapidly transported by running waters to other locations, then it is understandable that this fauna is relatively frequent in amber. Baltic amber has given us special insight into the various aquatic biotopes of the amber forests through the large number of aquatic insect inclusions it contains (Wichard & Weitschat 1996).

The imagoes of amphibious (merolimnetic) aquatic insects spend a relatively short period of their developmental cycle on land, e.g. stoneflies (Plecoptera) and mayflies (Ephemeroptera). The winged insects reproduce and spread during this fertile phase of their life. There is always a chance of them coming into contact with tree resins during their dispersion and swarming flights, as documented by the large number of aquatic Diptera (e.g. Chironomidae) and Trichoptera preserved in amber.

Many aquatic insects leave their aquatic habitat for the first time as larvae rather than imagoes, in order to hatch outside the water at elevated, dry locations, e.g. dragonflies and damselflies (Odonata). The larvae of several holometabolic aquatic insects go on land to pupate and finally hatch as imagoes, e.g. alderflies (Megaloptera), spongillaflies (Planipennia) and water beetles (Coleoptera). Consequently, it is no surprise that larvae of these aquatic animals are occasionally found in amber. In contrast, their pupae lead a hidden lifestyle and are out of reach of the resin.

However, several very rare and isolated finds contradict these explanations, such as inclusions of nymphs of water boatman (Heteroptera: Corixidae) and rheophilous mayflies (Ephemeroptera: Heptageniidae) (Wichard & Weitschat 1996). The larval stages of these animal groups normally do not leave their aquatic habitat. The same is true of amphipods (Crustacea: Amphipoda) and ostracods (Crustacea: Ostracoda), which can nonetheless survive in wet moss (Chap. 2.21). The occurrence of these animal groups in Baltic amber is difficult to interpret and may be the result of bodies of water running dry.

In addition, a few caddisfly larvae (Trichoptera) also occur in Baltic amber (WICHARD 2000, 2001). They all belong to the suborder Integripalpia, whose larvae always bear cases. However, none of the fossil larvae in amber have been preserved with cases, having probably left them beforehand. Extant larvae initially retreat into their cases when disturbed. If they are deprived of water for an extended period, so that the residual water in the case drains or evaporates, the caddisfly larvae change their behavioral strategy, leaving the case to crawl around in an unfamiliarly dry environment. The Integripalpia larvae in amber thus lead to the assumption that the bodies of water they inhabited ran dry, and that they were unable to synchronously adapt their life cycles.

1.8.2 Aquatic biotopes of the "amber forest"

The range of the aquatic insect fauna in Baltic amber supports far-reaching conclusions about the landscape of the Eocene "amber forest". It bears witness to the existence of numerous running and standing waters.

The study of the rich caddisfly fauna of Baltic amber is a good example for analyzing paleoecological conditions (ULMER 1912). Presupposing fossil and recent genera and families display similar behavior in adapting to their habitats (principle of actuality), ULMER reconstructed a detailed landscape from the range of the trichopteran fauna. A large majority of the roughly 5,000 identified Trichoptera prefer fast-flowing waters, while a much smaller portion is characteristic of standing waters and a few species display no preference at all. ULMER concluded that the "catchment area" of the resin-producing trees was a mountainous landscape with swift mountain streams flowing down into the valley, eventually into larger rivers and through the lowlands.

This theory is supported by our knowledge of the habits of other aquatic insects occurring in amber (WICHARD & WEITSCHAT 1996). Mayfly larvae (Ephemeroptera) and particularly stonefly larvae (Plecoptera) indicate a mountainous landscape with numerous, clear, rapid and oxygen-rich running waters inhabited by larvae.

A number of other amphibious insect groups preserved in amber provide additional information on paleoecological conditions. Water bugs and the various water beetles indicate ponds with dense vegetation. The marsh beetle family (Scirtidae), which is represented by a large number of species and individuals in Baltic amber, suggests the presence of small puddles in the vicinity of the resin-producing trees, e.g. moory hollows in swamp forests or puddles littered with leaves. The occurrence of predaceous alderfly larvae (Megaloptera) is an example of the hyporheic habitat of sandy, muddy, standing and slowly running waters (WICHARD 1997).

With flies and midges, the amphibious Diptera in amber represent a species-diverse and abundant insect group that inhabits a wide variety of biotopes and enables detailed descriptions of amber forest biotopes when coupled with precise knowledge of the biology and ecology of the taxa. Both Diptera groups include numerous species whose larvae and pupae inhabit lakes, ponds and puddles, or preferred running waters. SEREDSZUS & WICHARD (2002) pointed out the special paleoecological importance of true midges (Chironomidae). True midges are among the most common inclusions in Baltic amber and thousands of them have been preserved. Most of the specimens are Orthocladiinae species characteristic of running waters. A smaller proportion comprises species of the subfamily Chironominae, which today live in stagnant, often oxygen-poor bodies of water, e.g. on the bottom of large lakes. The great abundance of these animals in Baltic amber further indicates that their swarming flights took place when the amber trees were in the phase of active amber production.

1.9 Biogeographical aspects

Baltic amber inclusions have been passed down to us as a taphocoenosis. The question of whether this fossil community was also a biocoenosis has been the subject of intense debate since the beginning of amber research.

The "amber forest" contained genera and subfamilies, and even entire families in rare instances, that still live virtually unchanged today in the Palearctic or Holarctic regions, while other taxa are currently restricted to the subtropics or tropics, principally of the Indo-Malayan and Australian region, but also from Africa and tropical America. From today's standpoint, the amber fauna and flora appear to be a mixture of exotic plants and animals from a wide range of climates. A variety of attempts have been made to explain this phenomenon.

HEER (1865) suspected that the amber forest had different altitudes and assumed that temperate forms inhabited the mountainous elevations, while thermophilic forms occupied the lowlands (cf. SCHUBERT 1961). WHEELER (1914) believed climate changes were a more probable explanation. According to him, the actual amber age was preceded by a warmer period, whose subtropical-tropical elements were the last remaining evidence of former floras and faunas. ABEL (1935) compared the amber forest to the jungles and pine forests of Florida, while ANDER (1942) compared it to the considerably more humid forests of Southeast Asia. Both regions are mixed in terms of plant and animal biogeography. ANDER discussed another possibility that present-day tropical species were strongly eurythermic in the past, i.e. adapted to a broader temperature range. In conclusion, the assumption that Baltic amber inclusions do not stem from a single biotope, but rather document a wide variety of biocoenoses, would appear to be justified.

In order to clarify this problem, WEITSCHAT (1997) went back to criteria previously established by ANDRÉE (1937, 1951), who had proposed investigating "whether a single amber specimen contains organisms that indicate different regions when compared to present-day floral and faunal regions". ANDRÉE did indeed frequently find "tropical" and "temperate" organisms in the same piece of amber.

Animals and plants enclosed by the same resin flow must have lived at the same time, on the same tree and under the same climatic conditions. When applied to the problem under discussion, this can only mean that the faunas and floras of the Eocene amber forest were complex and cannot be assessed according to current biogeographical criteria.

This theory is supported by the drastic geographical and climatic changes that have occurred on earth over the last 50 million years. Paleoclimatic data show that the Late Mesozoic and Paleogene were characterized by very warm climatic conditions extending all the way into the high paleolatitudes. The Earth had no polar icecaps. The temperature differences in the N-S direction were therefore less distinct and subtropical-tropical life forms were spread over a very large area in the Eocene. It must be assumed that individual species had a much greater N-S distribution than they do today. At the end of the Eocene, roughly 40 million years ago, plate-tectonic events triggered drastic global cooling considered to be the most significant of the Cenozoic Era. As a result, the climatic zones shifted towards the equator and the differences between them became more distinct. Subtropical-tropical faunas and floras attempted to retreat to the South and disappeared from higher paleolatitudes.

During this period of worldwide cooling, all life forms were exposed to exceptional selective pressure. A great many genera and species became extinct. Faunas and floras dwindled and numerous genera only survived in isolated regions, living on as "relict species". Others adapted to the changed conditions and climates and colonized abandoned biotopes. The current biogeographical regions did not develop until the end of the Tertiary and into the Pleistocene.

Only from today's biogeographical point of view is the composition of flora and fauna preserved in Baltic amber a "curious mixture of temperate, subtropical and tropical life forms". In the Paleogene amber forest, they formed a biocoenosis that functioned for several million years in a vast region of the former northern European mainland characterized by stable, subtropical-tropical climates.

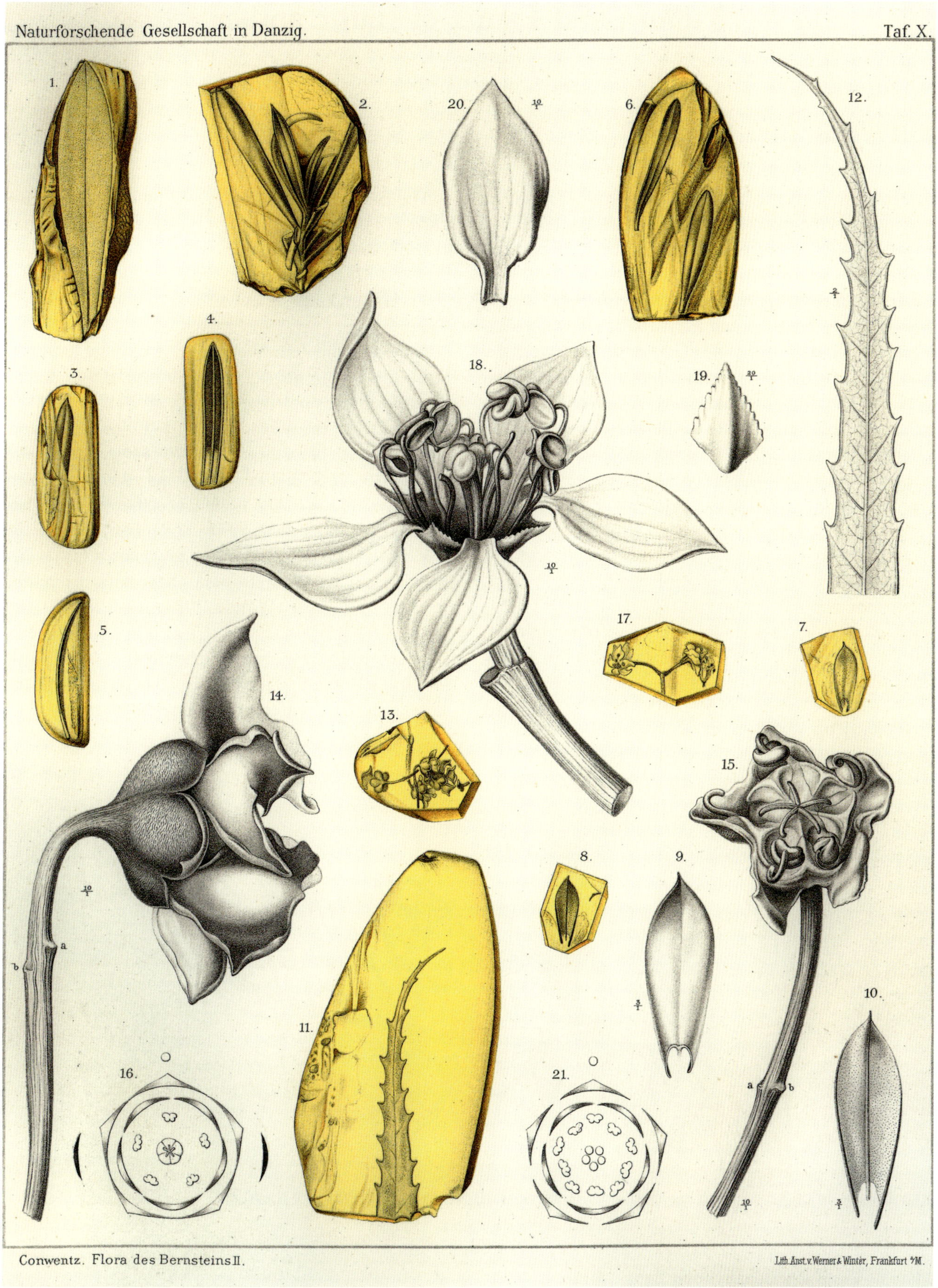

Fig. 24: Plate X from CONWENTZ, H.: Die Flora des Bernsteins. – 2. Die Angiospermen des Bernsteins. – Danzig 1886.

2 Flora and Fauna of Baltic amber

2.01 Ferns – Pteridophyta

The first comprehensive work on plant remains in amber was published by GOEPPERT & BERENDT in 1845 and encompassed 125 pages and seven plates. It covered fungi, algae and lichens, as well as mosses, ferns and flowering plants. *Sporotrichites heterospermus* GOEPPERT & BERENDT, 1845 and *Peziza* (*Pezizites*) *candida* (GOEPPERT & BERENDT, 1845) GOEPPERT, 1853 were the first fungi described from amber. Since then, there have been reports of both fossil species that have long been extinct, as well as species that still survive today. These groups were summarized by SPAHR (1993b) in a systematic catalog. *Licea* sp. (Liceaceae) and *Stemonitis splendens* cf. *succini* (Stemonitaceae) were the first slime fungi (Myxomycetes) to be recorded (PIELINSKA 1990, DOMKE 1952). Although GOEPPERT & BERENDT initially questioned the occurrence of algae and lichens in 1845, EHRENBERG (1848), SCHUMANN (1863) and ZANON (1929) soon after published studies of algae (Phycophyta), particularly the diatoms (Diatomeae). Confirmed reports of lichens then followed in publications by GOEPPERT (1852), CASPARY & KLEBS (1907), MÄGDEFRAU (1957) and PIELINSKA (1990).

The Baltic amber flora covered by GOEPPERT & BERENDT (1845) mainly included mosses, ferns and flowering plants. The first mosses of the genera *Jungermannites* and *Muscites* described in this early work today appear in updated reviews by GROLLE and FRAHM (Chap. 2.02).

Although the fern-like plants, which include true ferns (Filicatae), are extremely rare in Baltic amber, they were already noted by GOEPPERT & BERENDT (1845). Two species of the polypody family (Filicatae: Polypodiaceae) have been described: *Pecopteris humboldtiana* GOEPPERT & BERENDT, 1845 and *Alethopteris serrata* CASPARY, 1881. The frond piece shown here (Plate 1, Fig. 25), which displays some similarity to *Pecopteris humboldtiana*, is one of the few known fern inclusions in Baltic amber. Fern spores have been given little attention up to now (WETZEL 1953). Like today's forests, the Eocene amber forest had a diverse range of fern-like plants in the shrub and herbaceous layer. However, it is evident that flowering plants were dominant, particularly the coniferous and deciduous trees.

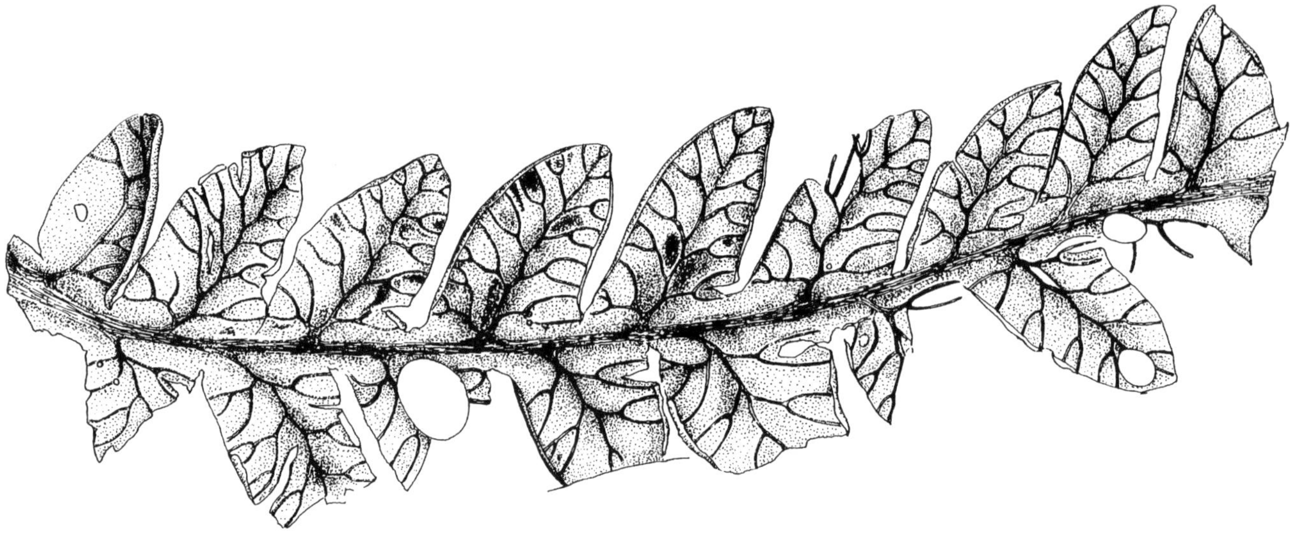

Fig. 25: Frond of ? *Pecopteris humboldtiana* GOEPPERT & BERENDT, 1845 (Polypodiaceae).

Plate 1: Ferns (Pteridophyta) in Baltic amber.

Frond of ? *Pecopteris humboldtiana* GOEPPERT & BERENDT, 1845 (Polypodiaceae). ×20.

2.02 Mosses and liverworts – Bryophyta

Bryophytes and fern-like plants are among the oldest of the terrestrial plants. They have been confirmed as far back as the Devonian Period 350 million years ago and found in all subsequent geological formations. Of the fossil bryophytes, the Tertiary forms display the greatest structural similarity to recent mosses. Some of the moss species found from the Tertiary Period have apparently survived up until today.

With regard to the liverworts (Hepaticae), *Notoscyphus lutescens* is one of the rare species that occurs as a fossil in Baltic amber and also as a present-day plant in Southeast Asia, where it is primarily adapted to the moist soils of the subtropical to tropical highlands and lowlands (GROLLE 1988 a). The ecological requirements of congeneric, recent species indicate that many of the species found in amber probably preferred a moist tropical and subtropical climate. Examples include representatives of the genera *Cheilolejeunea*, *Nipponolejeunea* and *Trocholejeunea*. The genus *Cheilolejeunea* today occurs south of 40 degrees latitude in the hot, humid climate of the subtropics and tropics (GROLLE 1984). The Eocene species *Nipponolejeunea europaea* has present-day relatives that exist as epiphytes on bark in the subalpine coniferous forests (*N. pilifera* and *N. subalpina*) and the deciduous forest zone (*N. pilifera*) of Northeast Asia, particularly Japan (GROLLE 1981a). The genus *Trocholejeunea*, which occurs in Baltic amber as *T. contorta*, has two mesophytic, (tropical-) subtropical species that are currently native to Southeast Asia (GROLLE 1982).

GROLLE (1988 b) elaborated a preliminary list of 18 liverwort species reported from Baltic amber, several of which have also been recorded from the Bitterfeld deposits in Germany. All the liverworts found in the Bitterfeld deposits are also known from Baltic amber. Although the range of species in Bitterfeld amber is less diverse, it fits in well with the overall range of liverwort species found in Baltic amber.

FRAHM (1996 a, b, 1999, 2000 a, b, 2001 a, b) summarized the mosses (Musci) in Baltic and Bitterfeld amber. In addition to numerous specimens that are either unidentifiable or have only been determined down to the level of genus, current findings include 12 fossil species that are apparently extinct. Some other species occur today in Asia. Two of these species also occur in Europe as relicts of the Tertiary Period. *Haplocladium angustifolium* (Plate 2 h) is indigenous to three areas of the southern Alps. *Fabronia ciliaris* (Plate 2 g) is restricted to the Mediterranean region and warm locations in Central Europe.

Most mosses that occurred in the early Tertiary and continue to exist today as presumably identical species, or in the same genera, are native to Southeast Asia. These Asian plants include the genera *Mastopoma* and *Symphyodon*, as well as the species *Campylopodiella* cf. *himalayana* and *Merrilliobryum* cf. *fabronionides* (FRAHM 1996 b). The current distribution of many amber mosses indicates not only the great age of some species, but also the favorable tropical-subtropical climate that probably prevailed in the Eocene Epoch, endured the subsequent ages in Southeast Asia and thus ensured the survival of the Eocene mosses. It is remarkable to note that many of the insects of Baltic amber also appear to be very closely related to extant fauna in Southeast Asia.

Plate 2: Mosses and liverworts (Bryophyta) in Baltic amber.

a, c Liverwort: *Frullania schumannii* **a** Moss stem, × 60, **c** Moss branch, × 20.
b, d Liverwort: *Radula oblongifolia* **b** Stem apex, × 60, **d** Stem in ventral view, × 20.
e-f Mosses: **e** unknown moss, × 10, **f** *Muscites* cf. *tortifolius* CASPARY & KLEBS, 1907, × 15.
g Moss: **g** *Fabronia* cf. *ciliaris* (BRIDEL, 1812) BRIDEL, 1827, × 10.
h Moss: **h** *Haplocladium angustifolium* (HAMPE &C.MÜLLER, 1855) BROTHERUS, 1907, × 10.

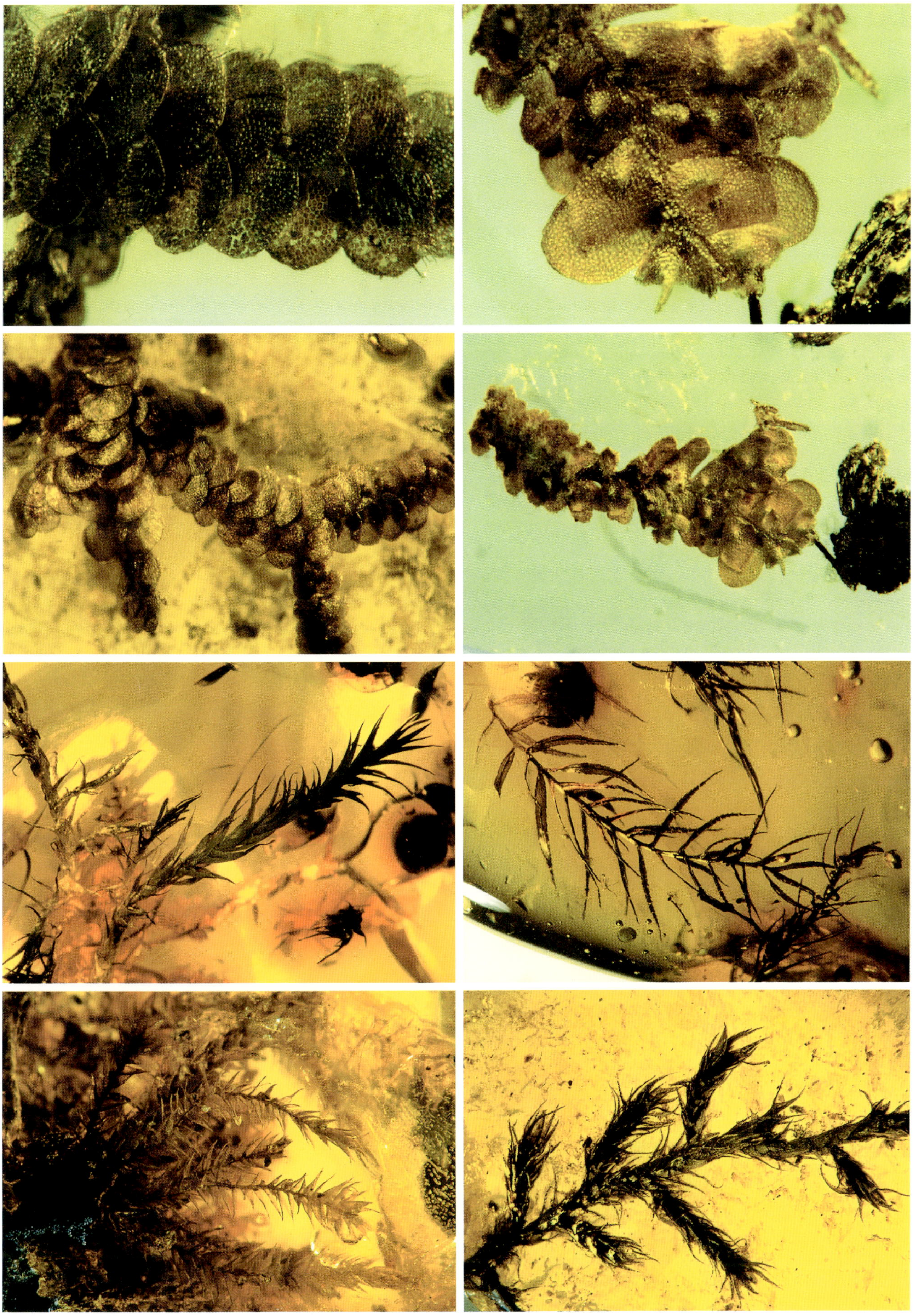

a b c d e f g h

2.03 Flowering plants: gymnosperms – Spermatophyta: Gymnospermae

Corticolous algae, lichens and mosses that were covered by dripping resin are regularly found in botanical inclusions. However, amber also contains fragments of ferns and flowering plants that obviously fell into the fossil resin, were light enough to be carried to it by the wind or were covered with resin as shreds on the ground. In the past, new species were described all too quickly on the basis of flowers, pollen, cones, seeds, fruits, leaves, needles and twigs, as well as bark fragments. The lack of knowledge of a plant's overall phenotype and the identification of individual fragments made duplicate descriptions unavoidable. Accordingly, there are a great number synonyms. Czeczott (1961) recognized and addressed this taxonomic problem in one of the first necessary revisions of the Baltic amber flora.

Three taxa of gymnosperms are discussed: Cycadinae (cycads), Coniferae (conifers) and Gnetinae (gnetophytes). The "very uncertain" report (Schuster 1931) of cycads (Cycadinae) is based on a roughly 1 cm-long impression of a leaf described as *Zamiophyllum sambiense* (Caspary, 1881) Caspary & Klebs, 1907. The impression, which can no longer be located, would likely prove to be an artifact. In a similar case, Caspary & Klebs (1907) claimed that gnetophytes (Gnetinae) were "not observed in amber", even though a flower stalk had already been described as *Patzea gnetoides* Caspary, 1872, "flower stalks whorled around the stems" as *Ephedra (Ephedrites) johnianus* (Goeppert & Berendt, 1845) Conwentz, 1886 and, finally, an infructescence as *E. mengeana* (Goeppert, 1883) Conwentz, 1886. Conwentz (1886) classified all three species of the genus *Patzea* in the Loranthaceae family, which belongs to the dicotyledon Angiospermae and not the Gymnospermae.

Conifers (Coniferae) are the only Gymnospermae in Baltic amber to date. This includes the families Cupressaceae, Taxodiaceae, Podocarpaceae and Pinaceae. There are also doubts about the Podocarpaceae, particularly since this family is only classified on the basis of one, small leaf described as *Podocarpites kowalewskii* Caspary & Klebs, 1907. On the other hand, however, *Podocarpus* has been confirmed as a fossil from the Eocene Epoch (Larsson 1978).

According to Czeczott (1961), the three remaining conifer families, Cupressaceae, Taxodiaceae and Pinaceae, include a total of 33 species, if uncertain taxa are disregarded and additional synonyms taken into account. This revision (Czeczott 1961), which has not yet been supplemented with new descriptions of conifers, includes 11 Pinaceae species in the genera *Pinus* (8), *Picea* (1) and *Abies* (2), four Taxodiaceae species in the genera *Glyptostrobus* (1) and *Sequoia* (3), 18 Cupressaceae species in the genera *Widdringtonia* (3), *Thuites* (4), *Libocedrus* (1), *Chamaecyparis* (4) and *Juniperus* (2), as well as uncertain species (4).

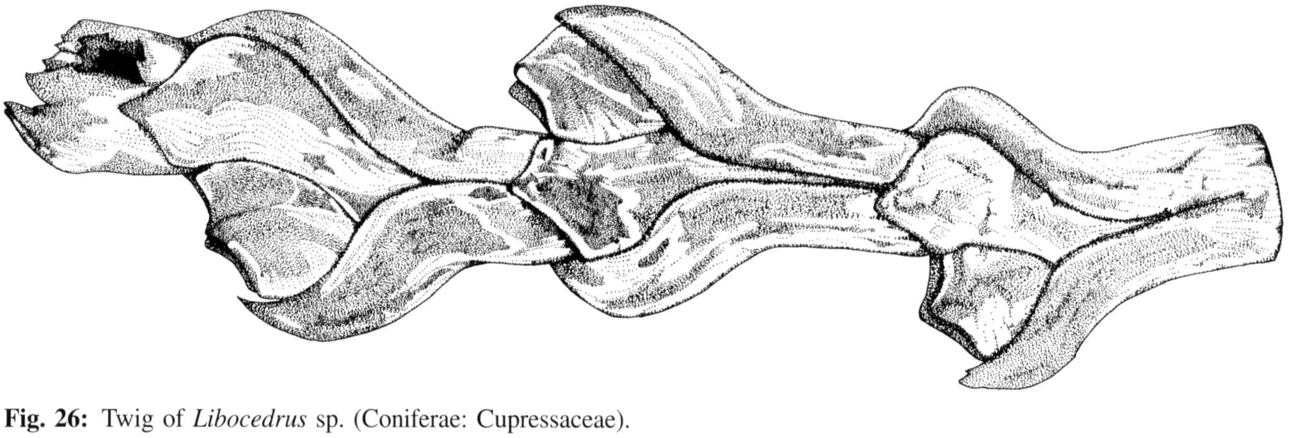

Fig. 26: Twig of *Libocedrus* sp. (Coniferae: Cupressaceae).

Plate 3: Flowering plants: gymnosperms (Spermatophyta: Gymnospermae) in Baltic amber.

a-b Cupressaceae (cypress family): **a** *Thuites* twig, ×8, **b** *Thuites* twig, ×10.
c-d Cupressaceae (cypress family): **c** *Thuites* flower, ×17, **d** *Libocedrus* twig, ×9.
e-f Cupressaceae (cypress family): **e** Twig, ×8, **f** Twig, ×10.
g-h Taxodiaceae (swamp cypress family): **g** Flower, ×12, **h** Flower, ×22.

a
b
c
d
e
f
g
h

2.04 Resin-producing conifers – Spermatophyta: Gymnospermae

The question of the identity of the resin supplier for Baltic amber has yet to be answered conclusively. GOEPPERT (1850, 1853, 1883) proposed several conifer species as candidates based on anatomical/histological studies of fossil wood. His student CONWENTZ made an attempt (1890) to find an answer in his "Monographie der baltischen Bernsteinbäume" (Monograph of Baltic amber trees), by suggesting the probability of the pine tree described by GOEPPERT & BERENDT, *Pinus succinifera* (GOEPPERT & BERENDT, 1845) CONWENTZ, 1890, being responsible for resin production. From a taxonomic standpoint, it is considered to be an aggregate species associated with numerous synonymous descriptions (cf. SPAHR 1993). CONWENTZ assumed one key factor to be the intensity of the resin production, without which extensive quantities of amber would be inexplicable. He used pathological findings to explain the increased resin flow, suggesting that wood-destroying insect infestation, parasitic fungi, lightening stroke, forest fires and storms were responsible for causing the symptoms of a disease called "succinosis".

In his paleohistological studies, SCHUBERT (1961) also firmly supports *Pinus succinifera* (GOEPPERT & BERENDT, 1845) CONWENTZ, 1890 as the resin supplier for amber. This species is characterized by intense cork production, but weak wood production. The late wood consists of just a few cell layers and thickening is minimal. The tracheids have extremely thin membranes. The resin-filled pockets, fissures and channels (Fig. 1), many of unusually large dimensions, indicate increased resin production.

According to SCHUBERT (1961), this pine was the dominant tree of Eocene amber forests: the pine-oak forest at fairly high elevations, the thick pine-sclerophyllous forest along rivers and the savanna forest with mixed pine-palm stands at relatively low, warm elevations. The drastic climate changes (humidity/temperature) resulting from the advancing Eocene ocean (Fig. 10) were associated with physiological changes, which SCHUBERT (1961) claimed to have caused increased resin flow.

In addition to pines, cedars *(Cedrus)* and araucarias *(Agathis)* have also been the subject of more recent discussion as potential resin suppliers for Baltic amber (KATINAS 1971, KRZEMINSKA & KRZEMINSKI 1992, KRUMBIEGEL & KRUMBIEGEL 1994). Their high, natural resin production would be a welcome argument for rejecting the hypothesis of "succinosis" (CONWENTZ 1890) or a metabolic disorder (SCHUBERT 1961). Resins from cedars and araucarias may be involved in the formation of copal and other types of amber (POINAR 1992, GRIMALDI 1996), but not Baltic amber, in which neither cedars nor araucarias have ever been recorded to date (SPAHR 1993 b, PIELINSKA 1997, TURKIN 1997). Furthermore, no agathic acid is found in Baltic amber.

The intensity of resin production in conifers is sure to vary from species to species and very probably corresponds to the climate to which a conifer species is adapted. The resin production of most conifers increases in a species-specific manner towards the equator, from temperate to subtropical and tropical climates. Consequently, what would oppose the assumption that natural and slightly elevated resin production was common in the tropical-subtropical amber forest? If the amber forest stretched from Central Europe all the way to the Ural Mountains and existed for a period of roughly 10 million years during the Eocene Epoch, then the production of resin must have been immeasurably high. Rivers may have flooded, washed away the soil and carried the resin and copal into the sea. Changing ocean currents may have repeatedly transported them to new beds and repositioned them until they eventually came to rest in today's deposits.

Plate 4: Flowering plants: gymnosperms (Spermatophyta: Gymnospermae) in Baltic Amber.

a-b Pinaceae (pine family): Needles **a** of an Abietoideae species, ×6, **b** of a Pinoideae species, ×3.
c-d Pinaceae (pine family): Different magnifications and views of pollen, **c** ×40, **d** ×180.
e-f Pinaceae (pine family): Full and partial view of male cones, **e** ×6, **f** ×18.
g-h Pinaceae (pine family): Male cones in different states of preservation, **g** ×9, **h** ×5.

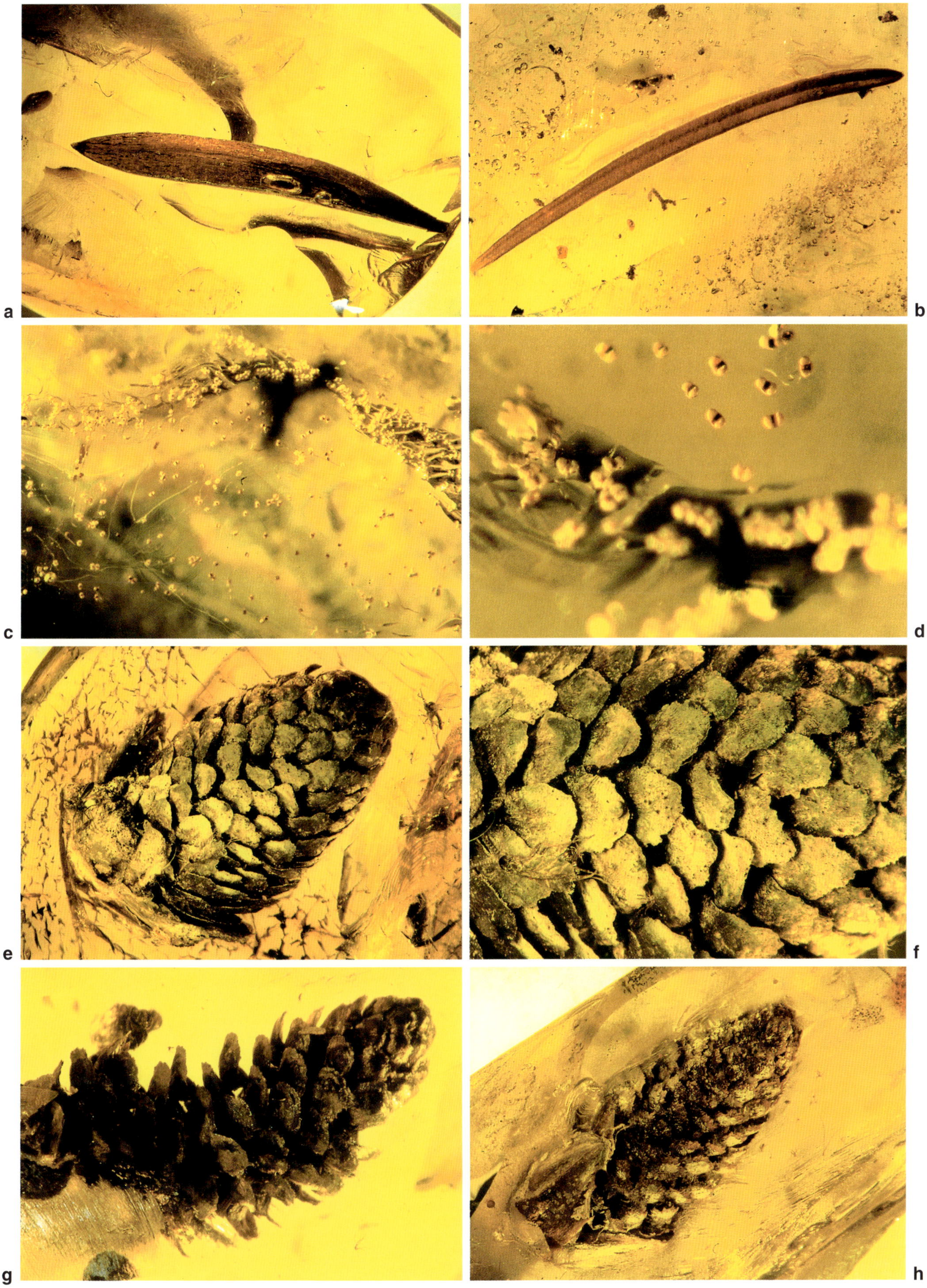

2.05 Flowering plants: angiosperms – Spermatophyta: Angiospermae

Flowering plants are quite numerous in Baltic amber compared to all other botanical inclusions. According to Czeczott (1961), 67 % of the Angiospermae were identified on the basis of flowers, fruits and seeds, and 33 % on the basis of leaves and twigs. They are distributed among 44 families, an overview of which is shown in Chapter 2.06, supplemented by the results of more recent studies (Willemstein 1978, Pielinska 1990). Tropical and subtropical plants are represented by about 10 to 12 families and thus make up approximately 25 %, although purely tropical families (Dipterocarpaceae, Melastomataceae and Musaceae) are apparently not among them. There are several families of deciduous plants from the temperate regions: Aceraceae, Betulaceae, Hamamelidaceae, Juglandaceae, Rosaceae, Saxifragaceae and Umbelliferae. However, nearly half the families cannot be clearly linked to climatic factors. Many display a cosmopolitan distribution.

Of the flowering plants, the monocotyledon palm trees suggest a subtropical to tropical climate. Overall, the inclusions in Baltic amber indicate a diverse flora, which was not uniform from the standpoint of the plant sociology and characterized by a more subtropical than tropical climate. The flora stretched far into the northern latitudes, where the present-day flora is determined by a temperate climate.

In this context, special emphasis must also be given to the dicotyledon family of the Fagaceae, which is represented in amber by the genera *Castanea, Dryophyllum, Fagus, Quercites* and *Quercus* (including *Tricolpopollenites*) (Goeppert 1845) and contains species characteristic of the flora in temperate and subtropical climates. Oak flowers and parts of the inflorescence are known to occur in all deposits of Baltic amber. Widely distributed and more common are the so-called “stellate hairs”, which develop on the flower and leaf buds of the oak (“down”) and are shed in great numbers year after year. They are found in nearly fifty percent of all Baltic amber pieces and considered to be an indication of a specimen’s authenticity. Unfortunately, no analytical studies are available to clarify either the origin of the fossil stellate hairs or their significance for amber.

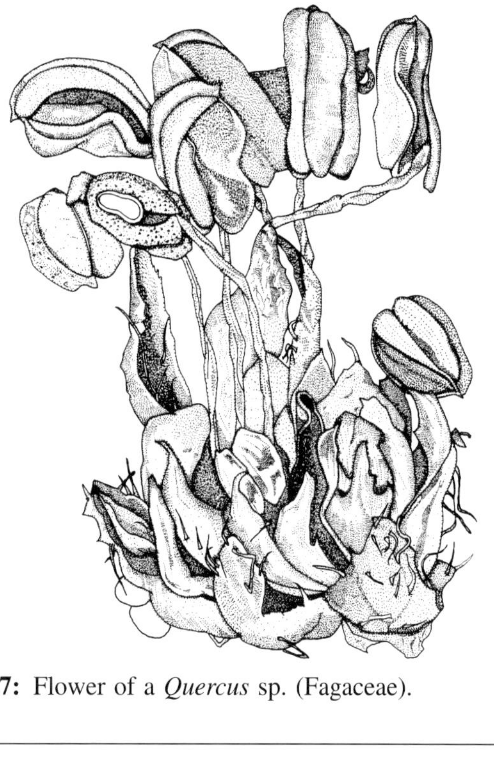

Fig. 27: Flower of a *Quercus* sp. (Fagaceae).

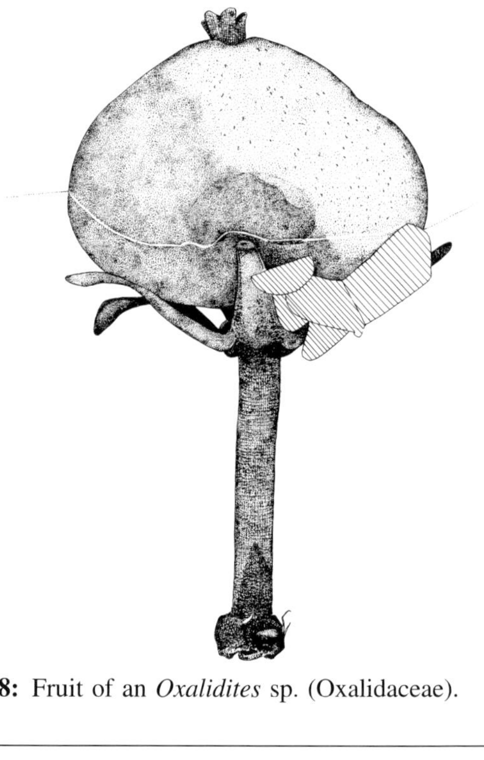

Fig. 28: Fruit of an *Oxalidites* sp. (Oxalidaceae).

Plate 5: Flowering plants: angiosperms (Spermatophyta: Angiospermae) in Baltic amber.

a-b Fagaceae (beech family): *Quercus* sp. (oak), **a** Inflorescence, × 8, **b** Bud, × 9.
c-d Fagaceae (beech family): *Quercus* sp. (oak), two flowers, **c** × 15, **b** × 18.
e-f **e** Dilleniaceae (dillenia family) × 8, **f** Myricaceae (bogmyrtle family): Leaves, × 7.
g-h **g** Unknown family, × 5, **h** Salicaceae (willow family): Leaves, × 22.

a

b

c

d

e

f

g

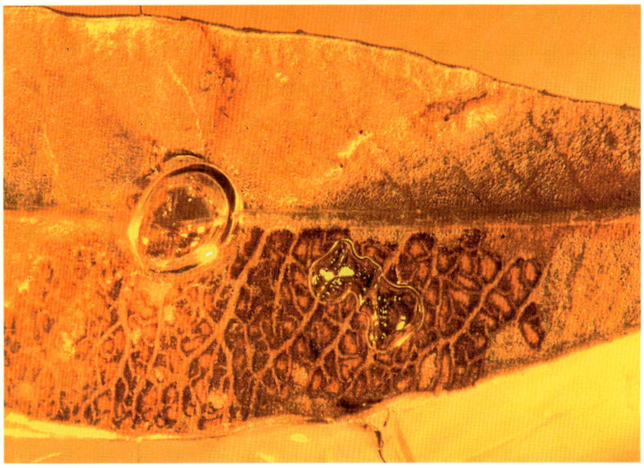
h

2.06 Systematic overview of the flowering plants in amber – Spermatophyta

GYMNOSPERMAE (gymnosperms)
Coniferae (conifers)

- **Cupressaceae** (cypress family)
- **Taxodiaceae** (swamp cypress family)
- **Pinaceae** (pine family)

ANGIOSPERMAE (angiosperms)
Monocotyledonae (monocotyledons)

- **Araceae** (arum family)
- **Commelinaceae** (spiderwort family)
- **Poaceae** (= **Gramineae**) (grass family)
- **Liliaceae** (lily family)
- **Najadaceae** (naiad family)
- **Palmeae** (palm family)

Dicotyledonae (dicotyledons)

- **Aceraceae** (maple family)
- **Apocynaceae** (periwinkle family)
- **Aquifoliaceae** (holly family)
- **Betulaceae** (birch family)
- **Campanulaceae** (bellflower family)
- **Caprifoliaceae** (honeysuckle family)
- **Celastraceae** (spindle tree family)
- **Chenopodiaceae** (goosefoot family)
- **Cistaceae** (rockrose family)
- **Clethraceae** (pepperbush family)
- **Asteraceae** (= **Compositae**) (composite family)
- **Connaraceae** (bean tree family)
- **Brassicaceae** (= **Cruciferae**) (crucifer family)
- **Dilleniaceae** (dillenia family)
- **Droseraceae** (sundew family)
- **Ericaceae** (heath family)
- **Euphorbiaceae** (spurge family)
- **Fagaceae** (beech family)
- **Geraniaceae** (geranium family)
- **Hamamelidaceae** (witch hazel family)
- **Hippocastanaceae** (horse chestnut family)
- **Lauraceae** (laurel family)
- **Loranthaceae** (mistletoe family)
- **Magnoliaceae** (magnolia family)
- **Myricaceae** (bogmyrtle family)
- **Myrsinaceae** (myrsine family)
- **Olacaceae** (tallow wood family)
- **Oleaceae** (olive family)
- **Oxalidaceae** (woodsorrel family)
- **Papilionaceae** (bean family)
- **Pentaphylaceae**
- **Pittosporaceae** (pittosporum family)
- **Polygonaceae** (buckwheat family)
- **Proteaceae** (silky oak family)
- **Pyrolaceae** (wintergreen family)
- **Rhamnaceae** (buckthorn family)
- **Rosaceae** (rose family)
- **Rubiaceae** (bedstraw family)
- **Salicaceae** (willow family)
- **Santalaceae** (sandalwood family)
- **Saxifragaceae** (saxifrage family)
- **Scrophulariaceae** (figwort family)
- **Theaceae** (tea plant family)
- **Thymelaeaceae** (daphne family)
- **Tiliaceae** (lime family)
- **Ulmaceae** (elm family)
- **Apiaceae** (= **Umbelliferae**) (parsley family)
- **Urticaceae** (nettle family)
- **Vitaceae** (grapevine family)

This overview of the flowering plant families (Spermatophyta) reported for Baltic amber is based on CZECZOTT (1961), WILLEMSTEIN (1978), PIELINSKA (1990) and SPAHR (1993b). Although it has been critically assessed, it is still of a preliminary nature. Several families have only been verified by the identification of pollen (WILLEMSTEIN 1978, WETZEL 1953): Chenopodiaceae, Asteraceae, Brassicaceae, Hippocastanaceae, Tiliaceae and Urticaceae. PIELINSKA (1990) reported the occurrence of the families Najadaceae, Droseraceae and Vitaceae without providing any further evidence. A revision of the flowering plants will give a detailed picture of the amber flora.

Plate 6: Flowering plants: angiosperms (Spermatophyta: Angiospermae) in Baltic amber.

a Rosaceae (rose family): Flower, × 12.
b Hamamelidaceae (witch hazel family): Flower, × 15.
c-d Clethraceae (pepperbush family): **c** Flower, × 8, **d** Bud, × 11.
e-f Theaceae (tea plant family): Flowers, **e** × 10, **f** × 8.
g-h Oxalidaceae (woodsorrel family): Fruits, × 12.

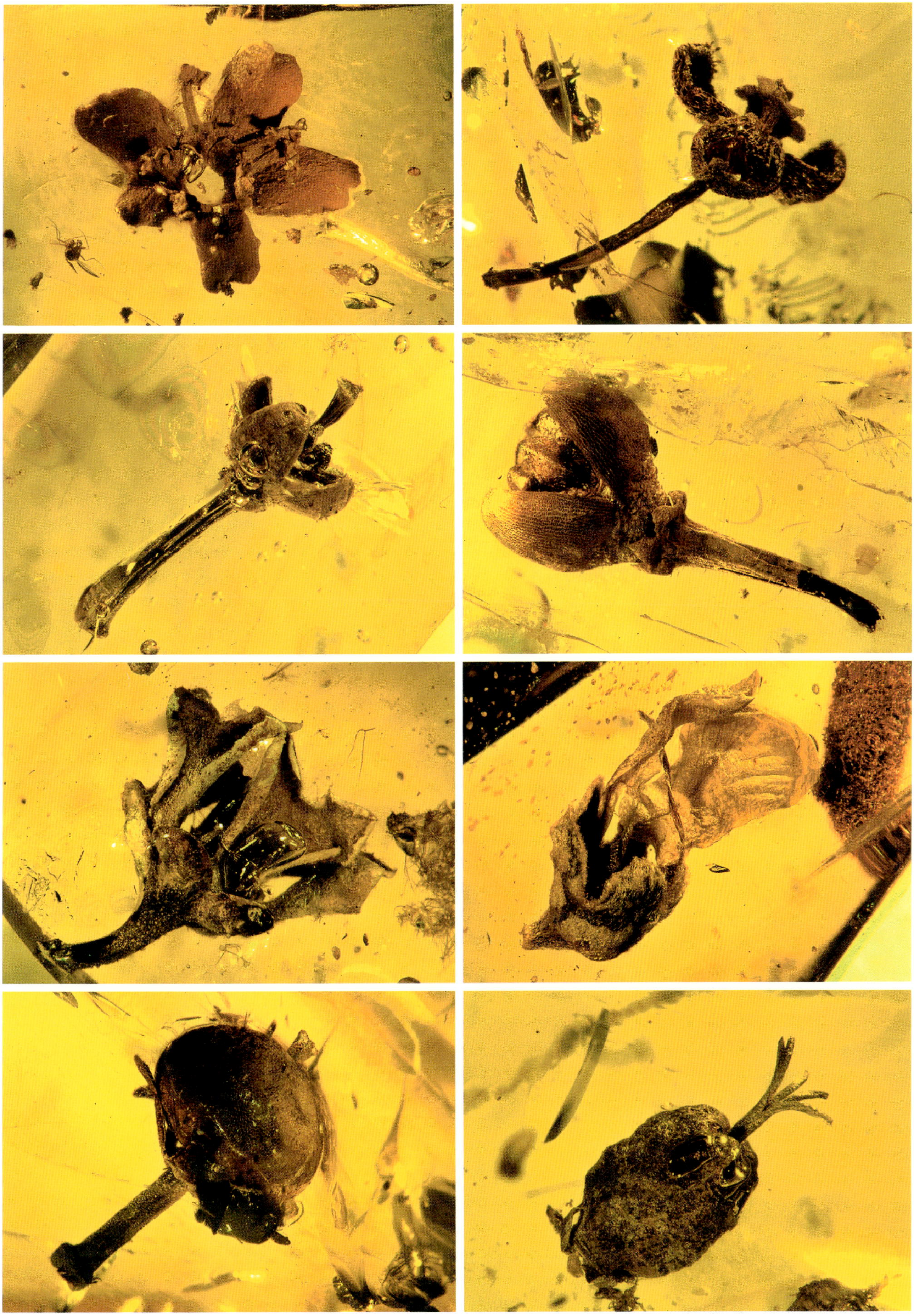

a b c d e f g h

2.07 Worms – Nematoda, Annelida

The body shape of worm-like animals is an indication of their adaptation to a subterranean or parasitic lifestyle. The worms that have been found to date in Baltic amber include the Nematoda (roundworms) of the phylum Nemathelminthes and the Oligochaeta (oligochaete worms) of the phylum Annelida.

Most nematodes are just a few millimeters long and live in the soil. The great number of species and individuals makes roundworms the largest group of soil animals. Despite this fact, soil-dwelling nematodes are rare in amber, because they live euedaphically, deep in the root area of plants and in the pore system of the soil. Most of the few nematodes found in amber are parasitic Mermithidae that infest insects. Poinar (1984), Schlee & Glöckner (1978) and Schlee (1990) all mention Mermithidae on midges, whose larvae live in water (Plate 7). Many recent Mermithidae are aquatic and lay egg masses in the muddy bottom of a body of water or on aquatic plants. After hatching, the larvae search for aquatic insects, bore into the insect's body and feed as endoparasites in the second larval stage. The hosts are often midge larvae (Chironomidae). A number of worms are carried phoretically to new biotopes by adult midges and, 40 to 50 million years ago, some apparently ended up being brought by chance into Baltic amber. Due to the difficulty of identification, the fossil species described by Duisburg (1860) and Menge (1866, 1872) are assigned collective generic names for fossil nematodes: *Heydenius* (Mermithidae) and *Oligoplectus* (Plectidae), as well as *Vetus* (fam. incert. sed.) for free-living nematodes that cannot be assigned to any family (cf. Spahr 1993 a).

The soil-dwelling worms also include the Oligochaeta of the family Enchytraeidae. They live in the leaf litter and also deep underground as substrate-feeders. Like earthworms, they come above ground as soon as living conditions in the soil deteriorate due to the weather. Predaceous flies (Dolichopodidae) take advantage of this opportunity to grab small Enchytraeidae with their proboscis, pull them out of the ground and then take the prey to a safe place, usually leaves. A worm can be torn in two when it is grabbed and extracted from a potentially fixed position in the ground.

Ulrich & Schmelz (2001) applied this observation of recent Dolichopodidae and Enchytraeidae to a scenario in Baltic amber. A small specimen contains both a 2.4 mm-long dolichopodid (*Gheynia bifurcata*) and a fragment of an enchytraeid worm (Plate 7 g, h). The worm is slightly arched and only its caudal end has been preserved. The head and front section are missing. Body fluids may have flowed out at the point of rupture and formed the visible halo around the wound. The fly and the worm are located in the same resin flow of a slightly rounded drop of amber, which is covered by a second resin flow that encases the two animals together. All signs indicate that the worm fragment got caught in the amber as prey.

This taphocoenosis confirms that Dolichopodidae already preyed on Enchytraeidae in the Eocene amber forest. It illustrates how soil-dwelling Oligochaeta may have gotten captured in amber. This scenario provides a plausible explanation for the rareness of Enchytraeidae which, as soil-dwellers, do not have an affinity for resin.

Plate 7: Worms (Nematoda, Annelida) in Baltic amber.

a, b Crane fly (Limoniinae) with a swarm of small roundworms (Nematoda?), **a** ×12, **b** ×25.
c Sciarid fly (Sciaridae) with roundworms (Nematoda, Mermithidae?), ×22.
d Free-living (?) roundworm (Nematoda), ×22.
e Midge (Chironomidae) with roundworm (Nematoda, Mermithidae?), ×20.
f Free-living (?) roundworm (Nematoda), ×25.
g Long-legged fly (Dolichopodidae): *Gheynia bifurcata* with Enchytraeidae, ×10.
h Oligochaete worm (Oligochaeta, Enchytraeidae): Fragment, front section severed, ×30.

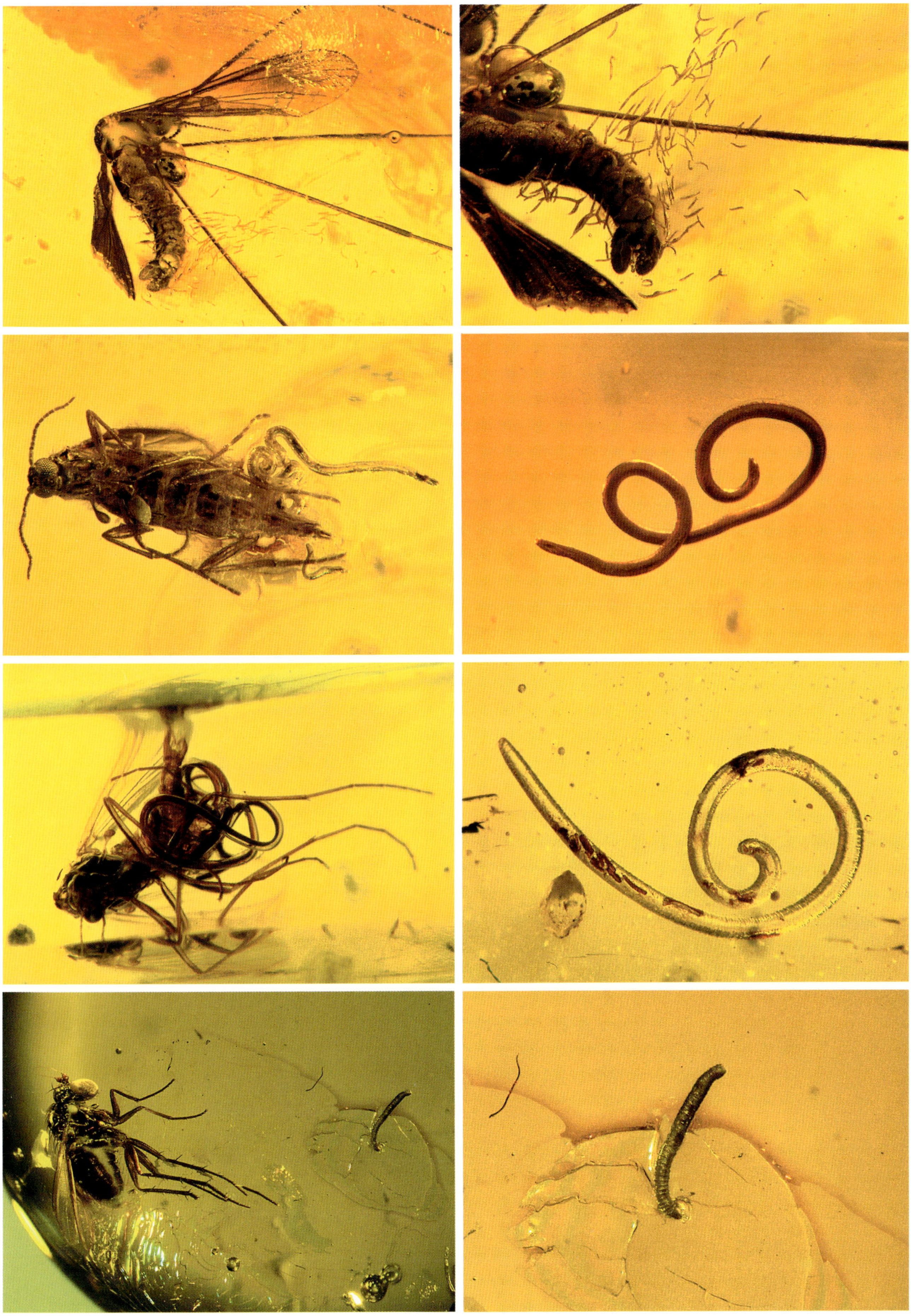
a
b
c
d
e
f
g
h

2.08 Snail shells in Baltic amber – Mollusca: Gastropoda

Of the mollusks (Mollusca), only snails (Gastropoda) with a shell have been reported from Baltic amber. KLEBS (1886 a, b), SANDBERGER (1887) and BACHOFEN-ECHT (1949) studied snails in amber. More recent works are not available, as only a few individual discoveries have been made in the past few years. Overviews have been elaborated by LARSSON (1978) and SPAHR (1993 a).

The terrestrial snails that have been reported are lung snails. According to SPAHR (1993a), the following species of Pulmonata have been described:

Clausiliidae
Balea antiqua KLEBS, 1886
Helicidae
Helix sp. BACHOFEN-ECHT, 1949
Parmacellidae
Parmacella succini KLEBS, 1886
Pupillidae
Vertigo hauchecornei KLEBS, 1886
Vertigo künowii KLEBS, 1886
Zonitidae
Hyalina alveolus SANDBERGER, 1887
Hyalina gedanensis (KLEBS, 1886)
Hyalina sp. KLEBS, 1886
Microcytis kaliellaformis KLEBS, 1886

The amber snails appear to be related to snails of the South and East Asian fauna, as well as to North American species. *Helix* is said to be closely related to African species (BACHOFEN-ECHT 1949). The tropical-subtropical species appear to dominate overall.

In addition to the pulmonate snails cited, KLEBS (1886 a, b) also described another species, *Electra kowalewskii* KLEBS, 1886, which he assigned to the terrestrial family (Cyclostomatidae) of prosobranch mollusks (Prosobranchia).

The snail shown in Plate 8 d appears to be the only known specimen in amber to extend its soft body far out of the shell. If this is not merely an artifact, the snail appears to be creeping on its foot over a resinous floor. No slime trails are evident. At the front, the right tentacle can be seen on the protruding head region. The snail did not become fully embedded until it was covered by a second gush of resin, which must have been of very low viscosity. While the tentacles normally retract at the slightest disturbance, the right tentacle appears to be extended in this case and there are no identifiable signs of retraction.

Plate 8: Snails (Mollusca: Gastropoda) in Baltic amber.

a-d Snail shells (Gastropoda: Pulmonata), **a** × 16, **b** × 15, **c** x 12, **d** × 15.

a

b

c

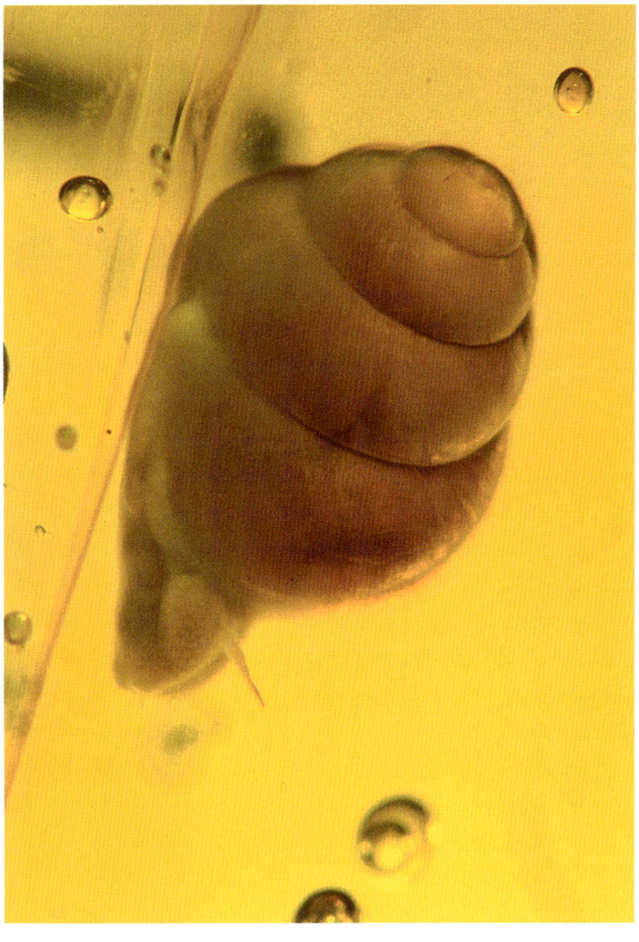
d

2.09 Scorpions in Baltic amber – Arachnida: Scorpiones

The arachnids (Arachnida) are represented in Baltic amber by scorpions (Scorpiones), pseudoscorpions (Pseudoscorpiones), harvestmen (Opiliones), mites (Acari) and, in particular, spiders (Araneae). The Arachnida have common features that distinguish them from other arthropods: the body is divided into the anterior prosoma and the posterior opistosoma. Arachnids do not have antennae. The prosoma is equipped with six pairs of appendages. The first pair is the chelicerae, which have evolved into mouthparts that engage in pincer-like fashion. The chelicerae are bodkin-shaped in some mites and have evolved into retractable poisonous claws in several harvestmen. The second pair comprises the powerful (in the Scorpiones and Pseudoscorpiones), frontal pedipalps, which are also used for feeding. Finally, arachnids have four pairs of walking legs with seven segments each.

Unfortunately, all that remains of the first specimen described by Holl (1829) as *Scorpio schweiggeri* is a rather rough sketch and the description itself, as the inclusion was already considered to have been lost in the 19th century. The description only allows the systematic classification of the species in the family Buthidae.

Menge (1869) described a new species as *Tityus eogenus* on the basis of two other specimens from Baltic amber. In contrast to *Scorpio schweiggeri,* which the field of science hardly took note of at the time, this discovery attracted a great deal of attention, particularly due to its taxonomic classification in a recent, Neotropical genus. However, as both of these specimens also disappeared shortly after being described, the question of their exact taxonomic classification remains unanswered. The relatively imprecise diagram and description only permit us to conclude that *Tityus eogenus* probably belongs to the family Buthidae.

In 1994, a fourth scorpion was found that is also from the family Buthidae (Lourenço & Weitschat 1996). Most of the features indicate that this new find is closely related to the present-day genus *Lychas*. Nevertheless, several primitive features made it necessary to a make a distinction and establish a new genus. The specimen, which was redescribed as *Palaeolychas balticus*, is a subadult male of only 1.3 cm in length that can be assigned to the group of "microscorpions". They are considered to be "primitive forms" and include various, extremely small species with a simply structured keel and body. This group includes the recent genera *Tityobuthus* Pocock and *Microcharmus* Lourenço, which are today endemic to Madagascar, as well as representatives of the genus *Charmus* Karsch, which are native to Sri Lanka and Thailand.

Representatives of the closely related and widely distributed genus *Lychas* occur in parts of Africa, on the Seychelles and Mauritius, in the Indo-Malayan region, the southern Palearctic regions (China, Nepal) and northern Australia. This new discovery shows that the scorpion fauna of Baltic amber was more closely related to the present-day fauna of the Old World. This conclusion is further supported by *Palaeolychas balticus*, another subtropical-tropical species of Baltic amber whose closest modern relatives are indigenous to the tropical forests of Southeast Asia and Australia.

Three additional discoveries of scorpion inclusions also belong to the family Buthidae and have been assigned to new genera and species: *Palaeotityobuthus longiaculeus*, *Palaeoprotobuthus pusillus* and *Palaeoakentrobuthus knodeli*. These new finds expand our knowledge of fossil scorpion fauna and indicate the remarkable biodiversity of scorpions in the Paleogene Baltic amber forest (Lourenco & Weitschat 2000).

Plate 9: Scorpions (Arachnida: Scorpiones) in Baltic amber.

Amber scorpion *Palaeolychas balticus* Lourenço & Weitschat, 1996 (Buthidae), holotype, × 12.

2.10 Pseudoscorpions – Arachnida: Pseudoscorpiones

Pseudoscorpions were given the vernacular name "book scorpions" towards the end of the Middle Ages, when scholars derived their knowledge of natural history from books rather than Nature. They could directly observe pseudoscorpions preying on booklice in books. The flat, short body with its two, strong pedipalp pincers attracted attention and was reminiscent of scorpions. In ignorance of their original habitat, the scholars called these animals "book scorpions". Their size of just 1 to 7 mm and lack of a sting on the tail are distinct features that make it necessary to classify these arachnids not with the Scorpiones, but in their own taxonomic group: Pseudoscorpiones.

Due to their epedaphic habits, pseudoscorpions are perhaps most accurately referred to as "moss scorpions". They live in stone fissures and moss, under tree bark and often in the surface litter of humus-rich forest floors. Their flat bodies are adapted to this habitat. The lateral position of the four pairs of legs and the far, forward extension of the pedipalps emphasize the dorsoventral flattening of the body. Pseudoscorpions are predaceous animals that hunt the surface litter for Collembola, mites and other small, soil-dwelling animals. Their biology and ecology have been comprehensively discussed by BEIER (1937, 1947, 1963), VACHON (1949), WEYGOLDT (1966, 1969) and LEGG & JONES (1988).

Based on the taxonomic work of BEIER (1963), the Pseudoscorpiones are divided into three suborders, all of which are represented in Baltic amber (Plate 10):

1. Chthoniinea, e.g. *Chthonius mengei*
2. Neobisiinea, e.g. *Neobisium extinctum*
3. Cheliferinea, e.g. *Cheiridium hartmanni, Electrochelifer balticus.*

To date, 25 fossil species of Pseudoscorpiones in Baltic amber have been described by KOCH & BERENDT (1854), MENGE (1855) and BEIER (1937, 1947, 1955). Most of them are related to contemporary species of Central Europe and are either bark-dwellers (Cheiridiidae, Chernetidae, Cheliferidae) or soil-dwellers (Chthoniidae, Neobisiidae). Other species in amber are related to present-day species of the Nearctic region (Pseudogarypidae: *Pseudogarypus extensus* BEIER, 1937, *P. hemprichii* (KOCH & BERENDT, 1854), *P. minor* BEIER, 1947) and the tropical to subtropical zones (Atemnidae: *Progonatemnus succineus* BEIER, 1955).

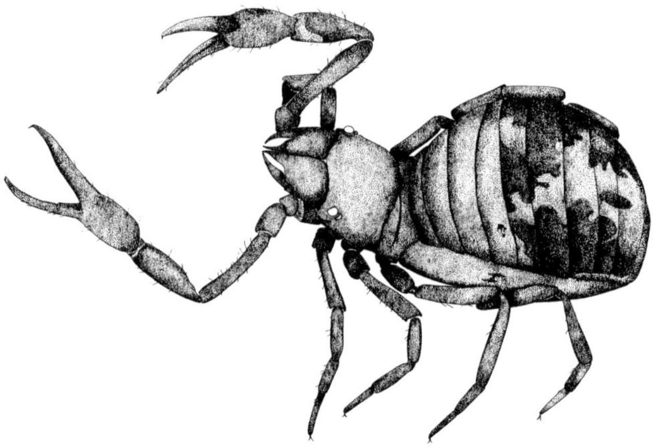

Fig. 29: *Neobisium extinctum* BEIER, 1955 (Neobisiidae), holotype.

Fig. 30: *Progonatemnus succineus* BEIER, 1955 (Atemnidae), holotype.

Plate 10: Pseudoscorpions (Arachnida: Pseudoscorpiones) in Baltic amber.

a Cheliferidae, × 10.
b Cheliferidae: Head with pedipalps, × 15.
c Neobisiidae: *Neobisium extinctum*, holotype, × 12.
d Garypidae, × 12.
e Atemnidae: *Progonatemnus succineus*, holotype, × 15.
f Cheliferidae: *Electrochelifer balticus*, holotype, × 12.
g Cheliferidae, × 10.
h Cheiridiidae: *Cheiridium hartmanni*, × 12.

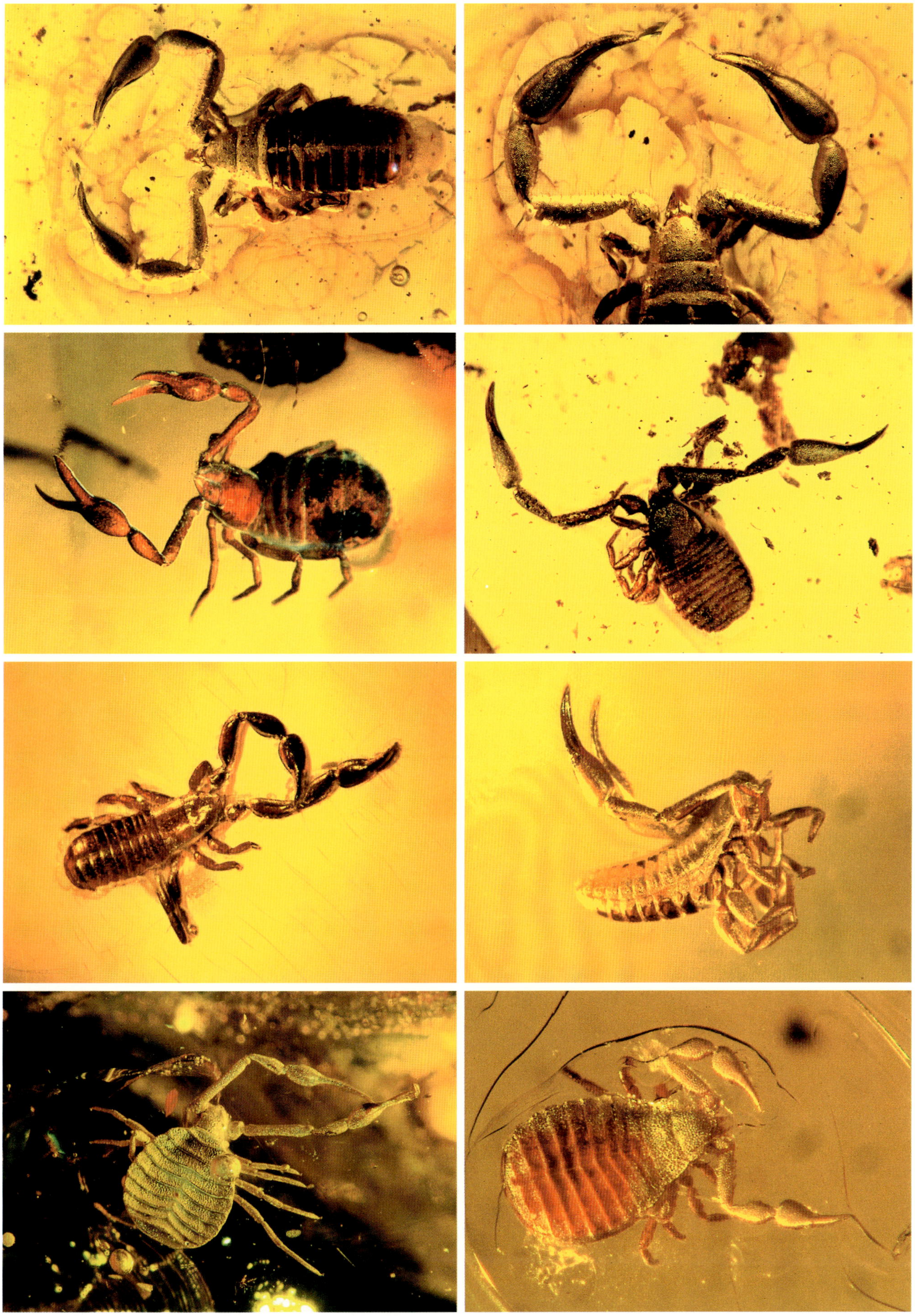

a b c d e f g h

2.11 Phoresy in pseudoscorpions – Arachnida: Pseudoscorpiones

The term phoresy refers to a mode of behavior in which small animals are transported by larger carrier-animals over long distances or extended periods of time to new habitats. The interacting animals come into contact when they both colonize the same habitat at least temporarily. Phoresy is a means for species to spread and propagate. This behavior is common among arthropods and also observable in Baltic amber among soil mites (Acari), particularly moss mites (Oribatida) (BEIER 1937, 1948, MUCHMORE 1971, VACHON 1949).

The transport hosts in Baltic amber, which phoretically transport pseudoscorpions as airborne passengers, include midges (Diptera: Tipulidae: Limoniinae), hymenopterans (Hymenoptera: Braconidae, Ichneumonidae) and caddisflies (Trichoptera) (MENGE 1855, BEIER 1937, 1948, SCHLEE & GLÖCKNER 1978, SCHAWALLER 1978):

1. Chernetidae (indet.) on Tipulidae: Limoniinae
2. *Oligochernes bachofeni* on Braconidae (Plate 11 a)
3. *Pycnochelifer kleemanni* on Braconidae
4. *Oligochelifer berendtii* on Ichneumonidae and Trichoptera.

The pedipalp pincers are effective grasping organs used by pseudoscorpions to grip the carrier-animals, preferably on the legs. If two pseudoscorpions take advantage of the same midge as a carrier-animal – something which occurs only rarely, but also in amber – they inevitably grasp opposing legs of the carrier-animal and thus unintentionally ensure the balance of the flying insect.

Long-legged harvestmen (Opiliones) are preferred carrier animals for pseudoscorpions (Fig. 31, Plate 11 b, c). The predaceous harvestmen search for food in the surface litter and encounter epedaphic pseudoscorpions in the process. Their passengers, who hold onto the thin legs with their pedipalps, are transported from one place to another in long, fast strides. The fact that both partners inhabit the surface litter or the bark of tree trunks ensures that pseudoscorpions will always have access to these habitats.

This phoretic relationship still helps pseudoscorpions spread today, just as it did back in the Eocene Epoch – even though the transfer sometimes came to an unexpected end as a fossil in amber (Plate 11).

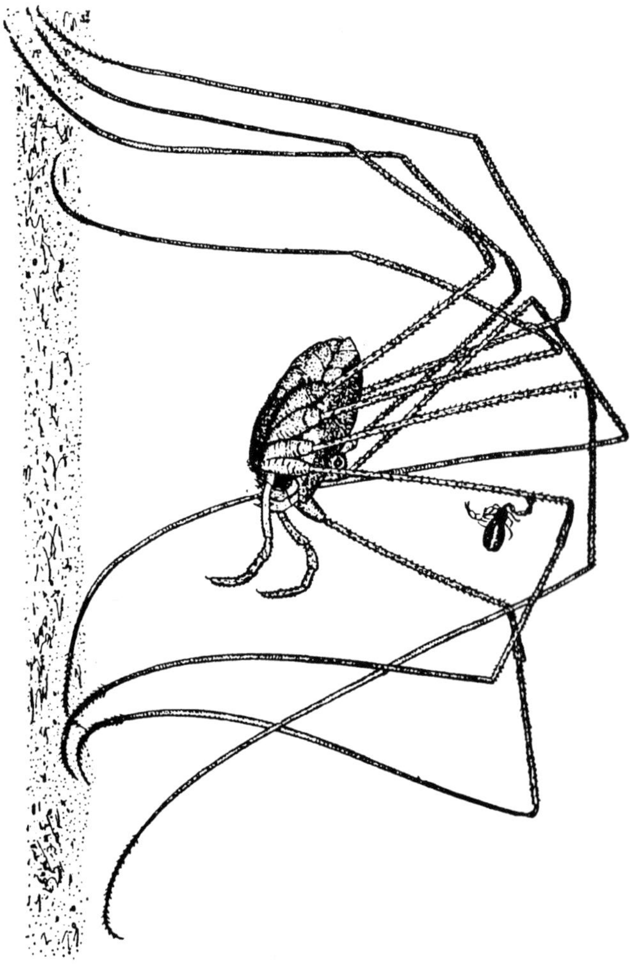

Fig. 31: Pseudoscorpion on the leg of a harvestman (Opiliones) (after CLOUDSLEY-THOMPSON 1988).

Plate 11: Phoresy among pseudoscorpions (Arachnida: Pseudoscorpiones) in Baltic amber.

a Pseudoscorpion *Oligochernes bachofeni* BEIER, 1937, holotype, on a braconid wasp (Braconidae), × 8.
b Pseudoscorpion clinging to the leg of a harvestman (Opiliones) (+ ground beetle), × 7.
c Pseudoscorpion with the left pedipalp on the tibia of a harvestman (magnification), × 25.

a

b

c

2.12 Harvestmen – Arachnida: Opiliones

Harvestmen or daddy-longlegs generally prefer the soil as a habitat (MARTENS 1978, EISENBEIS & WICHARD 1985). They move about in the upper soil layer of shady, damp forests, primarily in the leaf litter and under wood and stones. These predators capture prey using their chelicerae (mouthparts), pedipalps and front pair of walking legs. Their prey includes small soil animals, particularly snails and arthropods. Some species switch during the day to herbs, shrubs, bushes and trees. Only rarely do they venture out into open terrain. PFEIFER (1956) described the habits of harvestmen based on their distribution on the ground and in the herbaceous, shrub and tree layers. The Opiliones recorded in Baltic amber have yet to be subjected to corresponding paleobiological classification.

Despite their adapted and somewhat reclusive lifestyle, harvestmen are unusual animals that can easily be distinguished from all other arachnids not only as living specimens, but also as fossils in Baltic amber. Typical features of many Opiliones include the long legs they use to nimbly carry their short bodies over the ground. The prosoma and opistosoma are fused into a rounded body. The laterally oriented eyes are positioned centrally on a dorsal tubercle.

KOCH & BERENDT (1854) and MENGE (1854) in BERENDT, G. C. (1845-1856) described the harvestmen in Baltic amber. Only ROEWER (1939) has studied the Opiliones since then, even though new finds and interesting forms have come to light (Plate 12), which support far-reaching conclusions about the Eocene past of the Opiliones when compared to recent animals.

LARSSON (1978) points out that numerous loose legs from harvestmen are encased in amber. Many of them apparently managed to escape, leaving just a leg or two stuck to the resin.

Caddidae (Caddonidae)
Caddo dentipalpus (KOCH & BERENDT, 1854)
Gaggrellidae
Liobunum inclusum ROEWER, 1939
Liobunum longipes MENGE, 1854
Gonyleptidae
Gonyleptes nemastomoides KOCH & BERENDT, 1854
Nemastomatidae
Mitostoma denticulatum (KOCH & BERENDT, 1854)
Nemastoma clavigerum MENGE, 1854
Nemastoma incertum KOCH & BERENDT, 1854
Nemastoma succineum ROEWER, 1939
Nemastoma tuberculatum KOCH & BERENDT, 1854
Phalangiidae (Phalangodidae, Gyantidae)
Cheiromachus coriaceus MENGE, 1854
Dicranopalpus palmnickensis ROEWER, 1939
Dicranopalpus ramiger (KOCH & BERENDT, 1854)
Opilio corniger MENGE, 1854
Opilio ovalis KOCH & BERENDT, 1854
Opilio ramiger KOCH & BERENDT, 1854
?*Scotolemon nemastomoides* (KOCH & BERENDT, 1854)
Sabaconidae
Sabacon bachofeni ROEWER, 1939

Plate 12: Harvestmen (Arachnida: Opiliones) in Baltic amber.

a Phalangiidae: *Dicranopalpus* sp., ×4.
b Phalangiidae: *Dicranopalpus* sp., ×12.
c Phalangiidae: *Megabunus* (?) sp., ×8.
d Phalangiidae: Sclerosomatinae, ×10.
e Nemastomatidae, ×5.
f Nemastomatidae, ×10.
g Caddidae: *Caddo* sp., ×8.
h Sabaconidae: *Sabacon* sp., ×10.

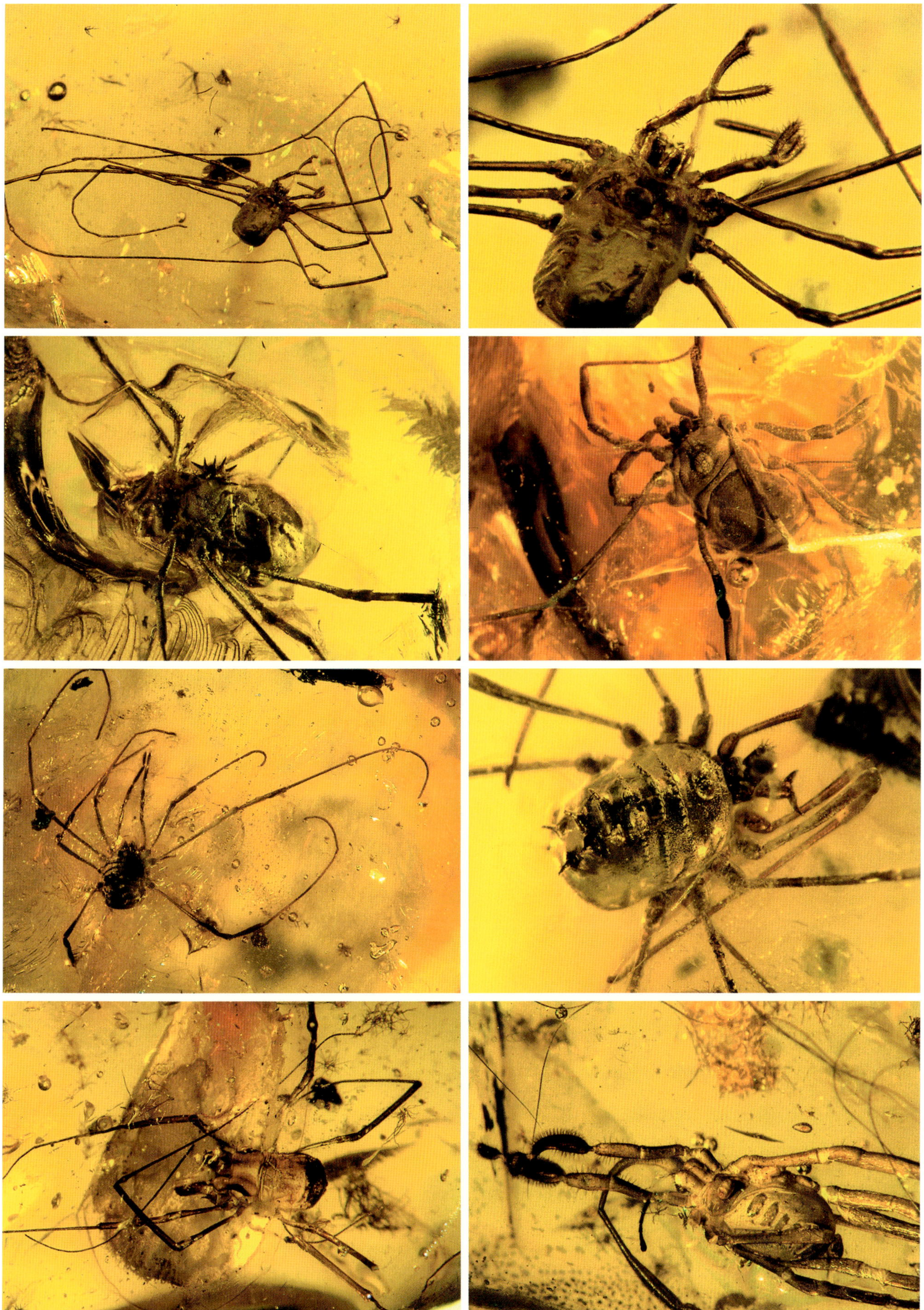
a
b
c
d
e
f
g
h

2.13 Mites – Arachnida: Acari

Mites (Acari = Acarina) are the most successful arachnids in terms of the number of species and their worldwide distribution. They populate marine and freshwater environments, parasitize plants and animals and inhabit the soil on all continents in a density that is only exceeded among the soil fauna by protozoa and roundworms. The diversity of mites in the Eocene Epoch was probably no different than today, as evidenced by the numerous mite inclusions found in many pieces of amber. Mite inclusions are often very small and easily overlooked, as the larger inclusions usually attract the most attention.

The order Acari comprises three suborders, Acariformes, Opilioacariformes and Parasitiformes (LINDQUIST 1984, EVANS 1992), and includes mites as well as ticks. The usually oval body is often thin, but always sclerotized if above-ground evaporation makes it necessary to minimize water loss. The sclerotized mites in the suborder Acariformes are the Oribatida (wood or moss mites), while those in the suborder Parasitiformes include the Ixodida (ticks) and Holothyrida, which live on moss and fern plants in the tropics. Although mites have four pairs of legs, they frequently walk on three pairs, using the first pair for tactile functions (Plate 13 e, g, h).

Ticks (Plate 13 a) were first reported in Baltic amber by WEIDNER (1964) in a description of the species *Ixodes succineus* (Plate 13 a). According to WEIDNER, this fossil species is closely related to the recent *Ixodes ricinus*, differing only slightly in terms of the shape of its coxal spines. WEIDNER (1964) suspects that, like recent species, the Eocene form fed on the blood of lizards and small mammals.

KOCH & BERENDT (1854) described 16 mite species. Another three were added by KARSCH (1884). The wood or moss mites from Baltic amber (Plate 13 c) were analyzed by SELLNICK (1919, 1927, 1931) and compiled into a list with 71 Oribatidae species. "Of the Oribatidae in amber that resemble present-day forms, many are so similar that you could easily mistake them for current forms, if you were unaware you were dealing with amber inclusions". SELLNICK (1931) thus described several amber moss mites as fossil forms of extant species. In his opinion, the evolution of the Oribatidae has been stagnant for the last 40 to 50 million years.

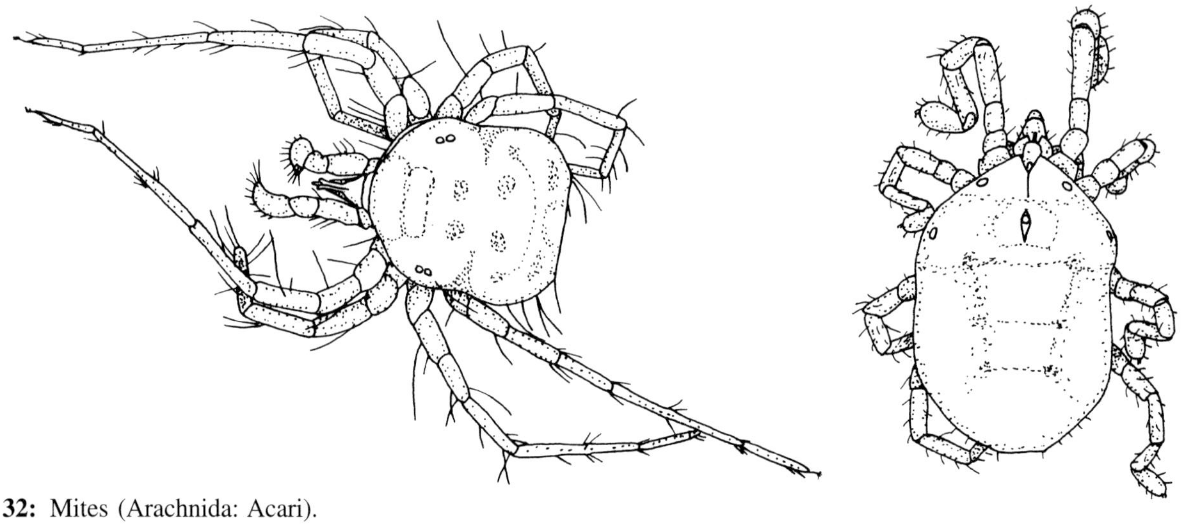

Fig. 32: Mites (Arachnida: Acari).

Plate 13: Ticks and mites (Arachnida: Acari) in Baltic amber.

- **a** Ixodidae (tick): *Ixodes* sp., × 25.
- **b** Bdellidae (snout mites), × 30.
- **c** Oribatidae (moss mites): *Liodes* sp., × 50.
- **d** Caeculidae (mites), × 30.
- **e** Parasitengona (mites), × 15.
- **f** Erythraeidae (mites): *Eatoniana* sp., × 16.
- **g** Parasitengona (mites), × 18.
- **h** Parasitengona (mites), × 18.

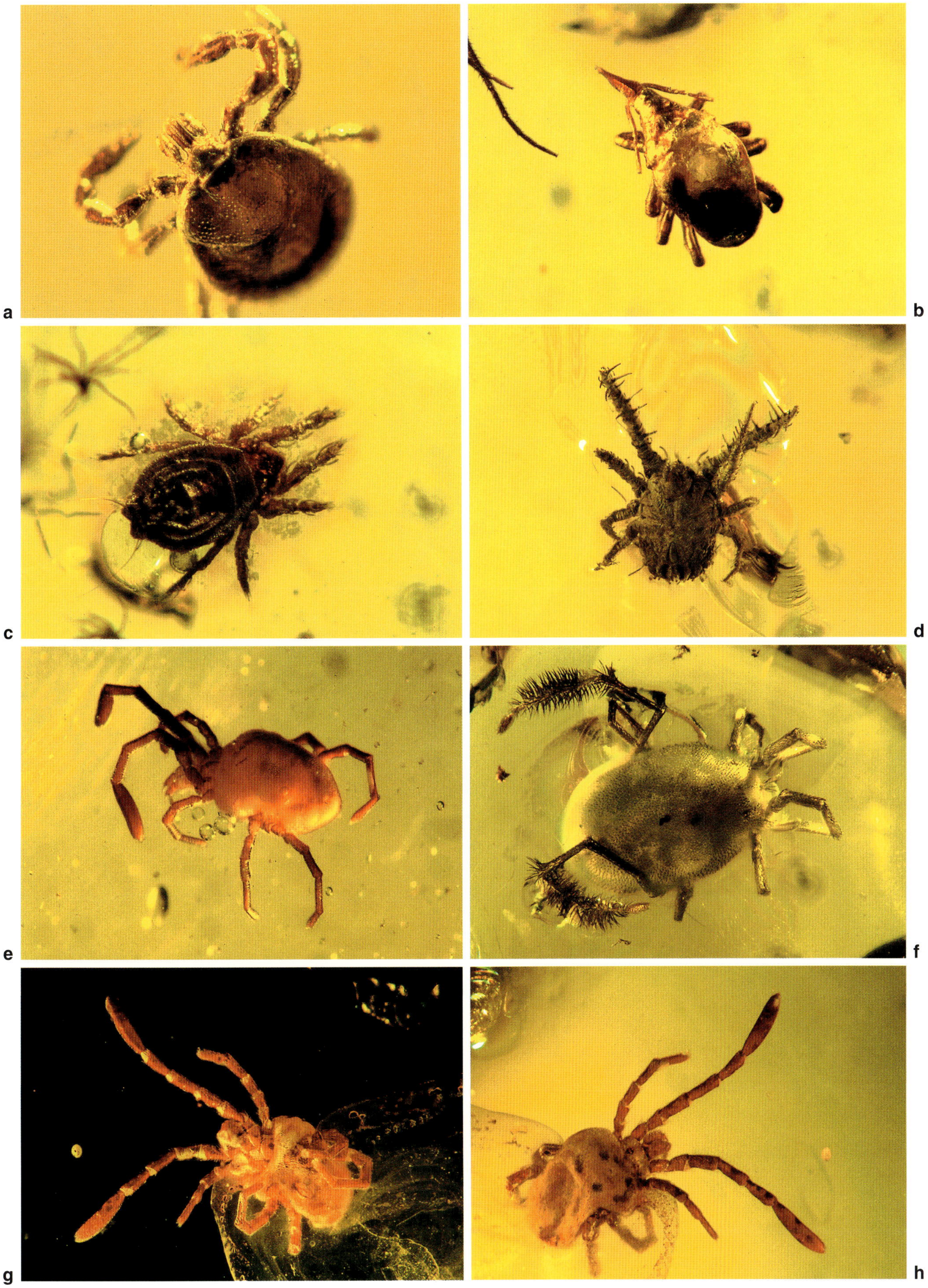
a
b
c
d
e
f
g
h

2.14 Parasitism and phoresy in mites – Arachnida: Acari

The characteristic habits of mites include parasitism and phoresy (KRANTZ 1978). The hosts or carrier animals for the parasitic or phoretic lifestyle with soil mites originate from the epedaphic soil fauna and are usually arthropods, such as midges, flies, spiders, harvestmen or woodlice. In the case of water mites (Hydrachnellae), the carriers are merolimnetic (amphibian) insects that live in water in the developmental stages and occasionally on land as imagines. Unequivocal indications of whether a case involves phoresy or parasitism are often lacking in amber. This question can only be answered very cautiously by way of comparison with recent examples. Phoresy (cf. Chap. 2.11) helps mites spread, in that migratory nymphs attach themselves to carrier animals and are carried away. All migratory nymphs have reduced mouthparts and none of them feed. Consequently, such forms cannot be taken into consideration as parasites (EISENBEIS & WICHARD 1987).

The mite eggs hatch into larvae, which are generally followed by three additional larval stages: protonymphs, deutonymphs and tritonymphs. These stages are not, however, fully developed in all taxonomic groups. The deutonymphs are the most important transport stage for mites and are referred to as migratory nymphs. They are phoretic and use special adhesive organs to cling to carrier animals. Deutonymphs in the family of tortoise mites, Uropodidae (Parasitiformes, Mesostigmata, Uropodina), secrete a substance from the anus that hardens to form a rigid stalk they use to attach to the carrier animal. Deutonymphs in the Anoetidae family (Acariformes, Sacroptiformes, Astigmata) use a sucker to attach to smooth surfaces. Both structures, whether stalked or suctioned, allow enough room for numerous migratory nymphs, more than 50 of which have been found on a checkered beetle (Cleridae) in Baltic amber (Plate 14 a, b).

Examples of parasitism in mites are provided by the family Erythraeidae (Acariformes, Trombidiformes, Prostigmata), which is represented in Baltic amber by the genus *Leptus*. The extant species of this genus prefer sclerotized areas on the hosts and apparently avoid the intersegmental membranes. The fossil nymphs sit in the pronotum region and penetrate the cuticula with their chelicerae. The hosts are midges (e.g. Mycetophilidae, Plate 14 c, and Limoniinae, Plate 14 d) and flies (e.g. Dolichopodidae, Plate 14 e). Present-day *Leptus* species have a broad host spectrum that encompasses the range of insects and arachnids living on the ground surface. The nymphs attach themselves by means of a sticky secretion from the gnathosoma and use their bodkin-like chelicerae to penetrate the cuticula. They remain in this position for five to eight days, sucking on the hemolymph of the host until their weight increases by a factor of 15 to 25, after which they let themselves drop and continue their development.

Plate 14: Parasitism and phoresy among mites (Arachnida: Acari) in Baltic amber.

a-b Uropodidae: Tortoise mites on a checkered beetle (Coleoptera, Cleridae), **a** × 8, **b** × 20.
c Erythraeidae: parasitic mite larva *Leptus* sp. on a fungus gnat (Diptera, Mycetophilidae), × 15.
d Erythraeidae: parasitic mite larva *Leptus* sp. on a crane fly (Diptera, Limoniinae), × 22.
e Erythraeidae: parasitic mite larva *Leptus* sp. on a long-legged fly (Diptera, Dolichopodidae), × 90.

2.15 Spiders in Baltic amber – Arachnida: Araneae

Spiders (Araneae) are just as common in Baltic amber as ants, beetles, midges and flies. Their frequency, which is estimated to be at least 8 % of all inclusions (KRZEMINSKA & KRZEMINSKI 1992), is coupled with a broad species diversity. The fossil species are divided among extinct, as well as surviving, present-day genera and families. Reliable identification is usually only possible with male specimens. Their external sex characteristics are good criteria for describing the species. However, disregard for this limitation has led to some incorrect descriptions and numerous synonyms.

The first fundamental studies of amber spiders were published by KOCH & BERENDT (1854), in BERENDT (1845-1856), and by PETRUNKEVITCH (1942, 1950, 1958) in a second phase of examination nearly 100 years later. Research on amber spiders has not stood still since then and was revived in a third phase by studies of Mexican (PETRUNKEVITCH 1963, 1971), Dominican (SCHAWALLER 1981 a, b, 1982, 1984, WUNDERLICH 1988) and Baltic amber (WUNDERLICH 1986, 1993 a, b).

The following list of recorded families is based on WUNDERLICH (1996), apart from a few changes: the families Agelenidae (funnel weavers) and Hahniidae (hahniids) were retained and not classified as subfamilies of other families. The 43 families are supplemented by the two questionable families Theraphosidae and Ctenidae. The list clearly shows that spider fauna is well-represented in Baltic amber. Taking into account the growing number of new descriptions of fossil species, a revision of the amber spiders may lead to the identification of even more families (WUNDERLICH 2001, in prep.).

Suborder MYGALOMORPHAE (Orthognatha)
- **Ctenizidae** (trapdoor spiders)
- **Dipluridae** (bird spiders)

Suborder ARANEOMORPHAE (Labidognatha)
- **Agelenidae** (funnel weavers)
- **Amaurobiidae** (hackledmesh weavers)
- **Anapidae** (anapid spiders)
- **Anyphaenidae** (ghost spiders)
- **Araneidae** (orbweavers)
- **Archaeidae** (dawn spiders)
- **Clubionidae** (sac spiders)
- **Corinnidae** (= **Myrmeciidae**) (antmimic spiders)
- **Cyatholipidae** (cyatholipid spiders)
- **Deinopidae** (net-casting spiders)
- **Dictynidae** (meshweb weavers)
- **Dysderidae** (dysderids)
- **Ephalmatoridae** (fossil only)
- **Gnaphosidae** (ground spiders)
- **Hahniidae** (hahniids)
- **Hersiliidae** (hersiliid spiders)
- **Leptonetidae** (leptonetid spiders)
- **Linyphiidae** (dwarf weavers, sheetweb weavers)
- **Liocranidae** (liocranid spiders)
- **Mimetidae** (pirate spiders)
- **Nesticidae** (cave cobweb spiders) (incl. *Acrometa*)
- **Oecobiidae** (flatmesh weavers)
- **Oonopidae** (dwarf sixeyed spiders)
- **Oxyopidae** (lynx spiders)
- **Philodromidae** (philodromid spiders)
- **Pholcidae** (daddylongleg spiders)
- **Pisauridae** (nursery web spiders)
- **Plectreuridae** (plectreurid spiders)
- **Salticidae** (jumping spiders)
- **Scytodidae** (spitting spiders)
- **Segestriidae** (segestriids)
- **Sparassidae** (= **Heteropodidae**) (hunting spiders)
- **Spatiatoridae** (fossil only)
- **Synotaxidae** (synotaxid spiders)
- **Telemidae** (telemid spiders)
- **Tetragnathidae** (longjawed orbweavers)
- **Theridiidae** (cobweb weavers)
- **Theridiosomatidae** (ray orbweavers)
- **Thomisidae** (crab spiders)
- **Uloboridae** (hackled orbweavers)
- **Zodariidae** (zodariid spiders)

Plate 15: Spiders (Arachnida: Araneae) in Baltic amber I.

a Dipluridae: bird spider, × 9.
b Oonopidae: *Orchestina tubulosa*, × 12.
c Scytodidae: *Scytodes weitschati* (holotype), × 10.
d Spatiatoridae: *Spatiator praeceps*, × 12.
e Archaeidae: *Archaea paradoxa*, × 12.
f Hersiliidae: hersiliid, × 8.
g Araneidae: orbweaver, × 8.
h Uloboridae: hackled orbweavers, × 9.

a
b
c
d
e
f
g
h

2.16 Frequency of amber spiders – Arachnida: Araneae

In view of the fact that the Baltic amber forest survived for 10 million years, it is likely that the abundance of the various species stabilized in ecological and ethological terms and that selection pressures became relatively moderate. Several of the euryecious species probably colonized the widest variety of environments in the Baltic amber forest (WUNDERLICH 1986). These biotopes primarily included the crevices and surface of tree bark. Small spiders with sufficient jumping capabilities had the highest probability of being captured in fossil resin. The bark-dwellers include the common dwarf sixeyed spiders of the genus *Orchestina* (Oonopidae). They are joined by the cobweb weavers (Theridiidae), including *Dipoena* and *Clya*, the web-spinning *Acrometa cristata* and *Anandrus inermis*, as well as the dwarf weaver (Linyphiidae) *Custodela cheiracantha*, all of which spun their nets near the trunk, between forked branches or possibly in the system of crevices in the bark. The lively behavior of some species may well have contributed to their frequency in amber. The preference of the *Orchestina* species and the Salticidae (jumping spiders) for jumping increased the probability of their capture, regardless of the question as to whether the spiders were also attracted by the odor and color of the resin.

The genera *Dipoena* and *Clya* of the family Theridiidae (cobweb weavers) and the genus *Orchestina* of the family Oonopidae (dwarf sixeyed spiders) display the greatest abundance of individuals in amber, followed in order of descending frequency by cave cobweb spiders (Nesticidae: *Acrometa*), synotaxids (Synotaxidae: *Anandrus*), dawn spiders (Archaeidae: *Archaea*), dwarf weavers (Linyphiidae: *Custodela*) and jumping spiders (Salticidae).

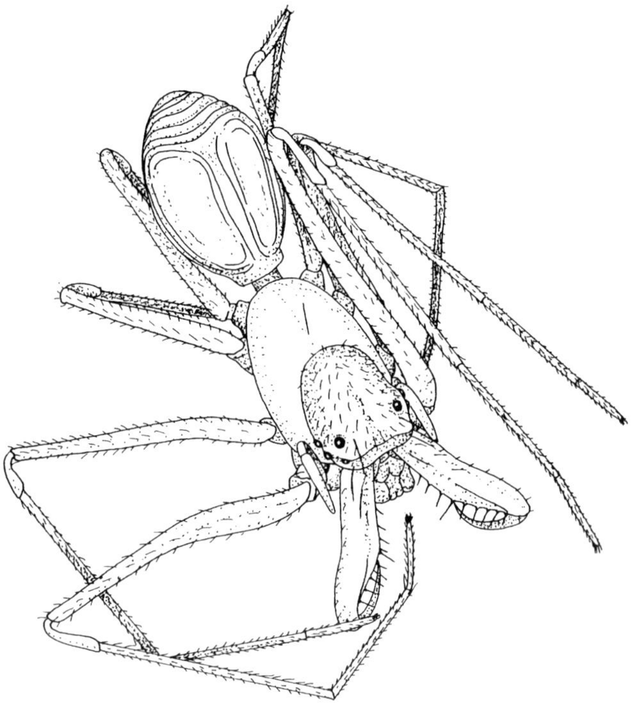

Fig. 33: Dawn spider (Archaeidae).

Plate 16: Spiders (Arachnida: Araneae) in Baltic amber II.

a Archaeidae (dawn spiders): *Archaea paradoxa*, ×35.
b Salticidae (jumping spiders): jumping spider, ×72.

a

b

2.17 Climatic environment of amber spiders – Arachnida: Araneae

The diversity and composition of the fossil spider fauna in Baltic amber give insight into the climatic conditions under which spiders lived in the Eocene amber forest. Impressions of this climate are partly based on the comparison of spider families recorded from Baltic amber with those that still occur today. The climate in their current areas of distribution provides information on the paleoclimate to which their Eocene predecessors were adapted.

According to WUNDERLICH (1986, 1996), the spiders that have been identified in Baltic amber and are today primarily distributed in the tropics include, among others, Ctenizidae: Ctenizinae, Dipluridae, Oecobiidae: Urocteinae, Hersiliidae, Archaeidae, Deinopidae, Anapidae: Anapinae, Theridiosomatidae, Cyatholipidae and Thomisidae: Stephanopinae. However, the spider fauna found in distinctly tropical Dominican amber comprises twice as many tropical families. Baltic amber particularly lacks the exclusively tropical families (subfamilies): Tetrablemmidae (Tetrablemmonae), Araneidae (Nephilinae), Selenopidae, Salticidae (Lyssomaninae). It must be considered that the range of the Hersiliidae stretches from the tropics all the way to the temperate zones. The distribution ranges of the Anapidae and Archaeidae are similarly indifferent.

Apart from cosmopolitan forms, the recent relatives of Baltic amber spiders are biogeographically concentrated in the Oriental and Ethiopian regions. Worthy of note are also the Archaeidae (dawn spiders) and Cyatholipidae (cyatholipids), whose closest relatives can be found in South Africa, Madagascar and Australia (Archaeidae), and in Jamaica, South Africa, New Zealand and Australia (Cyatholipidae), respectively. Only a few genera of spiders in Baltic amber have a purely Palearctic distribution today.

On the basis of numerous examples and comparisons, this Atlas of the Plants and Animals in Baltic Amber clearly shows that the various higher taxa occurring both in Baltic amber and today are indicative of a subtropical climate in the Eocene amber forest. This theory also concurs with the record of fossils whose recent relatives prefer either temperate or tropical climates, presupposing the subtropical amber forest was a tolerable climate for these plants and animals. The majority of plants and animals in Baltic amber are adapted to a subtropical climate and most closely related to the flora and fauna of Southeast Asia. ANDER (1942) already questioned and dismissed previous assumptions that Baltic amber primarily contained tropical faunal elements.

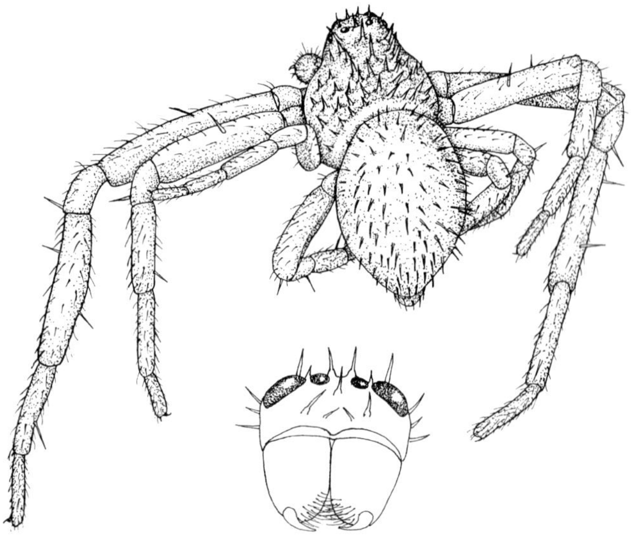

Fig. 34: Crab spider (Thomisidae), frontal view of the head.

Plate 17: Spiders (Arachnida: Araneae) in Baltic amber III.

a Anapidae: *Balticorma* sp., × 12.
b Mimetidae: pirate spider, × 5.
c Linyphiidae: dwarf weaver, × 6.
d Linyphiidae: dwarf weaver, × 6.
e Cyatholipidae: *Spinilipus teuberi*, holotype, × 6.
f Synotaxidae: *Anandrus* sp., × 8.
g Nesticidae: *Acrometa* sp., × 6.
h Nesticidae: *Acrometa* sp., × 6.

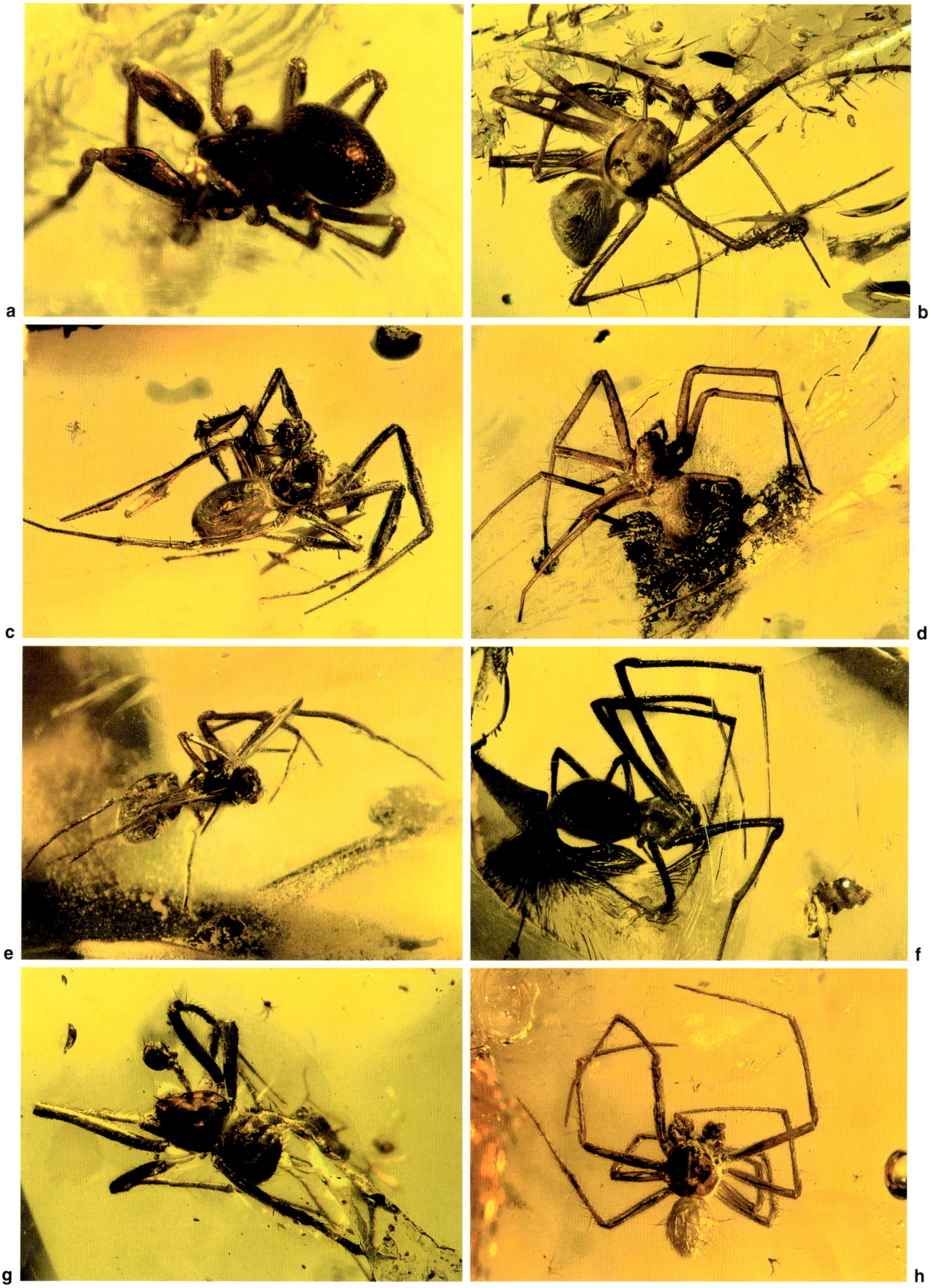
a
b
c
d
e
f
g
h

2.18 Biogeography of the amber spiders – Arachnida: Araneae

We know from comparative studies of the Baltic amber spiders and present-day spiders distributed worldwide (WUNDERLICH 1986) that roughly 75 % of all spider genera from Eocene amber have since become extinct. The remaining 25 % still occurring today supports further comparison and the paleobiogeographical breakdown of the spider fauna.

The present-day spiders that also occurred in the Palearctic region and Europe in the Eocene amber forest, include several cosmopolitan genera, e.g. *Orchestina* (Oonopidae), *Segestria* (Segestriidae), *Hyptiotes* (Uloboridae), *Argyrodes* (Theridiidae) and *Amaurobius* (Amaurobiidae), as well as a few additional genera that currently display a purely Palearctic (Agelenidae: *Mastigusa*) or Holarctic (Araneidae: *Zygiella*) distribution.

In most cases, however, the recent relatives of Baltic amber spiders are concentrated in the Oriental and Ethiopian regions, e.g. several species of jumping spiders (Salticidae). It is also remarkable to note the records of the Archaeidae (dawn spiders) and Cyatholipidae (cyatholipids), whose closest relatives can be found in the southern hemisphere: the Archaeidae in South Africa, Madagascar and Australia, and the Cyatholipidae in South Africa, New Zealand and Australia. Only one species of Cyatholipidae additionally occurs in Jamaica. Both families were apparently also distributed over the northern hemisphere in the Eocene Epoch and not restricted to Gondwanaland.

The three cyatholipid species recorded to date from Baltic amber and the Bitterfeld deposits *(Spinilipus kerneggeri, Succinilipus teuberi* and *Succinilipus saxoniensis)* represent the tropical spider fauna that may have retreated to the southern hemisphere in the wake of climatic changes. WUNDERLICH (1993) suggests another reason for their extinction in the northern hemisphere may have been competition with dwarf weavers (Linyphiidae: Erigoninae), which are unknown from the Tertiary, but today dominate the temperate zones of the northern hemisphere. While they hardly occur at all in the tropics and subtropics, their competitors, the cyatholipids, are very well-represented there.

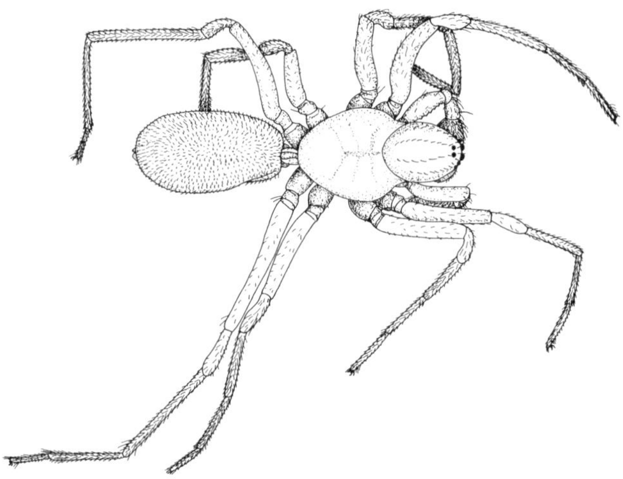

Fig. 35: *Spatiator praeceps* PERTRUNKEVITCH, 1942 (Spatiatoridae).

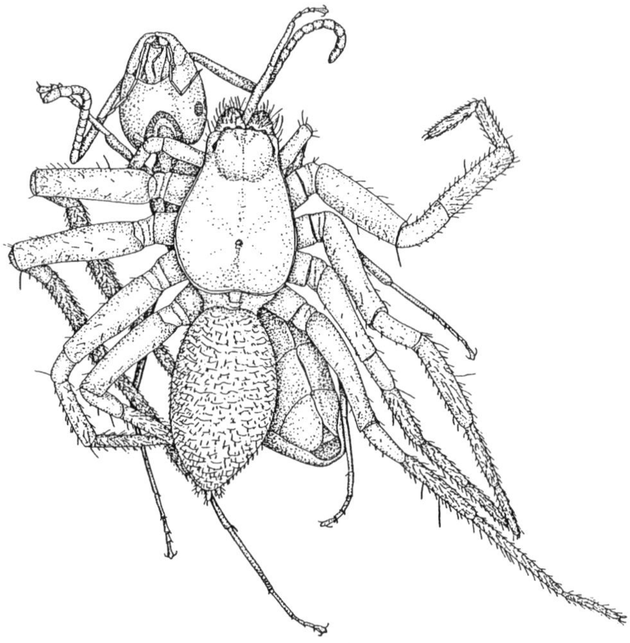

Fig. 36: Dysderid (Dysderidae) with ant prey.

Plate 18: Spiders (Arachnida: Araneae) in Baltic amber IV.

a Theridiidae: *Dipoena* sp., ×9.
b Theridiidae: *Eomysmena* sp., ×12.
c Theridiidae: cobweb weaver, ×8.
d Theridiidae: *Laseola* sp., ×10.
e Agelenidae: funnel weaver, ×8.
f Dictynidae: *Mastigusa* sp., ×8.
g Zodariidae: zodariid, ×6.
h Zodariidae: zodariid (subadult), ×9.

a
b
c
d
e
f
g
h

2.19 Spider webs in Baltic amber – Arachnida: Araneae

One of the most well-known characteristics of spiders is their ability to produce silk threads and spin them into webs. Not only the orbweavers (Araneidae), but also many other spider families create a great variety of webs, which they use for equally diverse capture methods. In contrast, cobweb weavers (Theridiidae) build webs of primitive structure, which are irregularly and loosely spun and usually fixed by a frame of tension threads. The vertically strung threads for capturing prey are provided with adhesive droplets and tear easily, as soon as the prey is ensnared. This method of capture is also used with the typical orb webs, whose spirals are equipped with either cribellate silk (Cribellatae) or adhesive droplets (Ecribellatae). The aggressively struggling animals come into contact with more and more sticky threads as they try to free themselves, until they are completely wrapped up and the spider's fatal bite finally puts and end to their struggle (Kullmann & Stern 1975, Foelix 1979). Ants (Fig. 37) and other prey in amber wound up in spider threads provide information on the method of capture used by Eocene web-spinning spiders.

Menge (1856) and Bachofen-Echt (1934) were the first to thoroughly document the occurrence of spider webs in amber. They referred to inclusions of webs and threads stemming from cobweb weavers (Theridiidae) and orbweavers (Araneidae). Additional webs and threads have also been recorded since then. The spider webs, with their delicate threads, silk (Plate 20) and adhesive droplets (Bachofen-Echt 1949), are hardly torn or stuck together in the amber and rather well-preserved in most cases. This indicates that the resin was of very low viscosity and only slowly soaked the spider webs at warm temperatures. The subtropical climate alone is not likely to have caused the low viscosity. There must have been a prevalence of low-viscosity resins that are not comparable to present-day resins.

The silk glands are located in the opisthosoma of the animals and terminate on the outside in fusulae on paired spinners. Some spiders in amber have threads hanging from the spinners, which can more likely be interpreted as a direct defense reaction to the approaching resin than the interruption of the web-spinning process.

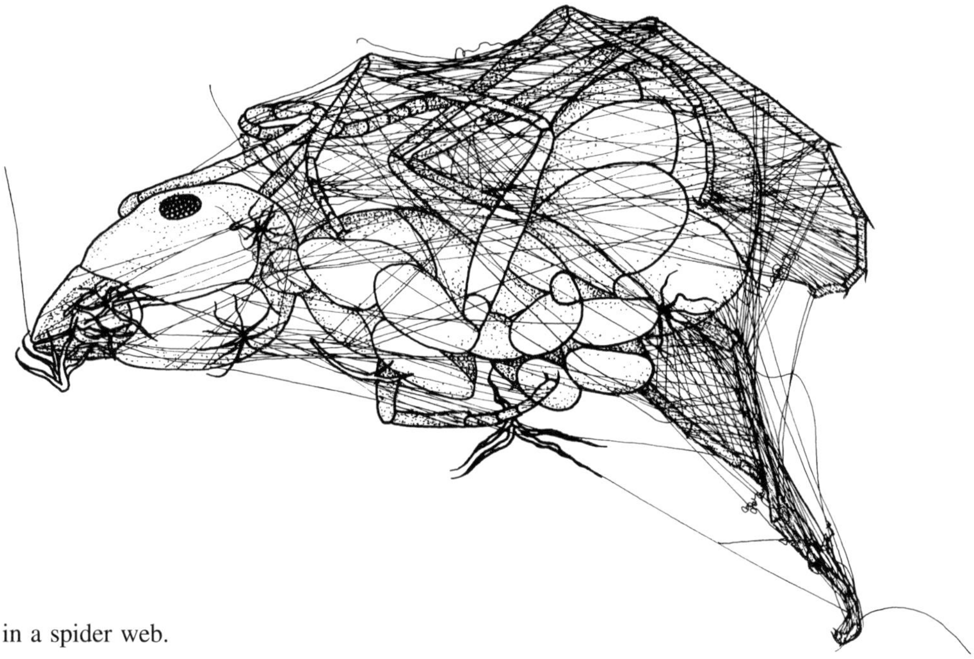

Fig. 37: Ant in a spider web.

Plate 19: Spiders (Arachnida: Araneae) in Baltic amber V.

a Sparassidae: hunting spider (juv.), × 12.
b Clubionidae: sac spider, × 8.
c Gnaphosidae: ground spider, × 10.
d Philodromidae: philodromid (juv.), × 18.
e Thomisidae: crab spider, × 9.
f Liocranidae: liocranid (juv.), × 5.
g Salticidae: jumping spider, × 32.
h Salticidae: jumping spider, × 25.

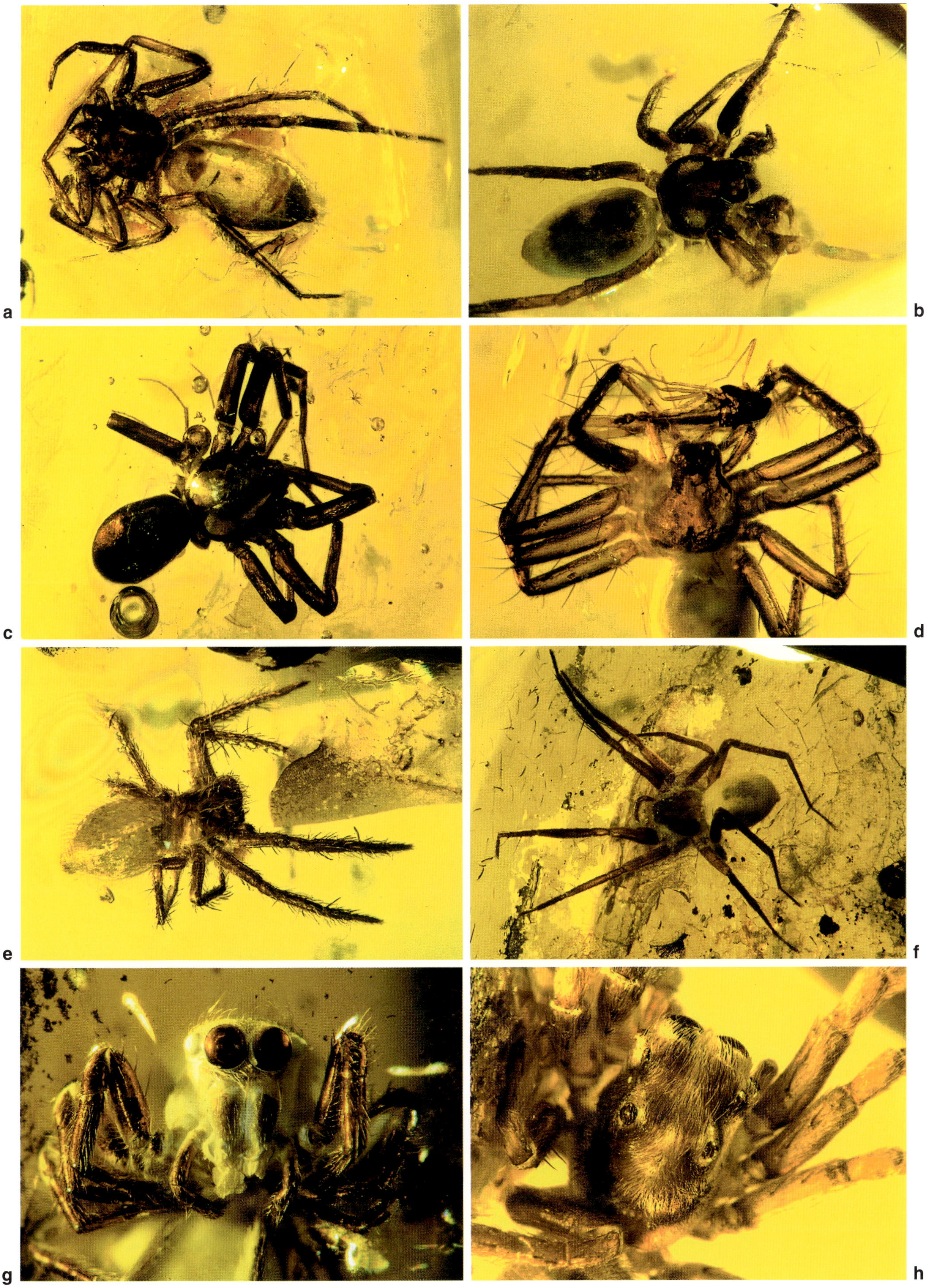
a
b
c
d
e
f
g
h

2.20 Brood care in amber spiders – Arachnida: Araneae

Spiders have a distinct sex dimorphism, where the females are always larger than the males. Apart from their smaller size, mature males can be distinguished from females by their enlarged palps. The female palps can be viewed as shortened extremities. The male palps, particularly the distal segments, are adapted to function as transfer organs for semen. For this purpose, they have a usually acuminate bulb on the distal segment. The tip of the bulb has an opening (i.e. embolus) for the spermophor, which spirals through the bulb and is closed at the opposite end. The male releases a drop of sperm from the genital pore on the abdomen and uses the palps to transfer the drop into the spermophor of the bulb. Once the spermophor is filled with sperm, the male goes in search of the female. In the courtship ritual, the male approaches the female extremely cautiously, so as not to be treated as prey. In order to improve communication, many males use "vibration instruments", which have been found to be similar in both fossil spiders in Baltic amber and present-day spiders, such as the stridulating organs in dwarf weavers (Linyphiidae), cobweb weavers (Theridiidae) and Spatiatoridae. Copulating fossil spiders have only rarely been found in amber. Mating behavior observed in fossil spiders in Baltic amber was first described for a dwarf sixeyed spider *Orchestina* sp. (WUNDERLICH 1986).

Many males copulate with other females as soon as the pedipalps are filled with sperm. They die very soon after copulation. The stronger females have a longer life-span, lay the eggs and build a sac. The females of some species care for their brood. WUNDERLICH (1986) describes an egg sac of a spider from Baltic amber with an outside diameter of 5 mm and a batch of 20 eggs with a diameter of 2.1 mm. The egg sac of a pirate spider (Mimetidae) is of similar size and shape (Plate 20e). Plate 20f shows an egg sac being carried by a daddylongleg spider (Pholcidae), where spiderlings close to hatching are visible inside the eggs. The other two photographs (Plate 20g, h) show further steps of the spiderlings as they leave the egg sac. Brood care among many spiders consists not only of constructing and guarding the egg sac, but also feeding the hatched spiderlings. Although no evidence is yet available, brood care of this kind can certainly also be expected to have been captured in Eocene amber.

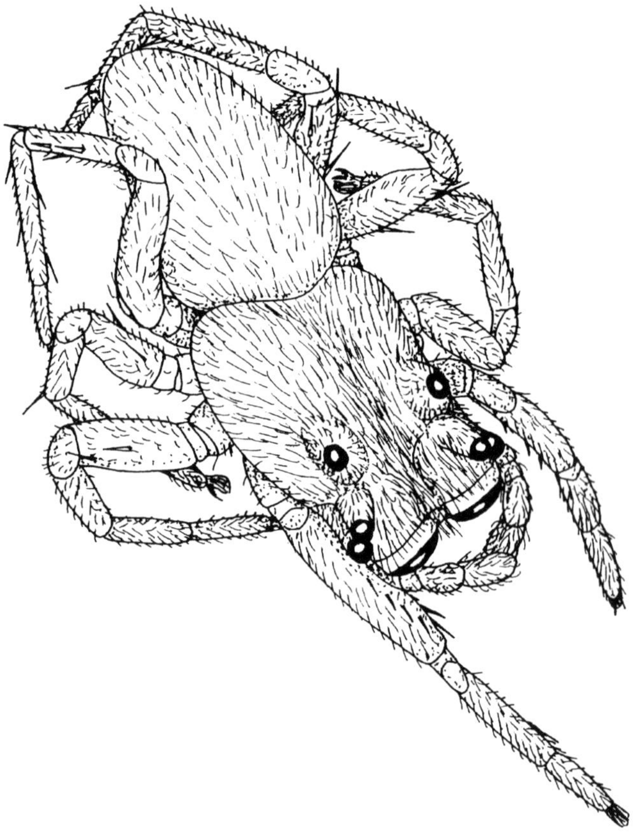

Fig. 38: Jumping spider (Salticidae).

Plate 20: Spiders (Arachnida: Araneae) in Baltic amber VI.

a Hersiliidae: hersiliid with threads, ×9.
b Magnification, ×25.
c Gnaphosidae: ground spider with threads, ×6.
d Agelenidae: funnel spider?, spider web, ×6
e Mimetidae: pirate spider, egg sac, ×8.
f Pholcidae: daddylongleg spider, egg sac with spiderlings, ×9.
g Web with spiderlings, ×12.
h Free-roaming spiderlings, ×10.

a
b
c
d
e
f
g
h

2.21 Crustaceans – Crustacea: Ostracoda, Isopoda, and Amphipoda

Crustaceans are represented in Baltic amber by woodlice (Isopoda), beach fleas (Amphipoda), and seed shrimps (Ostracoda). These groups represent originally marine crustaceans that gradually adapted to fresh water and terrestrial habitats.

The few ostracods found to date from the family Candodidae are indicative of aquatic habitats in the amber forest. They prefer to inhabit muddy sediments:

Candodidae: *Cyclocypris* sp. (PLATE 21 g)

Beach fleas still inhabit the ocean in various zones and depths, although several species today live in fresh water. *Gammarus pulex* and *G. roeseli* of the family Gammaridae are commonly encountered in European inland waters. Beach fleas do not go beyond a depth of two meters, preferring shallow, running and standing water. They are also found in swamps, bog pools and wet moss. The detailed study published by ZADDACH (1864) classifies amber beach fleas in a new genus: *Palaeogammarus* (Crangonycidae). Recently, two additional families were recorded:

Isaeidae (Plate 21 h)
Niphargidae
Crangonycidae
Palaeogammarus balticus LUCKS, 1927
Palaeogammarus danicus JUST, 1974
Palaeogammarus sambiensis ZADDACH, 1864

Woodlice also inhabit marine, freshwater and terrestrial habitats. A comparison of the biology of the oniscoids (Oniscoidea) with regard to their adaptation to moist and dry habitats results in a successive order of families (EISENBEIS & WICHARD 1987). The Ligiidae are amphibious woodlice occurring in the moist littoral zones of running waters and forest ponds. The Trichoniscidae prefer moist habitats in intact forest floors. The Oniscidae and Porcellionidae prefer moderately moist and dry biotopes under bark, rocks and leaf litter. The Armadillidiidae have become even further removed from aquatic biotopes. They look for dry and sunny sites. This range of very moist to sunny and dry biotopes is well represented by the fossil oniscoid fauna in Eocene Baltic amber (SPAHR 1993):

Armadillidiidae
Armadillidium pulchellum (ZENKER, 1798)
Ligiidae
Ligia sp. MENGE, 1856
Oniscidae
Oniscus cinvexus KOCH & BERENDT, 1854
Porcellionidae
Porcellio cyclocephalus MENGE, 1854
Porcellio granulatus MENGE, 1854
Porcellio notatus KOCH & BERENDT, 1854
Trachelipidae
Protracheoniscus politus (KOCH, 1841)
Trichoniscidae
Hyloniscus riparius (KOCH, 1838)
Trichoniscus asper MENGE, 1854

Three woodlouse specimens have been found in Bitterfeld amber and identified as extant species (SCHUMACHER & WENDT 1989): *Armadillidium pulchellum* (ZENKER, 1798), *Protracheoniscus politus* (KOCH, 1841) and *Hyloniscus riparius* (KOCH, 1838). Their identity with present-day species requires review, particularly as no reliable evidence is yet available to support the assignment of fossil specimens from Eocene Baltic amber and extant animals to the same species.

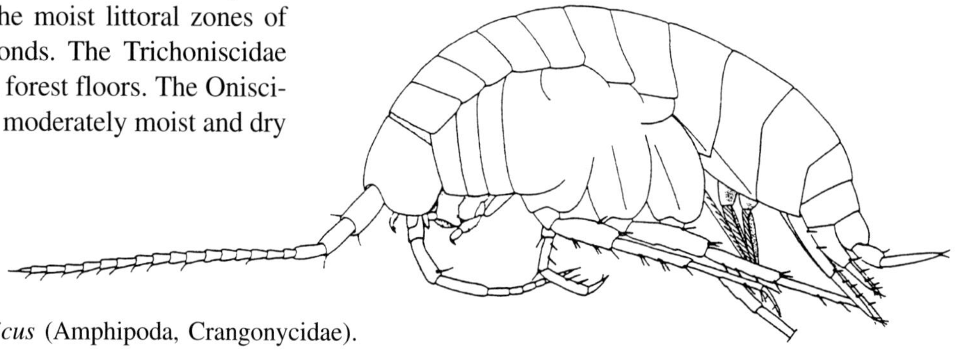

Fig. 39: *Palaeogammarus danicus* (Amphipoda, Crangonycidae).

Plate 21: Crustaceans (Crustacea: Ostracoda, Isopoda, and Amphipoda) in Baltic amber.

a-f Isopoda (woodlice): Oniscoidea, **a, b** × 10, **c, d** × 8, **e** × 9, **f** × 20.
g Ostracoda (mussle shrimps): Candodidae, *Cyclocypris* sp., with preserved soft parts, × 76.
h Amphipoda (beach fleas): ? Isaeidae, beach flea, × 35.

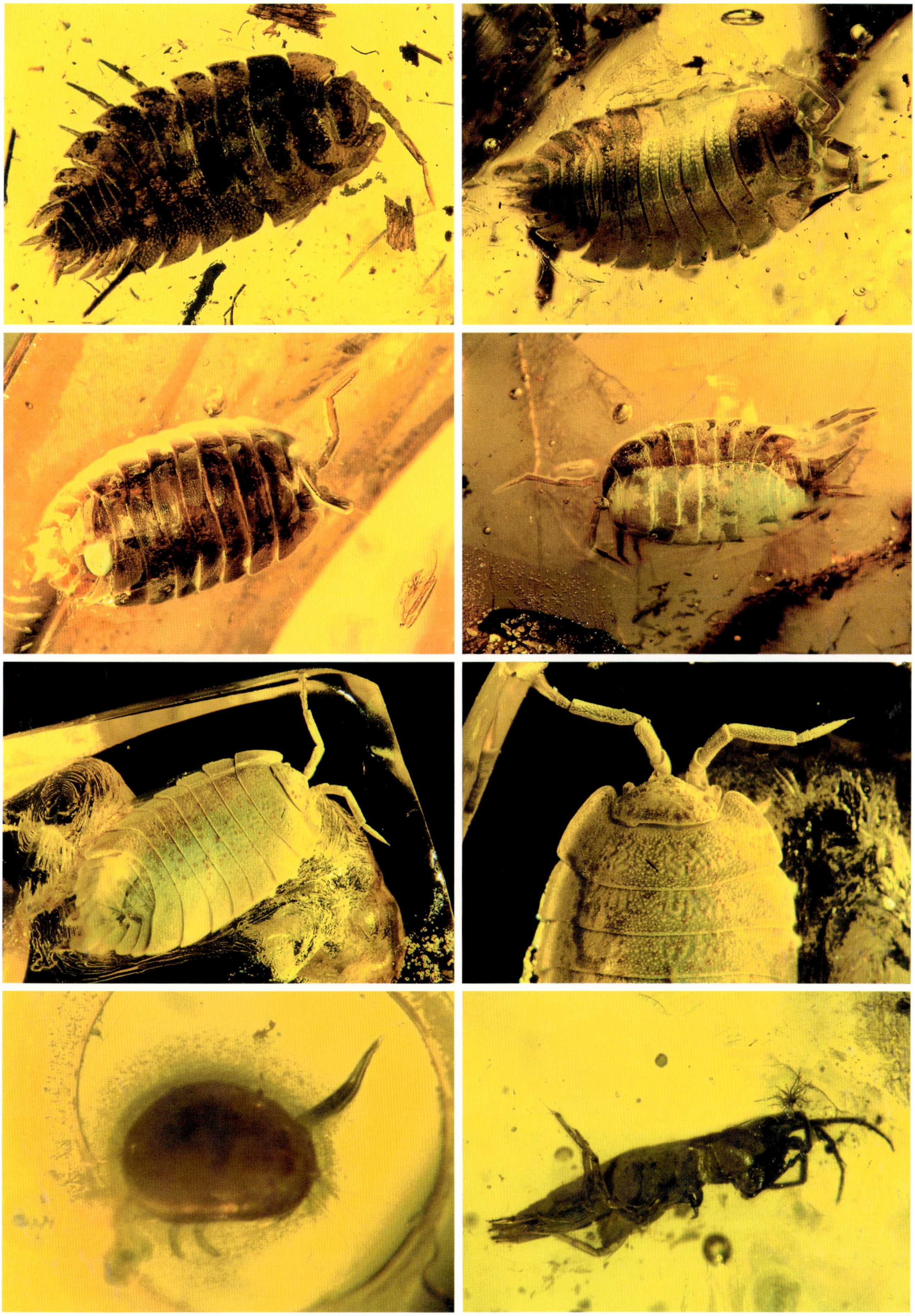

a b c d e f g h

2.22 Myriapods – Myriapoda: Chilopoda, Symphyla, and Pauropoda

Myriapods are arthropods and, together with the Insecta, form the systematic group Tracheata. The class Myriapoda includes the Chilopoda (centipedes), Symphyla (symphylids), Pauropoda (pauropods) and Diplopoda (millipedes). The monophyly of the Myriapoda is of some question (Ax 1984, Kraus 1995).

While the Chilopoda, Symphyla and Diplopoda have been recorded from Baltic amber, there were no records of the Pauropoda for quite some time. Only recently was a fossil pauropod found in Baltic amber, described as *Eopauropus balticus* and assigned to the family Pauropodidae (Scheller & Wunderlich 2001). Pauropods are eyeless, unpigmented soil animals with a maximum length of 2 mm. They live deep underground, avoid the surface and, for this reason alone, are hardly to be expected to occur in amber. Symphylids (Symphyla) are rare in amber (Plate 22 g, h) because, like pauropods, they are euedaphic animals (Eisenbeis & Wichard 1987), lacking eyes and having an unpigmented body in adaptation to life underground. With a maximum length of 8 mm and a diameter of approximately 0.5 mm, symphylids are adapted to the pore system of the soil. The agility they need to move about underground is provided by increased segmentation of the tergum, which has up to 24 tergites and thus exceeds the number of body segments.

Only a few, individual specimens from two genera have been reported from Baltic amber: *Scolopendrella* sp. (Bachofen-Echt 1942, 1949) and *Scutigerella* sp. (Kosmowska-Ceranowicz & Mierzejewski 1978).

Centipedes (Chilopoda) are encountered more frequently in Baltic amber (Plate 22 a-f). They are typical epedaphic soil-dwellers and search for prey on the surface. Their agility is due to 15 pairs of lateral walking legs and flat posture, which enables them to pursue prey more easily through the leaf litter, in crevices and under stones. Centipedes have well-developed eyes and also long antennae for tactile location of their prey.

Centipedes are represented in Baltic amber by the families Cryptopsidae *(Crytops)*, Geophilidae *(Geophilus)*, Lithobiidae *(Lithobius)*, Scolopendridae *(Scolopendra)* and Scutigeridae *(Scutigera)*. Bachofen-Echt (1942) identified the family Cryptopsidae with a description of the genus *Crytops* sp. Apart from several nomina nuda (Keilbach 1982), the following families have been reliably identified on the basis of their species:

Geophilidae
Geophilus brevicaudatus Menge, 1854
Lithobiidae
Lithobius longicornis Koch & Berendt, 1854
Lithobius maxillosus Koch & Berendt, 1854
Lithobius planatus Koch & Berendt, 1854
Scolopendridae
Scolopendra proavita Menge, 1854
Scutigeridae
Scutigera illigeri (Koch & Berendt, 1854)
Scutigera leachi (Koch & Berendt, 1854)

Plate 22: Centipedes and symphylids (Myriapoda: Chilopoda and Symphyla) in Baltic amber.

a Chilopoda, Lithobiidae: lithobiid, × 7.
b Chilopoda, Lithobiidae: lithobiid, × 8.
c Chilopoda, Geophilidae: geophilid, × 5.
d Chilopoda, Geophilidae: geophilid, × 5.
e Chilopoda, Scutigeridae: scutigerid, × 8.
f Chilopoda, Scutigeridae: scutigerid, × 18.
g Symphyla: symphylid, × 11.
h Symphyla: symphylid, × 11.

a

b

c

d

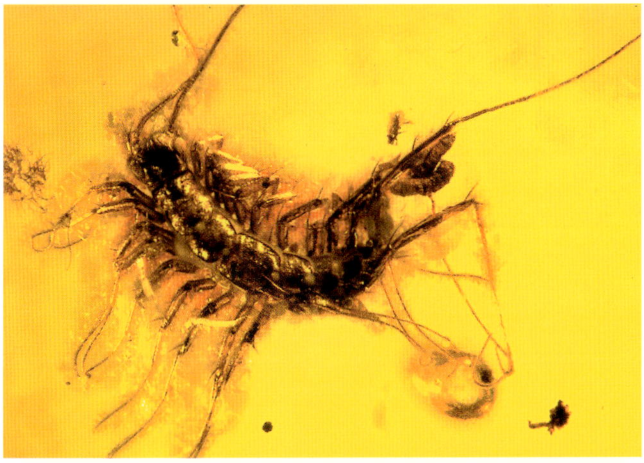
e

f

g

h

2.23 Myriapods – Myriapoda: Diplopoda

Millipedes or diplopods lead an epedaphic or hemiedaphic lifestyle. DUNGER (1974) and MANTON (1977) defined the following types:

1. Bulldozer type (e.g. Julidae)
2. Pill type (e.g. Glomeridae)
3. Wedge-pushing type (e.g. Polydesmidae)

All diplopods contribute to the process of decomposition by consuming fresh and decaying plant material, primarily from the leaf litter on the forest floor.

The hemiedaphic family Julidae inhabits the top soil, occasionally penetrating deeper into the ground in search of food. For this purpose, the species use their numerous serial legs to develop a propulsive power that forces the head and body-column into the soil in the manner of a bulldozer. If the ground is too firm, the Julidae become substrate-feeders like earthworms, gnawing their way through the soil. Unlike the bulldozer type, representatives of the pill type (Glomeridae) live epedaphically on the soil surface, where they consume plant material. They defend themselves against the increased danger of falling victim to predators by rolling up the body and releasing a secretion from their defensive glands. The wedge-pushing type is represented by the family Polydesmidae. The body of these species tapers towards the head and first body segments, while the subsequent segments are flat on top (tergites) and have lateral, wing-like extensions (paratergites). The wedge-shaped body enables the animals to push their way through the leaf litter or under stones. The dorsal surface, which is expanded by the paratergites, simultaneously prevents them from penetrating deeper layers, meaning that Polydesmidae species of the wedge-pushing type belong to the group of epedaphic diplopods. The pselaphognaths (Polyxenidae) are not typical soil-dwellers, although they also occur in the loose humus and under leaf litter. Some Polyxenidae can be found in weathered tree stumps and dead trees. As bark-dwellers, they inhabit the entire height of a tree and feed on algal growth.

The range of soil-dwelling forms observed among extant diplopods also appears to have occurred in the Eocene forest floor. All characteristic families have been recorded from Baltic amber. Diplopods are a good example of how individual organisms can provide pieces of information on the biology and ecology of the "amber forest". The diplopods are represented in Baltic amber by eight families, including the four mentioned as examples. Apart from several nomina nuda (KEILBACH 1982), they have been described down to genera and species (SPAHR 1993a):

Craspedosomatidae
Atractosoma sp.
Craspedosoma affine KOCH & BERENDT, 1854
Craspedosoma angulatum KOCH & BERENDT, 1854
Glomeridae
Glomeris denticulata MENGE, 1854
Julidae
Julus laevigatus KOCH & BERENDT, 1854
Nemasomatidae
Blaniulus sp.
Polydesmidae
Polydesmus sp.
Polyxenidae
Polyxenus conformis KOCH & BERENDT, 1854
Polyxenus ovalis KOCH & BERENDT, 1854
Polyzonidae
Polyzonium sp.
Synxenidae
Phryssonotus hystrix (MENGE 1854)

Plate 23: Millipedes (Myriapoda: Diplopoda) in Baltic amber.

a Julidae: Juliform millipede, × 10.
b Julidae: Juliform millipede, × 9.
c Polydesmidae: Flat millipede, × 10.
d Polydesmidae: Flat millipede (ventral), × 15.
e Polyzonidae: *Polyzonium* sp., × 12.
f Craspedosomatidae: *Craspedosoma* (?), × 10.
g Polyxenidae: *Polyxenus* sp., × 12.
h Synxenidae: *Phryssonotus* sp., × 17.

a

b

c

d

e

f

g

h

2.24 Diplurans and springtails – Insecta: Diplura and Collembola

The traditional division of the insects into the primary groups of wingless Apterygota and winged Pterygota is considered to be outdated. According to phylogenetic systematics, the insects are broken down into two monophyletic groups: the Entognatha (Diplura and Elliplura, the latter including the Protura and Collembola) and the Ectognatha (Archaeognatha and Dicondylia, the latter including the Zygentoma and Pterygota). Of the Entognatha, Diplura (diplurans), Collembola (springtails) and recently also Protura (proturans) have all been recorded from Baltic amber. Entognath insects are rare in amber, as they are usually euedaphic or epedaphic.

Proturans are strictly euedaphic, eyeless and unpigmented insects with a maximum length of 3 mm. They seldom come to the surface and can therefore be expected to occur only rarely in amber.

The Collembola are divided according to shape into the rounded symphypleonean and elongated arthropleonean springtails. Their species diversity and abundance in the soil is primarily attributable to the symphypleonean Onichiuridae, which are typical representatives of the euedaphon, being small, unpigmented, eyeless insects that have reduced legs and are adapted to humid conditions. No fossil species of this family have been recorded from Baltic amber. Epedaphic Collembola, including those in the herbaceous and shrub layers, as well as the bark-dwellers, are frequent in amber. In view of the fact that the taxonomic characteristics of the Collembola in amber are difficult to verify, and that the amber Collembola studied by Olfers (1907) have undergone major revision (Handschin 1926 a, b), only comparatively few species (19) remain of those previously described by Koch & Berendt (1854), Handschin (1926) and Stach (1922) (cf. Stach 1972, Lawrence 1985, Spahr 1990):

Entomobryidae
- *Entomobrya pilosa* (Koch & Berendt, 1854)
- *Lepidocyrtus ambricus* Handschin, 1926
- *(Paidium) crassicorne* Koch & Berendt, 1854
- *(Paidium) pyriforme* Koch & Berendt, 1854
- *Orchesella eocaena* Handschin, 1926

Hypogastruridae
- *Hypogastrura intermedia* Handschin, 1926
- *Hypogastrura protoviatica* Handschin, 1926

Isotomidae
- *Isotoma crassicornis* Handschin, 1926
- *Isotoma protocinerea* Handschin, 1926

Poduridae
- *Podura fuscata* Koch & Berendt, 1854
- *Podura pulchra* Koch & Berendt, 1854

Sminthuridae
- *Allacma pulmosa* Handschin, 1926
- *Allacma pulmosetosa* Handschin, 1926
- *Allacma setosa* Handschin, 1926
- *Sminthurus brevicornis* Koch & Berendt, 1854
- *Sminthurus longicornis* Koch & Berendt, 1854
- *Sminthurus ovatulus* Koch & Berendt, 1854
- *Sminthurus succineus* Stach, 1922

Tomoceridae
- *Tomocerus taeniatus* (Koch & Berendt, 1854)

The diplurans (Diplura) are epedaphic, predaceous animals that live in moss, under bark and stones, in the leaf litter and occasionally also deeper in the pore system of the soil. Silvestri (1912) described the fossil species *Campodea darwinii*. According to Silvestri (1912), another species *(Ocellia articulicornis)* described by Olfers (1907) is not a dipluran, but possibly a Dermaptera nymph. These discussions illustrate the difficulty associated with the taxonomic study of the entognath Collembola and Diplura in Baltic amber.

Plate 24: Diplurans and springtails (Insecta: Diplura and Collembola) in Baltic amber.

a-b Diplura: Dipluran, **a** ×13, **b** ×17.
c-e Collembola, Arthropleones: arthropleonean springtails, **c** ×20, **d** ×16, **e** ×15.
f-h Collembola, Symphypleones: symphypleonean springtails, ×25.

a

b

c

d

e

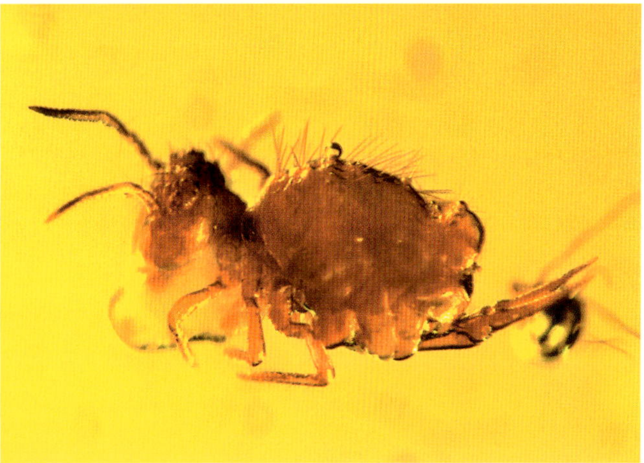
f

g

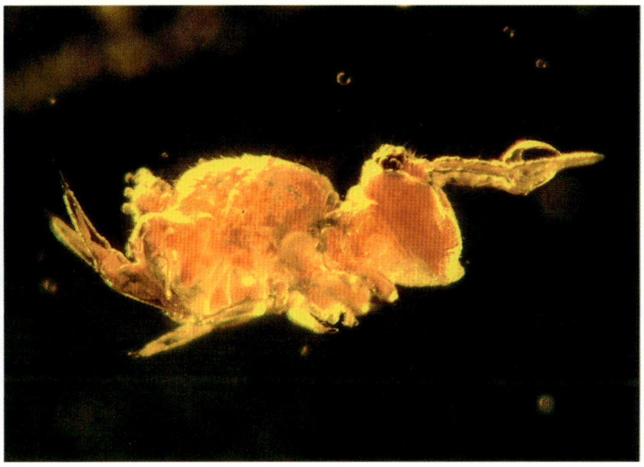
h

2.25 Bristletails and silverfish – Insecta: Archaeognatha and Zygentoma

Bristletails (Archaeognatha) and silverfish (Zygentoma) are the basal, primitively wingless orders of the Ectognatha and together sister to the Pterygota. Bristletails can be distinguished from silverfish by their large, compound eyes, which adjoin on the dorsal side of the head in many species (Plate 25 e). The tactile, labial palpi extend forward in the manner of legs (Fig. 41). The mesothorax, metathorax and all abdominal segments have distinct ventral styli (Plate 25 f). One or two coxal, eversible vesicles, which regulate water balance, are positioned medially in the direct vicinity of the styli. Silverfish have small, lateral eyes and inconspicuous styli on the last abdominal segments. They are relatively rare in amber, if for no other reason than that they are agile movers and can quickly avoid dangers.

Koch & Berendt (1854) and Menge (1854) in Berendt (1845-1856) provided excellent illustrations of bristletails and silverfish (Fig. 40, 41). The majority of the described species stem from their work. Olfers (1907) and Silvestri (1912) also contributed to our current knowledge of these groups. No more recent studies are available. Keilbach (1982) and Spahr (1990) took possible synonyms and nomina nuda into account:

ARCHAEOGNATHA

Machilidae

Machilis acuminata (Koch & Berendt, 1854)
Machilis albomaculata (Menge, 1854)
Machilis anguea (Koch & Berendt, 1854)
Machilis boops Olfers, 1907 (part.), Silvestri, 1912
Machilis (?) *caestifera* Silvestri, 1912
Machilis capito Olfers, 1907 (part.), Silvestri, 1912
Machilis confinis (Koch & Berendt, 1854)
Machilis corusca (Koch & Berendt, 1854)
Machilis diastatica (Olfers, 1907)
Machilis electra (Koch & Berendt, 1854)
Machilis imbricata (Koch & Berendt, 1854)
Machilis longipalpa (Koch & Berendt, 1854)
Machilis macrura (Menge, 1854)
Machilis palaemon Olfers, 1907 (part.), Silvestri, 1912
Machilis saliens (Menge, 1854)
Machilis seticornis (Koch & Berendt, 1854)
Praemachilis cineracea Olfers, 1907 (part.), Silvestri, 1912

ZYGENTOMA

Lepidotrichidae

Lepidothrix pilifera Menge, 1854

Lepismatidae

Lampropholis dubia (Koch & Berendt, 1854)

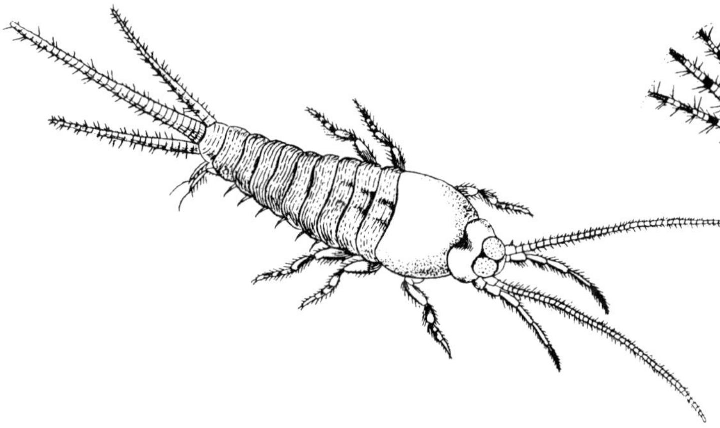

Fig. 40: *Machilis imbricata* (Koch & Berendt, 1854).

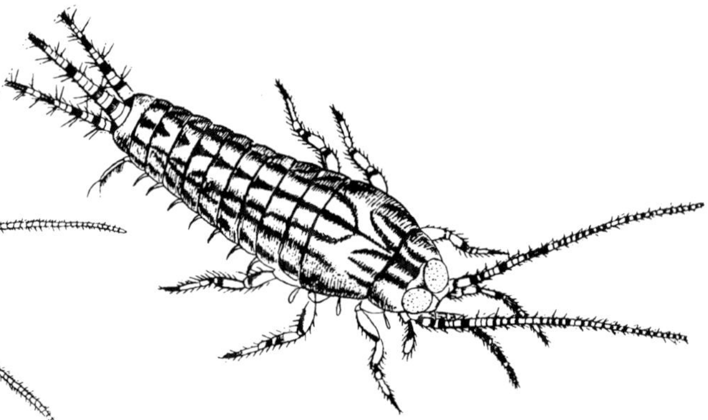

Fig. 41: *Machilis electra* (Koch & Berendt, 1854) (historical illustration from Koch & Berendt 1854).

Plate 25: Bristletails and silverfish (Insecta: Archaeognatha and Zygentoma) in Baltic amber.

a Zygentoma: Silverfish, × 11.
b Zygentoma: Silverfish, × 8.
c Archaeognatha: *Machilis* sp., × 5.
d Zygentoma: Silverfish, × 8.
e Archaeognatha: *Machilis* sp., head, × 12.
f Archaeognatha: *Machilis* sp., abdomen, with styli, × 20.
g Archaeognatha: *Machilis* sp., head, × 12.
h Archaeognatha: *Machilis* sp., abdomen, with styli, × 23.

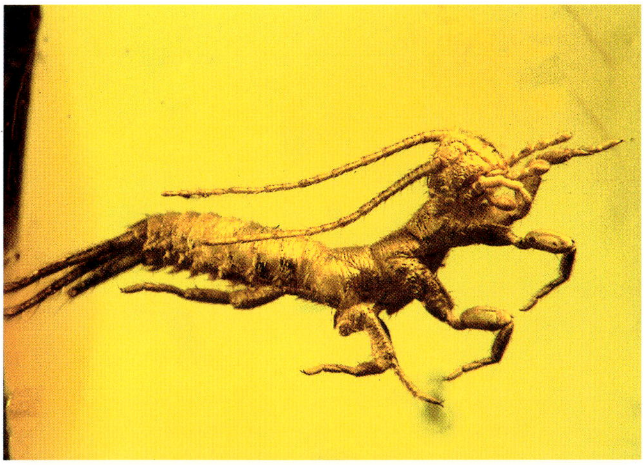
a

b

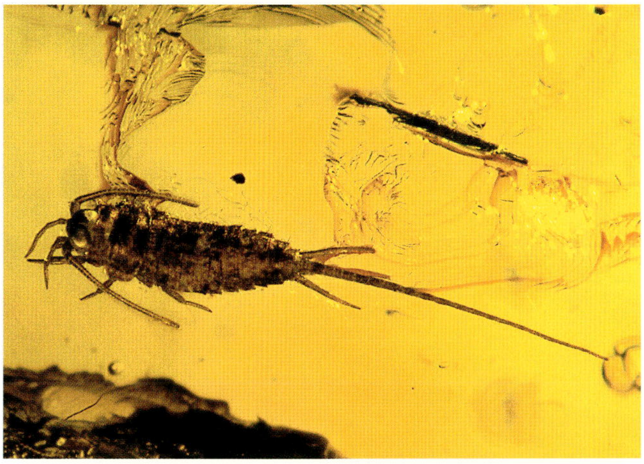
c

d

e

f

g

h

2.26 Mayflies – Insecta: Ephemeroptera

Mayflies are merolimnetic insects that first develop in water and then in a short phase on land. After mating, which occurs in flight in many species, most females drop their egg masses over water or shed them by briefly dipping the abdomen into the water. The eggs hatch into exclusively aquatic nymphs that gradually mature over many months by means of multiple molts. The last exuvia is shed by the subimagoes. At this winged developmental stage, the mayfly nymphs leaves its aquatic habitat and takes to the air. Final molting is followed by the emergence of the sexually mature mayfly, which does not feed with its reduced mouthparts and instead concentrates on mating and egg-laying during its brief lifetime of merely a few hours.

The mayflies preserved in amber are usually sexually mature imagoes and rarely winged subimagoes. Plate 27 b shows an imago shedding the exuvia of the winged subimago. The legs are spread and the wings still folded.

Mayflies normally do not leave the water as nymph, but rather as winged subimagoes, making records of nymphs in Baltic amber all the more surprising. *Succinogenia larssoni* is a fossil species described by DEMOULIN (1965) on the basis of a nymph from the collection of the Zoological Museum in Copenhagen. The dorsoventrally flattened body, scutiform head with dorsally oriented eyes and lateral, lamellar gills, are the most noticeable features of larvae belonging to the family Heptageniidae (Fig. 42). Since then, three additional heptageniid nymphs have been discovered (Plate 27 c). With their flattened, broad bodies, these nymphs represent a life form adapted to running water. They lie flat on firm substrates in a boundary layer that largely protects them from the rushing current, which would otherwise carry them off into open water (WICHARD et. al. 1995).

It is difficult to explain how these aquatic nymphs became trapped in amber. There is no evidence to date of aquatic animals being enclosed by resin in water and preserved as amber inclusions. If aquatic insects are embedded in amber, they must have previously left the water as nymphs or imagoes in accordance with their amphibious life cycle and dispersal behavior.

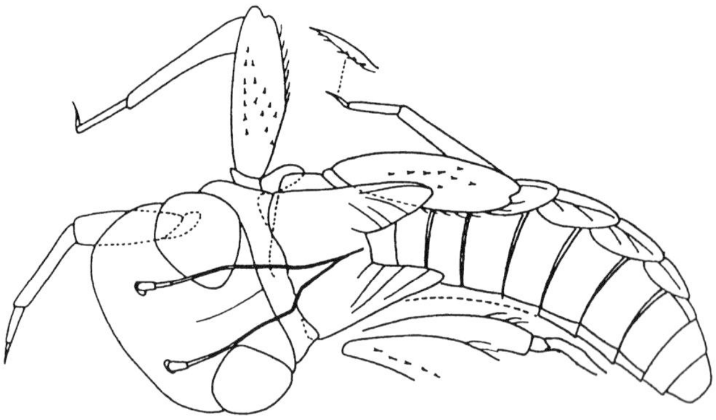

Fig. 42: Fossil nymph of *Succinogenia larssoni* DEMOULIN, 1965 (Heptageniidae).

Plate 26: Mayflies (Insecta: Ephemeroptera) in Baltic amber I.

a Heptageniidae: compound eyes of a stream mayfly, × 50.

b Heptageniidae: full view of a stream mayfly, × 8.

a

b

2.27 Mayfly species in Baltic amber – Insecta: Ephemeroptera

DEMOULIN monographed the mayflies of Baltic amber in a series of studies (1954, 1955, 1956, 1965, 1968 a, b, 1970 a, b) that built upon the pioneering work of HAGEN (1856) and PICTET (1856). He described a total of 25 fossil species. According to DEMOULIN, the extant species *Heptagenia (Kageronia) fuscogrisea* has also been recorded from Baltic amber. In addition STANICZEK & BECHLY (2001) described a new species of the family Baetiscidae.

The list of mayflies is based on the classification of the Ephemeroptera by HUBBARD (1990):

Ameletopsidae
Balticophlebia henningi DEMOULIN, 1968
Ametropodidae
Brevitibia intricans DEMOULIN, 1968
Baetidae
Baetis gigantea HAGEN, 1856
Baetis grossa HAGEN, 1856
Siphlonuridae
Metretopus henningseni DEMOULIN, 1965
Metretopus trinervis DEMOULIN, 1968
Siphloplecton jaegeri DEMOULIN, 1968
Siphloplecton macrops (PICTET, 1856)
Baltameletus oligocaenicus DEMOULIN, 1968
Siphlonurus dubiosus DEMOULIN, 1968
Oligoneuriidae
Cronicus anomalus (PICTET, 1856)
Cronicus major DEMOULIN, 1968
Arthropleidae
Electrogenia dewalschei DEMOULIN, 1956
Heptageniidae
Cinygma? baltica DEMOULIN, 1968
Heptagenia atypica DEMOULIN, 1968
Heptagenia bachofeni DEMOULIN, 1968
Heptagenia (Kageronia) fuscogrisea
Heptagenia gleissi DEMOULIN, 1968
Heptagenia ligata DEMOULIN, 1968
Heptagenia senex DEMOULIN, 1968
Rhithrogena sepulta DEMOULIN, 1968
Succinogenia larssoni DEMOULIN, 1965
Leptophlebiidae
Blasturophlebia hirsuta DEMOULIN, 1968
Xenophlebia aenigmatica DEMOULIN, 1968
Paraleptophlebia pisca (PICTET, 1856)
Ephemerellidae
Timpanoga viscata (DEMOULIN, 1968)
Baetiscidae
Baetisca velteni STANICZEK & BECHLI, 2001

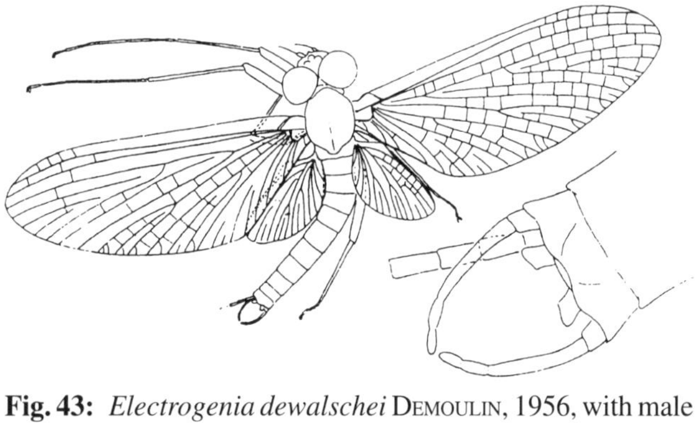

Fig. 43: *Electrogenia dewalschei* DEMOULIN, 1956, with male genitalia depicted.

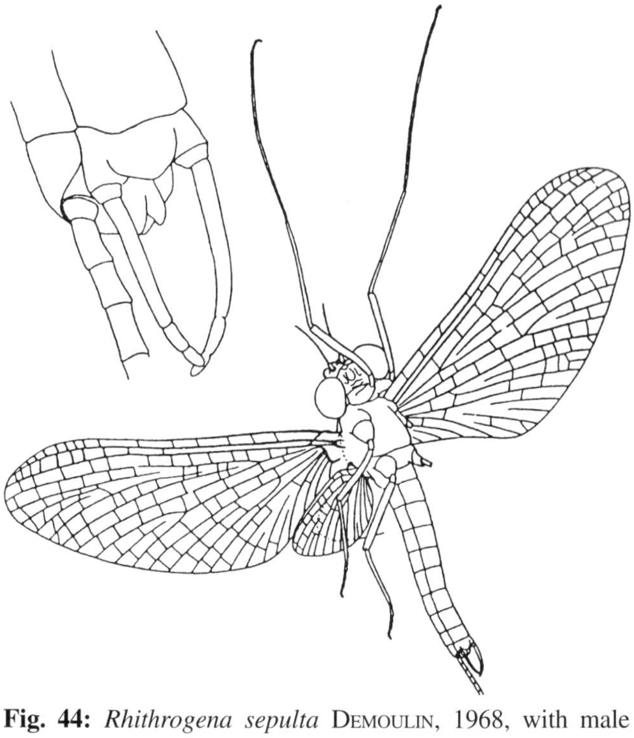

Fig. 44: *Rhithrogena sepulta* DEMOULIN, 1968, with male genitalia depicted.

Plate 27: Mayflies (Insecta: Ephemeroptera) in Baltic amber II.

a Siphlonuridae: primitive minnow mayfly, ×8.
b Imago shedding the subimago exuvia, ×6.
c Heptageniidae: flatheaded mayfly or stream mayfly nymph, ×10.

a

b

c

2.28 Dragonflies and damselflies – Insecta: Odonata

Dragonflies and damselflies are undoubtedly among the rarest insects in amber and have only been reported to date from Baltic and Dominican amber (LARSSON 1978, SCHLEE & GLÖCKNER 1978, SCHLEE 1990, SPAHR 1992). Their size and outstanding flying capabilities minimize their risk of getting caught in the viscous resin of trees, which they do occasionally land on. Nonetheless, several wing fragments found in amber (Plate 28) bear witness to the presence of dragonflies and damselflies. They clearly show that, like other large insects, the Odonata were also strong enough to free themselves from the resin, albeit in an injured state. Consequently, fully preserved dragonflies and damselflies are rarities in Baltic amber. The existing specimens include a Zygoptera of the family Agrionidae illustrated by BACHOFEN-ECHT (1949). Two other amber damselflies, which are relatively well-preserved, are the only ones to have been examined up to now. PFAU (1975) suggests that the two Zygoptera are closely related to the Platycnemididae *Platycnemis*? *antiqua* (PICTET & HAGEN, 1856) had been described 120 years earlier.

As merolimnetic insects, dragonfly and damselfly nymphs require an aquatic environment, which they leave for the first time as fully developed nymphs to molt on land into adults. Prior to molting, they climb up littoral plants or tree trunks, where they rest in a fixed position. The exuviae remain behind in this position and decay only very gradually. As a result, exuviae were captured much more frequently than nymphs by the flowing resin in the amber forest. It is also possible that the exuviae were torn away by the wind and blown onto the resin. The first exuviae recorded from Baltic amber were described by HAGEN (1848, 1856). WEIDNER (1958) studied a nymph exuvia in which the two front legs, the dorsal side of the head, the thorax of the right half of the body and the wing pads were preserved, thus permitting assignment of the exuvia to the family Platycnemididae.

As first representative of Epallagidae *Litheuphaea ludwigi* BECHLY, 1998 is described from Baltic amber. Taking into account the growing number of new descriptions of fossil Odonata, a revision of the dragonflies and damselflies (BECHLY, in prep. 2001 a, b) and the odonate larvae in Baltic amber (BECHLY & WICHARD, in prep. 2001) may lead to the identification of even more families.

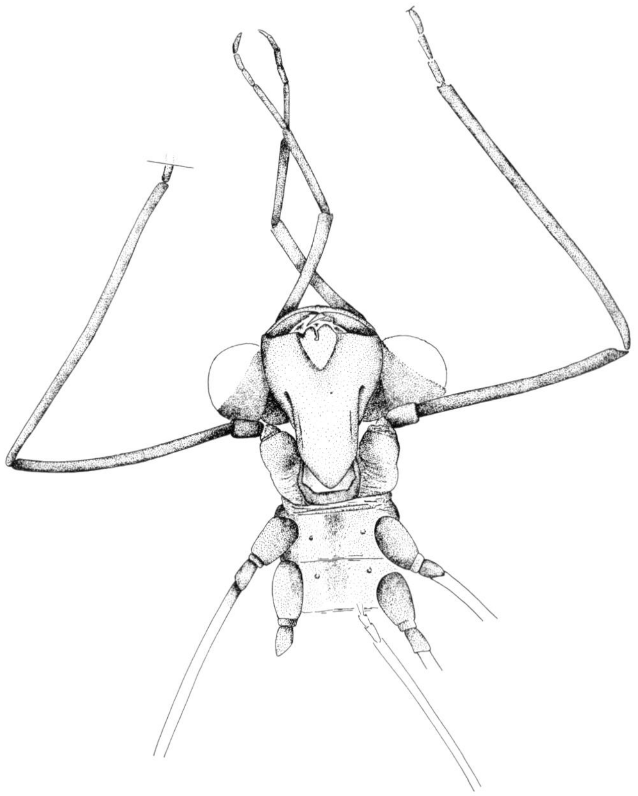

Fig. 45: Head of a Dicteriadidae (= Heliocharitidae) larvae in Baltic amber, from ventral view with mask (labium), antennae, and leg (after BECHLY & WICHARD, in prep. 2001).

Plate 28: Damselflies (Insecta: Odonata) in Baltic amber.

a Zygoptera, Coenagrionoidea: anterior region of a damselfly, ×8.
b Zygoptera, Coenagrionoidea: wings of a damselfly, ×4.5.
c Zygoptera, Coenagrionoidea: right forewing and hind wing of a damselfly, ×7.

a

b

c

2.29 Stoneflies – Insecta: Plecoptera

Although stoneflies (Plecoptera) are not frequent in Baltic amber, they have been found quite regularly. Apart from one exception, the 14 species described to date belong to genera and families that are still widely distributed in Europe today. Only *Megaleuctra neavei* Ricker, 1936 represents a currently North American genus that was verified to have a Holarctic distribution in the Tertiary (Illies 1967). In addition to Baltic amber, stoneflies have also been found in Oligocene Dominican and Upper Cretaceous Siberian amber (Wichard & Weitschat 1996).

While fully developed larvae are occasionally found in amber, the exuviae and particularly the winged imagoes are most common. The mature larvae leave the water and often climb up nearby tree trunks to shed their last exuvia and become winged imagoes. As stoneflies develop only a low affinity for flying and usually walk rather than fly in the event of danger, in search of food or to mate, their dispersal is minimal and often limited to the immediately vicinity of their original aquatic habitat. The stoneflies embedded in tree resins and preserved as amber inclusions provide not only information on the stonefly fauna of their Eocene past, but also evidence of the immediate biotope.

According to these indications, the "amber forest" was networked with running waters, from shallow streams to rapidly flowing rivers. In view of their limited flying activity, the stonefly exuviae imply that the resin-producing trees grew right up to the banks of the rivers and streams. The water, which the sensitive larvae prefer today and presumably preferred back then, was clear and certainly oxygen-rich. Stonefly larvae prefer running waters with a moderate to strong current, which are concentrated in mountainous regions. The hills and mountains dominating the landscape provided for the necessary current.

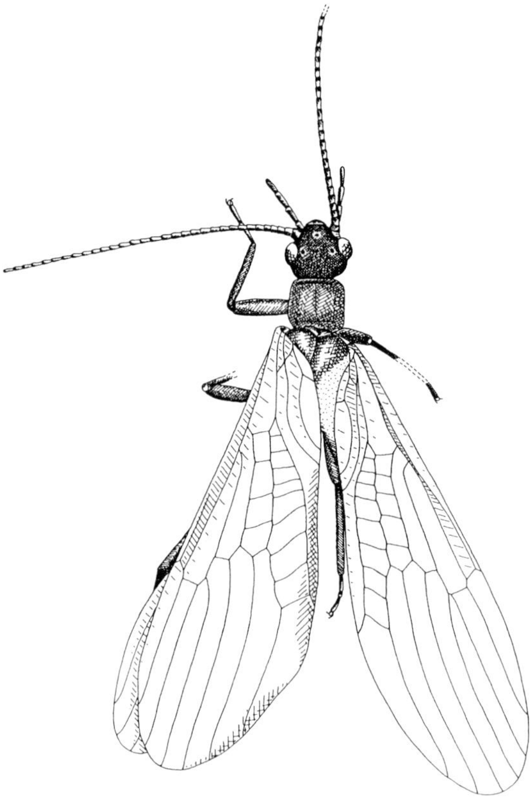

Fig. 46: Rolled-wing stonefly (Plecoptera, Leuctridae) (from the illustration collection of J. Illies).

Plate 29: Stoneflies (Insecta: Plecoptera) in Baltic amber I.

a Perlidae: common stonefly, × 8.
b Perlidae: common stonefly, nymph, × 5.
c Perlidae: common stonefly, nymph, × 25.
d Perlidae: common stonefly, nymph with filamentous gills, × 25.
e Perlidae: common stonefly, × 12.
f Perlidae: common stonefly, head and eyes, ommatids and antennae, × 25.
g Perlodidae: perlodid stonefly, × 9.
h Perlodidae: perlodid stonefly, exuvia, × 15.

2.30 Stonefly species in Baltic amber – Insecta: Plecoptera

Our current knowledge of stonefly species in Baltic amber is based on studies from the year 1856. The only other more recent publication is from Ricker (1935): *Megaleuctra*, which is today distributed in North America, has also been recorded there from the Miocene and occurs in Europe in Eocene Baltic amber.

This list of amber stoneflies, with 14 described species, stems from the systematics of the Plecoptera according to Illies (1965):

Leuctridae

Leuctra fusca Pictet, 1856
Leuctra gracilis Pictet, 1856
Leuctra linearis Hagen, 1856
Leuctra minuscula Hagen, 1856
Megaleuctra neavei Ricker, 1936

Nemouridae

Nemoura affinis Berendt, 1856
Nemoura lata Hagen, 1856
Nemoura ocularis Pictet, 1856
Nemoura puncticollis Hagen, 1856

Perlidae

Perla prisca Pictet, 1856

Perlodidae

Isoperla succinica (Hagen, 1856)
Perlodes resinata (Hagen, 1856)

Taeniopterygidae

Taeniopteryx ciliata Pictet, 1856
Taeniopteryx elongata Hagen, 1856

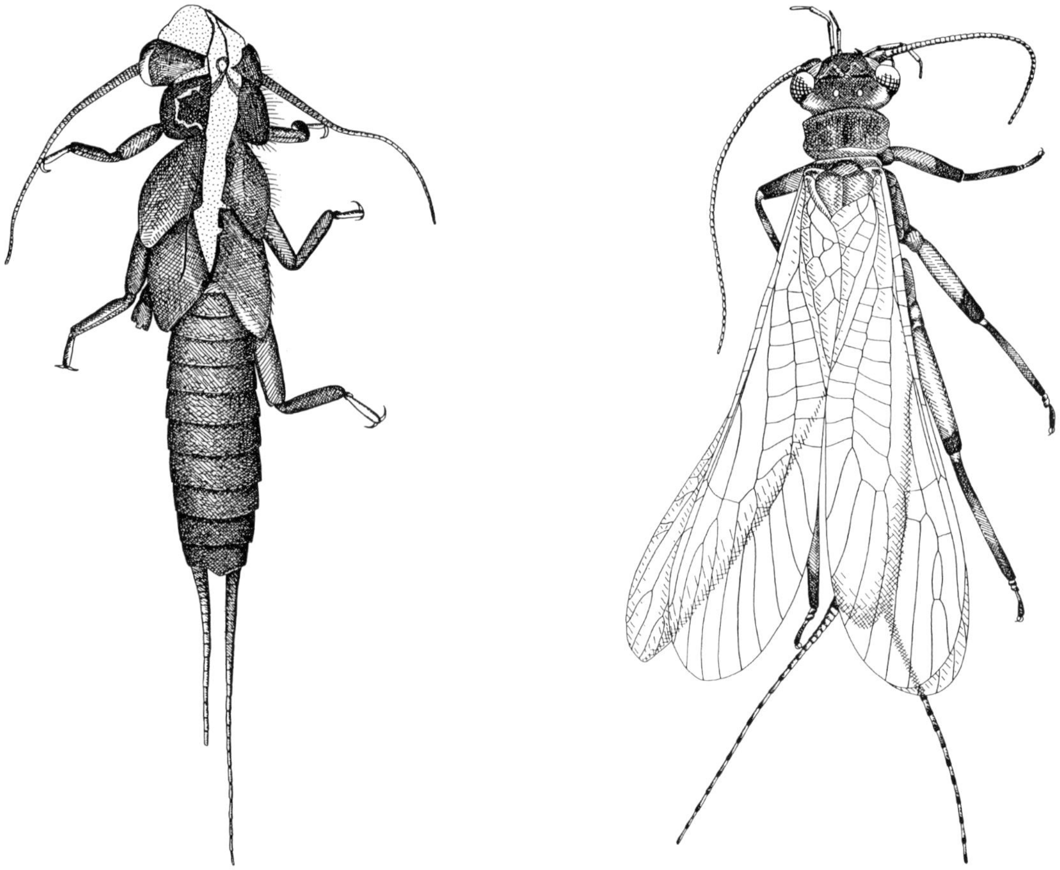

Fig. 47: Common stonefly (Plecoptera, Perlidae), exuvia and imago (from the illustration collection of J. Illies).

Plate 30: Stoneflies (Insecta: Plecoptera) in Baltic amber II.

a Leuctridae: rolled-winged stonefly, female with eggs, ×8.
b Leuctridae: rolled-winged stonefly larva, ×10.
c Leuctridae: rolled-winged stonefly, ×9.
d Leuctridae: rolled-winged stonefly, ×9.
e Taeniopterygidae: winter stonefly, ×6.
f Leuctridae: rolled-winged stonefly, ×8.
g Nemouridae: spring stonefly, ×9.
h Nemouridae: spring stonefly, ×10.

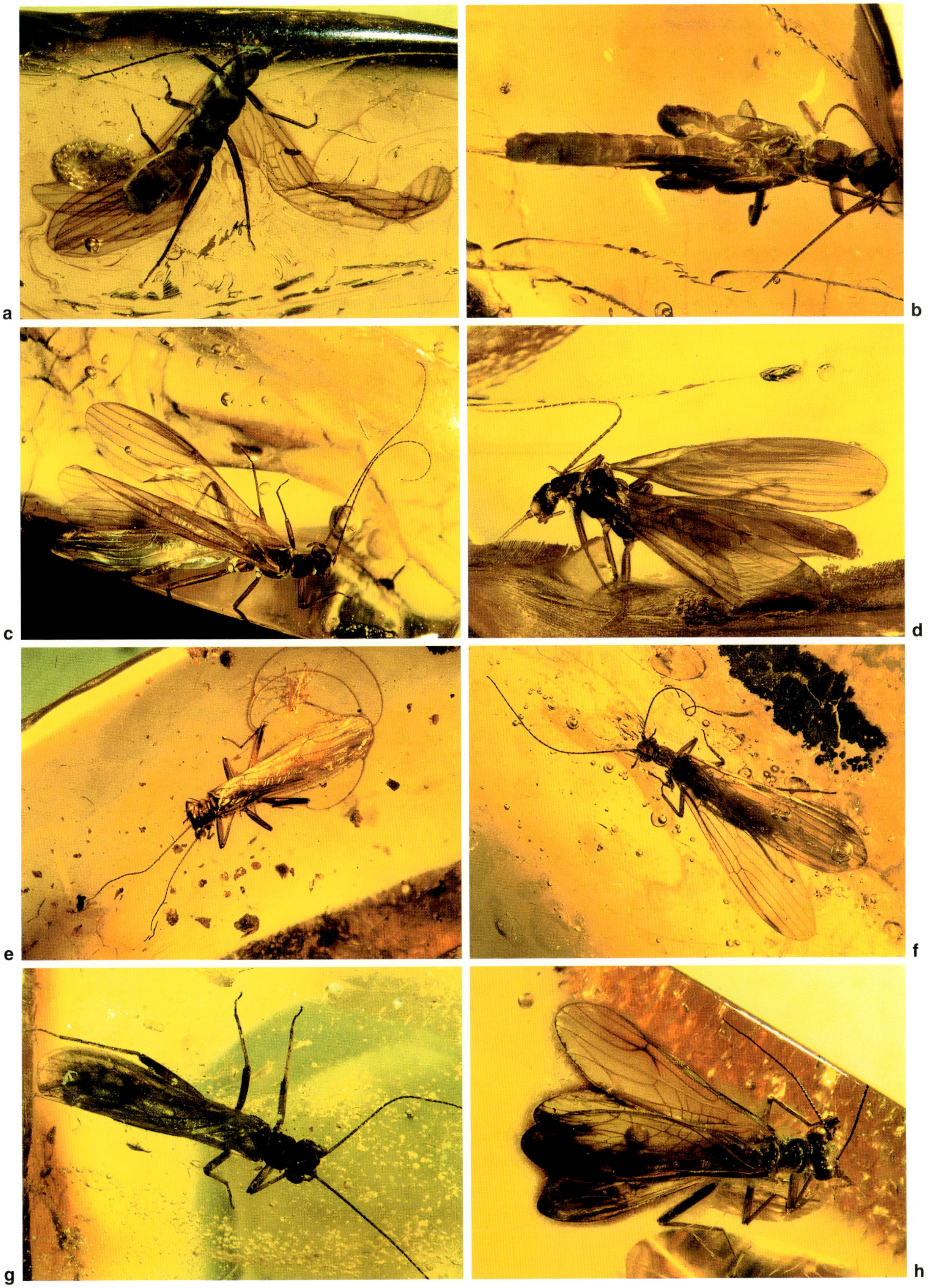
a
b
c
d
e
f
g
h

2.31 Web spinners – Insecta: Embiidina

Embiids (web spinners) are distributed not in the Baltic region, but rather in southern Europe, the subtropics and the tropics. They reach a maximum length of 20 mm and display distinct sexual dimorphism. While the males of most species are winged, the females are always wingless. The males are nocturnal, becoming active at twilight. During the day, they keep their wings folded flat on top of the abdomen.

The common name, "web spinners", refers to the ability of the embiids to produce silk from silk glands located on the first and second tarsal segments of the forelegs. They usually spin flat or tunnel-like galleries under stones. To do so, they rapidly move the forelegs back and forth along with the silk, attaching it as they progress. The insects are protected by these galleries and usually only leave them at night.

Two-hundred extant species divided into eight families have been described, compared to only one fossil species from Baltic amber: *Electroembia antiqua* (Pictet, 1854) of the family Embiidae. The description of this species is based on a wingless male, which was later redescribed by Ross (1956) and given the genus name *Electroembia*. Bachofen-Echt (1949) also illustrated a wingless male that is possibly conspecific with *E. antiqua*.

Plate 31 e, f shows two winged males. They are one of the rarities in Baltic amber and apparently represent further fossil Embiidina species. These winged males differ in form from the previously described species *E. antiqua* (Pictet, 1854), whose males are apterous.

Fig. 48: Web spinner *Electroembia antiqua* (Pictet, 1854) (Embiidina: Embiidae).

Plate 31: Web spinners (Insecta: Embiidina) in Baltic amber.

a-c *Electroembia* sp. (Embiidae), with head and swollen forelegs, **a** ×6, **b** ×15, **c** ×6.
d *Electroembia antiqua* (Pictet, 1854) (Embiidae), ×8.
e-f Winged males of Embiidina species, **e** ×5, **f** ×6.
g-h Wingless females of Embiidina species, **g** ×6, **h** ×6.

2.32 Earwigs – Insecta: Dermaptera

Earwigs (Dermaptera) are nocturnal, epedaphic insects primarily inhabiting the leaf litter. The nymphs and imagoes are considered to be omnivores that feed on fresh plant material, debris, mosses, lichens and fungal hyphae, but also hunt small insects, caterpillars, maggots, woodlice, millipedes and spiders. The feeding preferences of earwigs appear to be family-specific. While the Forficulidae mainly consume plant material, the primitive forms, particularly the Labiduridae, are predaceous. They raise the abdomen and use the cerci to grab their prey (Plate 32 b), which they immediately transfer to the biting mouthparts.

In the adults, the folded hind wings are located underneath the short, but very sclerotized tegmina (or hemelytra). For flight, the tegmina are pulled up on the sides, allowing the stable hind wings to unfold (Plate 32 a). The unfolding mechanism and the stability of the wings require static elements consisting of elastic joints, various bracing veins and radial folding lines (Kleinow 1966). These structures form an impressive wing pattern that can be observed in extant species, as well as in fossil earwigs in Baltic (Plate 32 a) and Dominican amber (Schlee 1980).

As earwigs seldom fly and instead move about on the ground, the hind wings are usually folded and hidden under the tegmina. The short tegmina are not much longer than the subsequent segments of the abdomen. The adaptation of the tegmina to the body segmentation makes these slender insects very agile and enables them to maneuver through the pore system of the soil. Earwigs would not be adapted to a terrestrial lifestyle, if the rigid tegmina covered the entire length of the abdomen, as they do in many beetles.

Fossil earwigs have been found in Burmese, Dominican, Mexican and Baltic amber, including its Bitterfeld deposits. The taxonomy of fossil earwigs from Baltic amber was last covered by Burr (1911). Imagoes are described from four species of the genus *Forficula*. The larvae are classified in the genera *Forficula*, *Labidura* and *Pygidicrana*.

Forficulidae

Forficula baltica Burr, 1911
Forficula klebsi Burr, 1911
Forficula praecursor Burr, 1911
Forficula ? pristina Burr, 1911

Labiduridae

Labidura ? sp.

Pygidicranidae

Pygidicrana ? sp.

Plate 32: Earwigs (Insecta: Dermaptera) in Baltic amber.

a Earwig imago with the left tegmen raised and the hind wing extended, × 11.
b Earwig nymph – as if in predatory position – with raised abdomen and cerci, × 20.

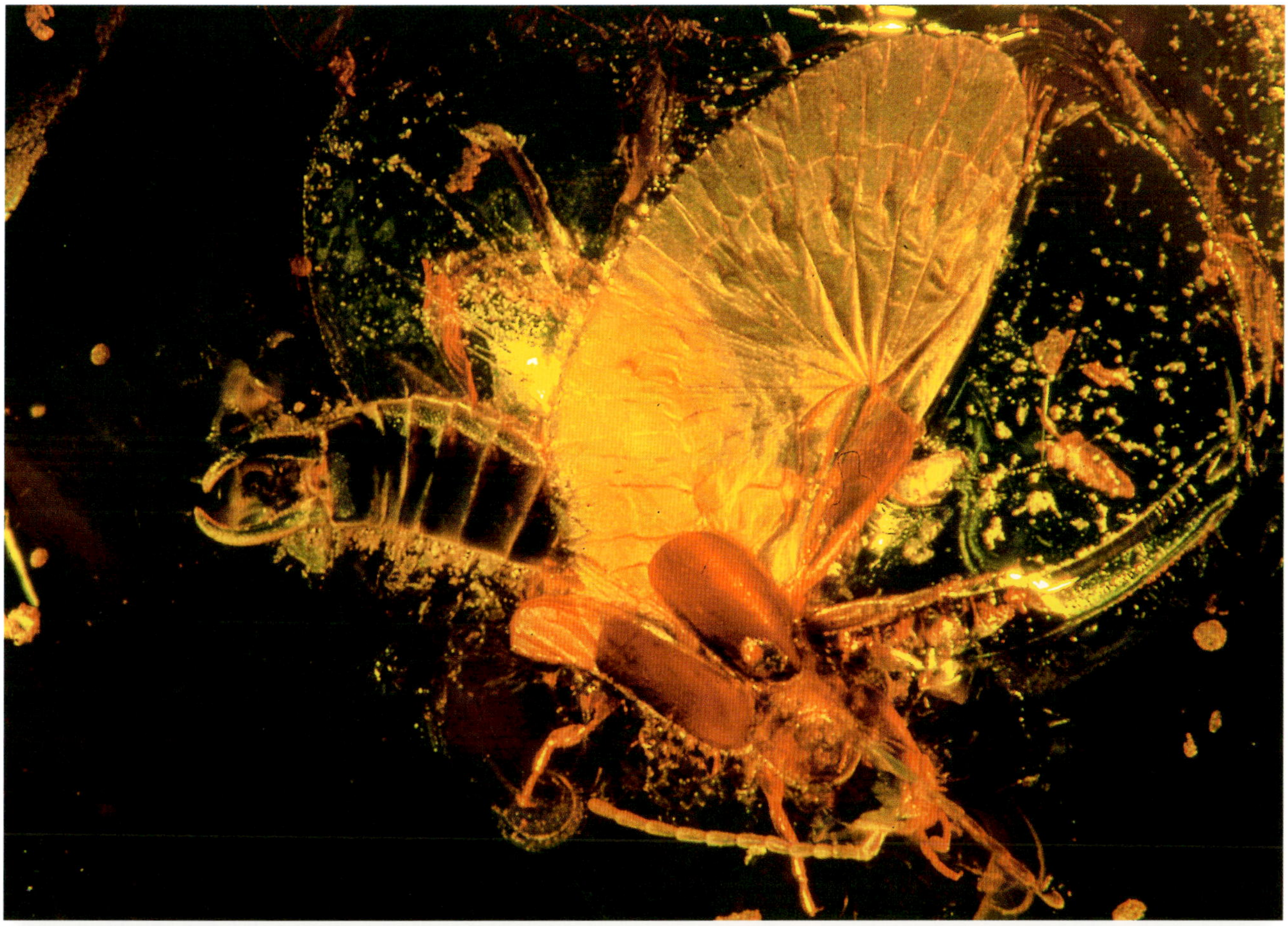

a

b

2.33 Mantids – Insecta: Mantodea

Praying mantids are not as harmless as their name suggests. They are diurnal, predaceous animals that hold their forelegs in an upraised position, as if in prayer, while lying in wait for prey. The coxae of the forelegs are elongated and the femorae equipped with two, parallel, longitudinal rows of pronounced spines. The tibiae, which also bear spines, snap against the femorae like a switchblade to capture prey. The very mobile head is equipped with a dominant pair of eyes for locating prey by sensing its motion. Once the praying mantid is within sufficiently close range, it triggers its rapid capture mechanism (Beier 1968, Hevers & Liske 1991).

The discovery of praying mantids in Baltic amber is a reliable indication of a dry, warm and at least subtropical climate in areas of the Eocene "amber forest". More than 20 specimens (primarily nymphs) from four families are known from Baltic amber: Liturgusidae, Chaeteessidae, Mantoididae and Mantidae (Ehrmann 1999). Most of the mantids belong to the family Mantidae. This is still the largest family today and includes *Mantis religiosa*, which occurs in central and particularly southern Europe (Brechtel et al. 1996).

Two praying mantids from the families Mantidae and Chaeteessidae have been recorded to date from the Bitterfeld deposits. If these deposits were to be mined more extensively, a greater number of mantids could be expected to be found, including the two, rare families reported from Baltic amber. However, the identification of the two praying mantids provides further evidence that Bitterfeld and Baltic amber are of identical origin (Weitschat 1997). The original dating of Bitterfeld amber in the Lower Miocene (Barthel & Hetzer 1982, Fuhrmann & Borsdorf 1986, Krumbiegel & Kosmowska-Ceranowicz 1989) contradicts global climate development, which indicates significantly lower temperatures starting in the Oligocene (Fig. 18). These conditions would have made it impossible for the Mantodea and other subtropical faunal elements to survive. Amber from Lower Miocene sediments was certainly redeposited several times and is at least in its third deposit (cf. Chap. 1.4.4). Contrary to initial assumptions, the Bitterfeld deposits are not located where Bitterfeld amber originally formed.

Fig. 49: Praying mantid of the family Chaeteessidae (Mantodea) with head and raptorial legs.

Plate 33: Mantids (Mantodea) in Baltic amber.

a Praying mantid (Chaeteessidae), × 8.
b Praying mantid (Mantoididae), × 10.
c Praying mantid (Mantidae), × 12.

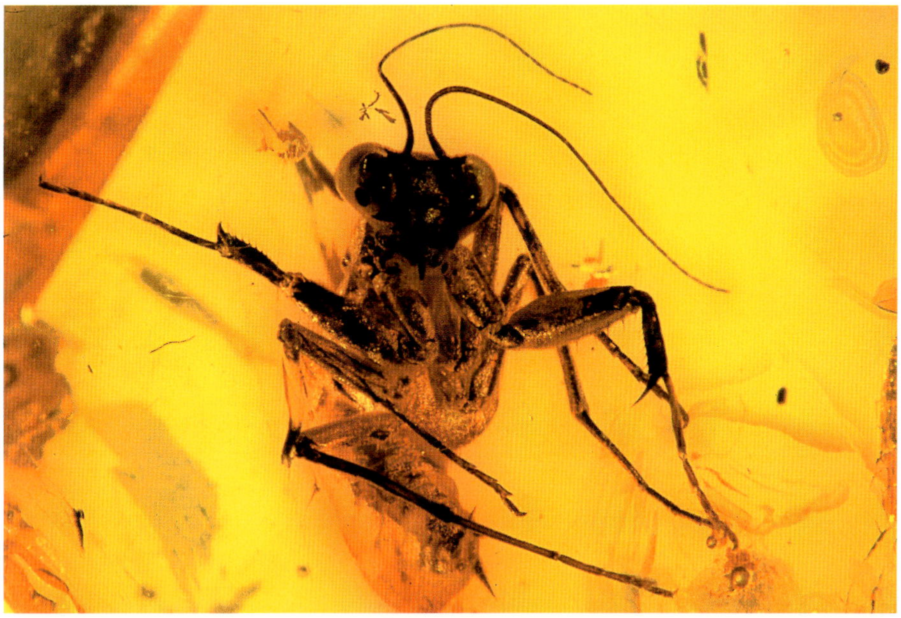
a

b

c

2.34 Cockroaches – Insecta: Blattaria

Cockroaches are among the larger animals found regularly, but not frequently in Baltic amber. Males are more common than females. Wingless nymphs are just as numerous as imagoes. GERMAR (1813), BERENDT (1836), GERMAR & BERENDT (1856), GIEBEL (1856), SHELFORD (1910, 1911) and PITON (1940) covered the Blattaria. SHELFORD primarily examined the KLEBS collection and that of the British Museum in London, describing numerous fossil species he exclusively classified in present-day genera. This group of fossil cockroaches includes the extant species *Euthyrrhapha pacifica*, which is today distributed in the southern hemisphere in South America, Africa, Madagascar and Polynesia. The cockroach family Blattellidae (Phyllodromiidae) has the most fossil representatives in Baltic amber. This family mainly occurs in tropical and subtropical regions (Australia, South America).

Blattellidae

Blatella baltica (SHELFORD, 1910)
Blatella furcifera (SHELFORD, 1910)
Blatella klebsi (SHELFORD, 1910)
Blatella latissima (SHELFORD, 1910)
Blatella praecursor (SHELFORD, 1910)
Blatella pristina (SHELFORD, 1910)
Blatella tenacula (SHELFORD, 1910)
Blatella woodwardi (SHELFORD, 1910)
Blatella yolanda (SHELFORD, 1910)
Ceratinoptera cruenta SHELFORD, 1910
Ceratinoptera (Blatta) didyma (GERMAR & BERENDT, 1856)
Ceratinoptera klebsi SHELFORD, 1910
Ceratinoptera miocenica SHELFORD, 1911
Ceratinoptera soror SHELFORD, 1910
Ischnoptera (Blatta) gedanensis (GERMAR & BERENDT, 1856)
Ischnoptera klebsi SHELFORD, 1910
Ischnoptera perplexa SHELFORD, 1910
Margattea (Blatella) germari (SHELFORD, 1910)
Margattea (Blatella) lorenmeyeri (SHELFORD, 1910)
Pseudophyllodromia succinica SHELFORD, 1911
Symploce (Blatella) antiqua (SHELFORD, 1910)
Temnopteryx klebsi SHELFORD, 1910

Blattidae

Blatta baltica GERMAR & BERENDT, 1856
Blatta berendti GIEBEL, 1856
Blattina succinea GERMAR, 1813
Periplaneta succinica SHELFORD, 1910
Polyzosteria parvula GERMAR & BERENDT, 1856
Polyzosteria tricuspidata (BERENDT, 1836)

Ectobiidae

Ectobius balticus (GERMAR & BERENDT, 1856)
Ectobius inclusus SHELFORD, 1910
Hololampra succini PITON, 1940

Euthyrrhaphidae

Euthyrrhapha pacifica (COQUEBERT)
Holocompsa fossilis SHELFORD, 1910

Nyctiboridae

? *Nyctibora succinica* SHELFORD, 1910

Perisphaeriidae

Larve

Polyphagidae

Polyphaga fossilis SHELFORD, 1910

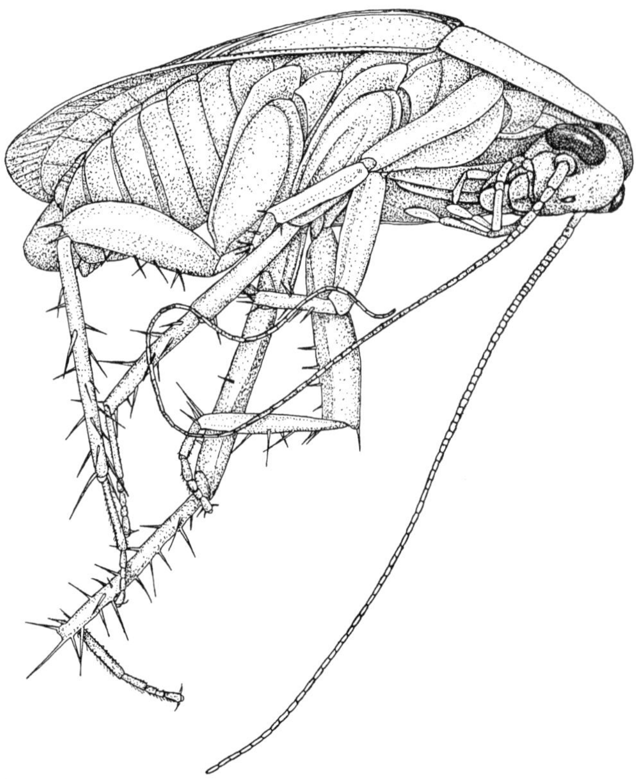

Fig. 50: Cockroach imago.

Plate 34: Cockroaches and cockroach nymphs (Insecta: Blattaria) in Baltic amber.

a ×8, **b** ×6, **c** ×6, **d** ×5, **e** ×4, **f** ×5, **g** ×3, **h** ×7.

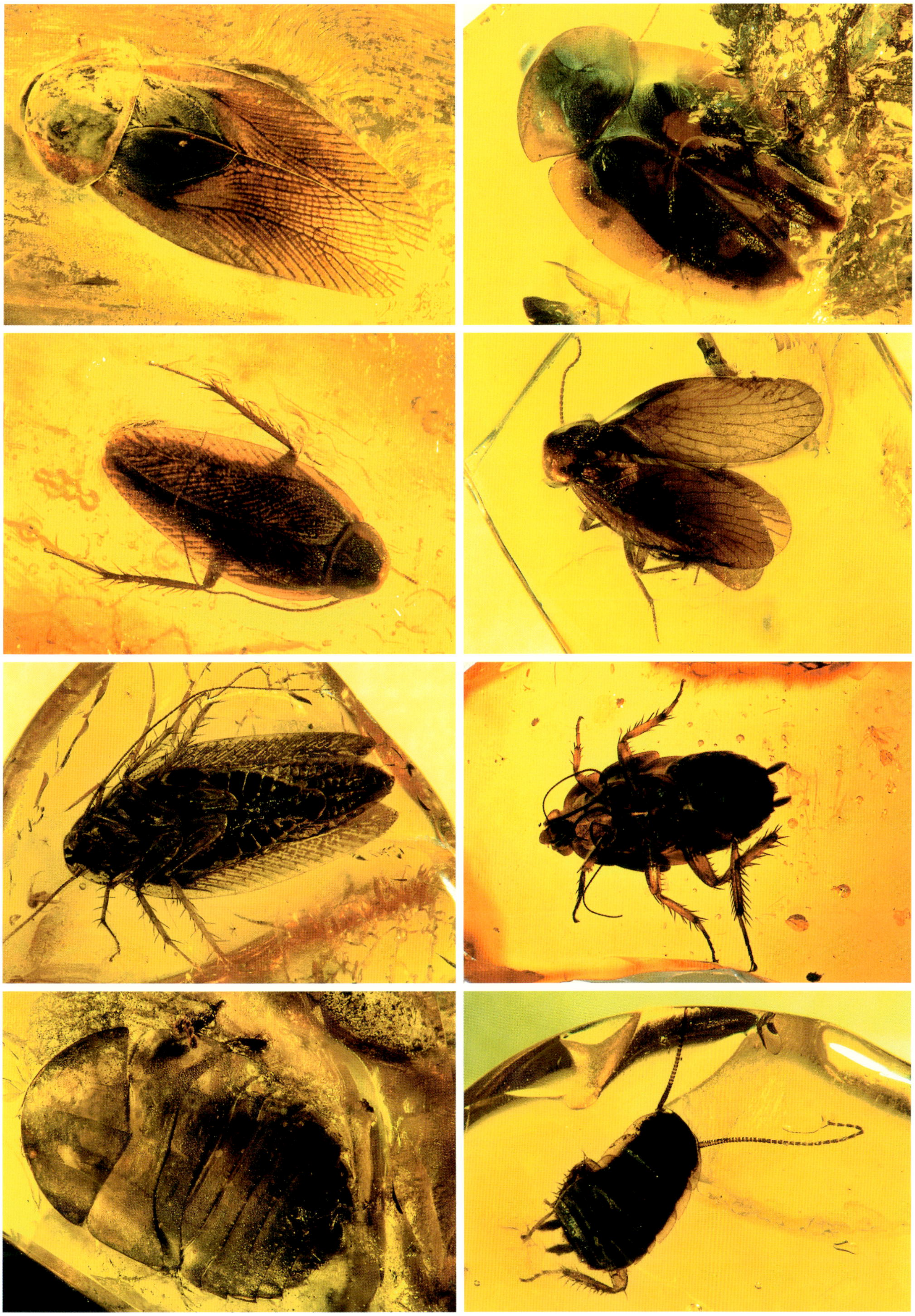
a b c d e f g h

2.35 Termites – Insecta: Isoptera

As subtropical insects, termites are typical members of the Baltic amber fauna. They are more frequent than Dermaptera (earwigs), Mantodea (mantids), Blattaria (cockroaches), Phasmatodea (walking sticks) and Orthoptera (grasshoppers, crickets, katydids, etc.). The last comprehensive revision prepared by WEIDNER (1955) recognizes eight, easily distinguishable species in three families. WEIDNER (1955) also included an identification key for these fossil species:

Termopsidae

Archotermopsis tornquisti ROSEN, 1913
Termopsis bremii (HEER, 1849)

Kalotermitidae

Electrotermes affinis (HAGEN, 1856)
Electrotermes girardi (GIEBEL, 1856)
Proelectrotermes berendti (PICTET, 1856)

Rhinotermitidae

Parastylotermes robustus (ROSEN, 1913)
Reticulitermes antiquus (GERMAR, 1813)
Reticulitermes minimus (SNYDER, 1928)

NATHAN SENDEL (1686-1757) provided the first illustration of a termite from Baltic amber. According to GERMAR (1813), the illustration in SENDEL's "Historia Succinorum" from 1742 represents *Reticulitermes antiquus*. Various authors have covered fossil termites since then (GERMAR 1813, HEER 1849, GIEBEL 1856, PICTET-BARABAN & HAGEN 1856, HAGEN 1858, ROSEN 1913, SNYDER 1928 and KRISHNA 1961).

The most common termite species in Baltic amber is *Reticulitermes antiquus* (Rhinotermitidae). According to WEIDNER (1955), extant species of this genus mainly inhabit dead pinewood. As termites form large swarms, it is easily conceivable that they were driven against the resin-producing trees during swarming flight, or caught in the resin thereafter while searching for suitable nesting sites, and then covered by a fresh flow of resin. Consequently, only winged termites are usually preserved, including specimens that shed their wings shortly after swarming and left them lying off to the side in the amber. Workers, soldiers and nymphs are very rare, because they primarily inhabit the trunks of trees. HAGEN (1858) was the first to report a nymph from Baltic amber. Two more nymphs and a soldier are shown in Plate 35 g, h.

The composition of the termite fauna in Baltic amber is apparently determined by ecological factors (WEIDNER 1955). Termites from the families Termitidae and Mastotermitidae, which today have a largely tropical and subtropical distribution, have not been found in amber. In contrast, the families Termopsidae and Rhinotermitidae, which currently inhabit warm climates and temperate elevations, have been recorded from amber. For example, *Archotermopsis wroughtoni* occurs at up to 2,700 meters in the northwestern Himalayas. The Termopsidae are also strictly pinewood termites. The only family with no basis for comparison is Kalotermitidae. The extinction of the three amber termite species may be related to climate change and the extinction of their preferred tree species.

Plate 35: Termites (Insecta: Isoptera) in Baltic amber.

a *Termopsis bremii* (Termopsidae), ×4.
b *Termopsis bremii* (Termopsidae), ×5.
c *Reticulitermes antiquus* (Rhinotermitidae), ×6.
d *Reticulitermes antiquus* (Rhinotermitidae), ×6.
e *Reticulitermes minimus* (Rhinotermitidae), ×8.
f *Electrotermes affinis* (Kalotermitidae), ×7.
g Termite larvae, ×8.
h Termite soldier, ×9.

a
b
c
d
e
f
g
h

2.36 Walking sticks – Insecta: Phasmatodea

The herbivorous Phasmatodea (stick-insects and walking-leaves) are distributed in tropical and subtropical areas. Most of the species resemble parts of plants like branches and leaves. The order is subdivided into three suborders. The Timematodea, which have only three tarsomeres, include only a few species distributed in the Southwestern USA. The Areolatae and Anareolatae include the majority of the genera and species. They have five tarsomeres. The Areolatae are characterized by a small, sunken, triangular area on the tibiae, which is absent in the Anareolatae.

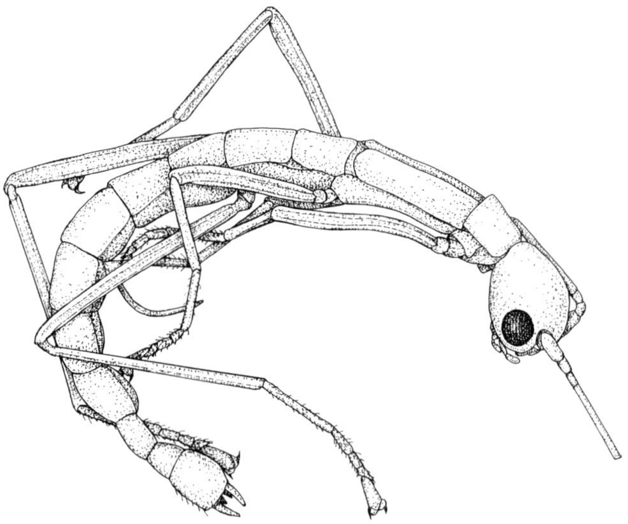

Fig. 51: Nymph of a walking stick (Phasmatodea: Archipseudophasmatidae).

In Recent phasmids the wings are developed or completely absent; except for the walking-leaves (Phylliidae) the forewings (tegmina) are always reduced and shorter than the hindwings (alae).

The phasmids found in Baltic Amber belong mainly to the extinct Archipseudophasmatidae, a family of the Areolatae, which are characterized by a strongly elongated third antennomere and fully developed tegmina. Members of Pseudophasmatidae or the suborder Anareolatae are very rare in Baltic Amber, both are only known from nymphs.

The nymphs are encountered most frequently in Baltic amber. Four species have been described to date (cf. ZOMPRO, 2001):

Archipseudophasmatidae

Pseudoperla gracilipes PICTET & BERENDT, 1854
Balticophasma lineata PICTET & BERENDT, 1854
Archipseudophasma phoenix ZOMPRO, 2001

Pseudophasmatidae

Electrobaculum gracile SHAROV, 1968

Several of the amber inclusions normally treated as phasmids do not belong to this order. In all probability they belong to a new orthopterean order. One genus and species is described to date:

Raptophasma kerneggeri ZOMPRO, 2001 (Plate 63 f)

Plate 36: Walking sticks (Insecta: Phasmatodea) in Baltic amber.

a *Balticophasma lineata* (PICTET & BERENDT, 1854), × 9.
b *Archipseudophasma* sp., nymph, × 8.
c *Balticophasma* sp., nymph, × 9.
d *Balticophasma* sp., nymph, × 8.
e *Balticophasma* sp., nymph, × 10.
f *Raptophasma kerneggeri*, nymph, × 10.
g *Balticophasma* sp., nymph, × 7.
h *Pseudoperla* sp., nymph, × 8.

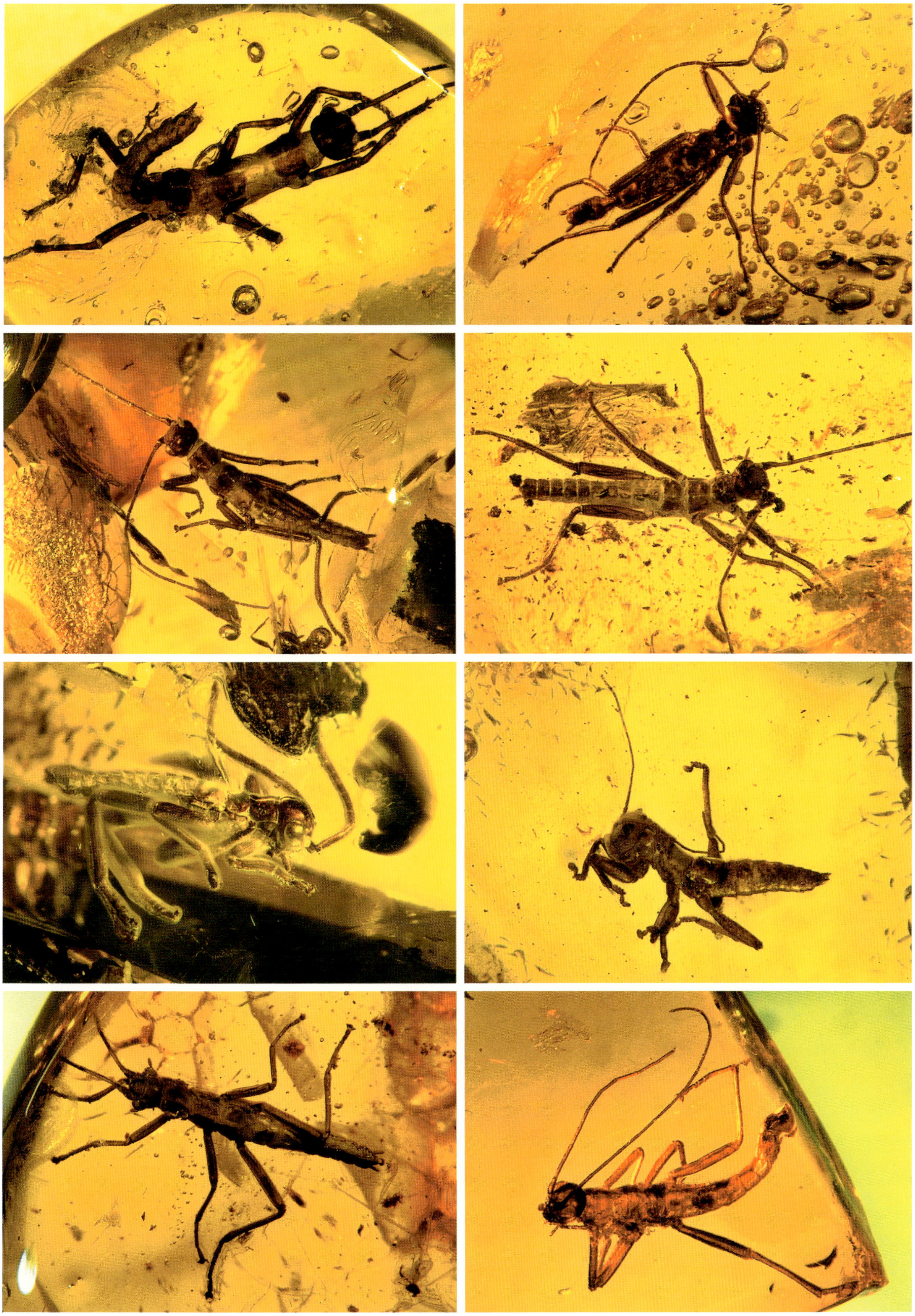

a b c d e f g h

2.37 Orthopterans – Insecta: Orthoptera

The orthopterans (Orthoptera = Saltatoria) include crickets, katydids, grasshoppers and locusts. The common feature of all Orthoptera is the development of the hind legs into elongate jumping legs. The jumping muscle is located in the swollen femur. The wings lie flat or rooflike over the abdomen when the insect is at rest. In nymphs, the wing pads of the front wings are often covered by the fanned pads of the back wings. Most of the Saltatoria can produce sounds by rubbing together the stridulating organs located at the base of the two front wings (Ensifera) or on the hind femorae and front wings (Caelifera).

The orthopterans are divided into two taxonomic groups: long-horned orthopterans (Ensifera), including katydids (Tettigonioidea), camel crickets (Gryllacridoidea) and crickets (Grylloidea), and short-horned orthopterans (Caelifera), including field grasshoppers (Acridoidea). The majority of orthopterans found in Baltic amber are long-horned orthopterans (Ensifera). Only two field grasshopper species (Caelifera: Acridoidea) of the family Tetrigidae have been recorded from Baltic amber.

Weidner (1956) covered the Orthoptera of Baltic amber. The comparison he made to contemporary relatives of the amber Orthoptera revealed the existence of relict groups in the humid, Indo-Malayan jungle. They are forest-dwellers that scale trees in search of insect prey, or wingless grasshoppers (Rhaphidophora) that occur in tropical Asia and in caves in the Mediterranean. Weidner concluded that Ander (1942) came closest to the truth when he suggested that the amber forest had a warm, temperate climate characterized by high humidity and minimal temperature fluctuations.

Ensifera

Tettigonioidea: Tettigonidae

Eomortoniellus handlirschi Zeuner, 1936
Lipotactes (?) *bispinatus* Weidner, 1956
Lipotactes martynovi Zeuner, 1936

Gryllacridoidea: Rhaphidophoridae

Rhaphidophora antiqua Chopard, 1936
Rhaphidophora tachycinoides Chopard, 1936
Rhaphidophora zeuneri Chopard, 1936
Protroglophillus sukatshevae Gorochov, 1989

Grylloidea: Gryllidae

Acheta (?) sp. (Chopard, 1936)
Heterotypus septentrionalis Chopard, 1936
Madasumma europensis Chopard, 1936
Stenogryllodes brevipalpis Chopard, 1936
Trichogryllus macrocercus (Germar & Berendt, 1856)

Caelifera

Tetrigoidea: Tetrigidae

Acrydium (s.l.) *bachofeni* Zeuner, 1937
Succinotettix chopardi Piton, 1918

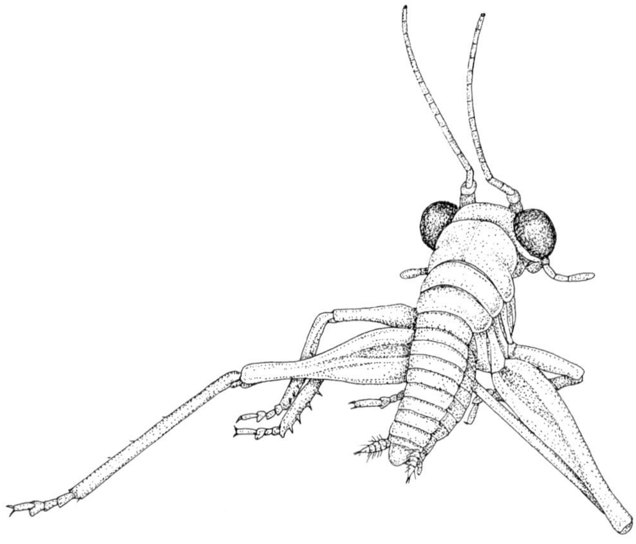

Fig. 52: Cricket nymph.

Plate 37: Orthopterans (Insecta: Orthoptera) in Baltic amber.

a-h Nymphs of long-horned orthopterans (Ensifera), **a** ×5, **b** ×4, **c** ×4, **d** ×5, **e** ×4, **f** ×7, **g** ×5, **h** ×5.

a b c d e f g h

2.38 Psocids – Insecta: Psocoptera

The information on psocids (barklice or booklice) from Baltic amber is based on the comprehensive monograph of the fossil Copeognatha and their phylogeny by ENDERLEIN (1911 a). This pioneering work was later supplemented by ROESLER (1943). Another new description originates from the Bitterfeld deposits: *Embidopsocus saxonicus* GÜNTHER, 1989. The phylogenetic relationships within the Copeognatha or Psocoptera have since been revised in parts; several new families were established (cf. WEIDNER 1972). A total of 30 species are described from Baltic amber, these being divided into roughly 20 genera and the following 11 families:

ATROPIDA
- **Amphientomidae**
- **Elipsocidae**
- **Liposcelidae**
- **Philotarsidae**
- **Sphaeropsocidae**
- **Trogiidae**

PSOCIDA
- **Archipsocidae**
- **Caeciliidae**
- **Epipsocidae**
- **Psocidae**
- **Trichopsocidae**

ENDERLEIN (1911 a) concluded that several species are very frequent in amber, while others are found only occasionally. The frequencies of amber Psocoptera do not correspond to those of extant species, but rather reinforce the impression that species expected to be rare are frequent, and vice versa. Species that developed on the resin-producing trees and were more easily captured by the flowing resin are overrepresented. It is noticeable that frequent species are also represented in amber by nymphs in every instance: e.g. *Psocidus multiplex* ROESLER, 1943 (= *Copostigma affinis* (PICTET & HAGEN, 1856)). According to this interpretation, species that were not directly dependent on the resin-producing trees visited them only randomly and are therefore rare in amber.

The phylogenetically younger Psocida (Isotectomera) are generally more species-diverse and abundant than the older Atropida (Heterotectomera). This distribution pattern agrees with that of extant tropical fauna. ENDERLEIN (1911 a) therefore suggested that the ground plan of the modern Psocoptera was already evolved in the Mesozoic and remained constant beyond the Tertiary (Eocene).

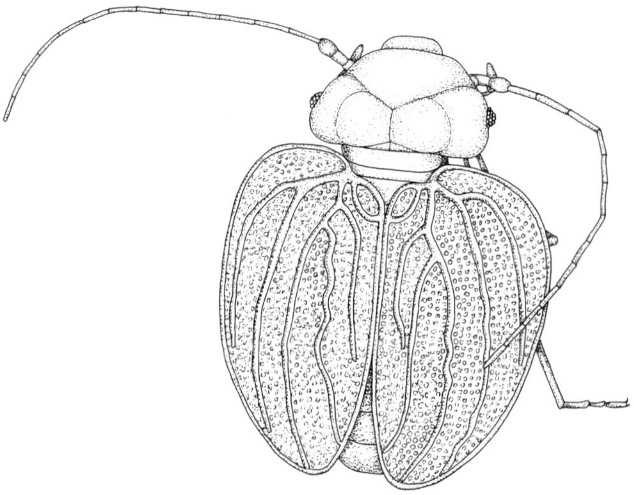

Fig. 53: *Sphaeropsocus* cf. *kuenowi* HAGEN, 1882, psocid (Psocoptera: Sphaeropsocidae).

Plate 38: Psocids (Insecta: Psocoptera) in Baltic amber.

a *Archipsocus puber* HAGEN, 1882, × 25.
b *Archipsocus puber* HAGEN, 1882, × 20.
c *Sphaeropsocus kuenowi* HAGEN, 1882, × 25.
d *Sphaeropsocus kuenowi* HAGEN, 1882, × 35.
e *Psocidus affinis* (PICTET & HAGEN, 1856), × 10.
f *Psocidus* sp., × 12.
g *Epipsocus ciliatus* (PICTET & HAGEN, 1856), × 9.
h *Epipsocus ciliatus* (PICTET & HAGEN, 1856), × 11.

a

b

c

d

e

f

g

h

2.39 Thrips – Insecta: Thysanoptera

The Thysanoptera are commonly known as thrips. They are small (1 to 2 mm), slender and somewhat flattened insects with piercing-sucking mouthparts used to suck nectar or the juices from pollen, leaves, stems, bark and fungal hyphae. The nymphs and imagoes live in the leaf litter, in the herbaceous and shrub layers, and on trees. Some thrips are predaceous on aphids, scales and mites.

The last segment of the foot bears the arolium, which can be expanded as necessary to form a large adhesive organ. The wings are typically slender with only a few veins or none at all. The wings have a fringe of long hairs that significantly increase the surface area of the wing. The immatures are wingless, although the wing pads are already present in the nymphs. Except for the species with reduced wings, the Thysanoptera occur as aerial plankton and occasionally form large swarms.

The wingless immatures may have ended up in the resin by actively coming into contact with it, or simply being covered by a resin flow. The winged adults belong to the aerial plankton, i.e. they drift passively on the wind. It is probable that they were blown onto the trees and into the resin. Flightless species have apparently not been preserved in Baltic amber (cf. Larsson 1978).

The Thysanoptera of Baltic amber were covered by Bagnall (1914, 1924, 1929), Priesner (1924, 1929) and Stannard (1956). The fossil species are listed in Jacot-Guillarmod (1970) and Spahr (1992). Schliephake (1990, 1993, 1997, 1999 a, b, 2000) expanded the list of Tertiary Thysanoptera. Apart from undetermined immatures, the roughly 70 species from Baltic amber are divided among the following six families:

Terebrantia
- **Aeolothripidae**
- **Heterothripidae**
- **Merothripidae**
- **Opadothripidae**
- **Thripidae**

Tubulifera
- **Phlaeothripidae**

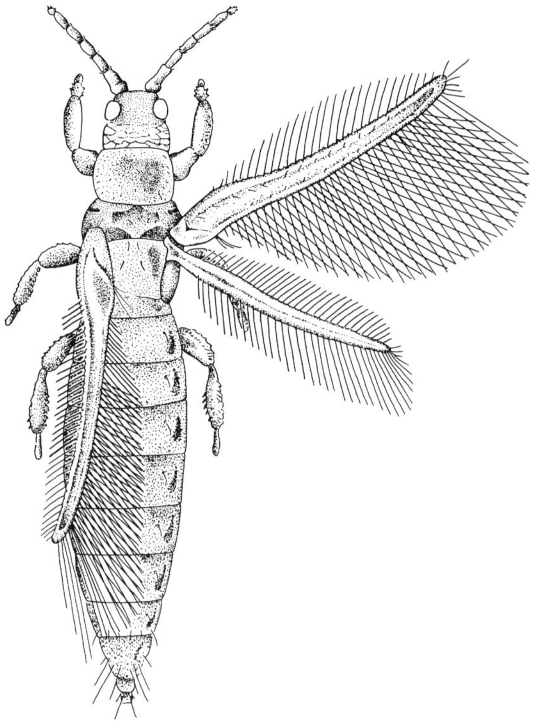

Fig. 54: An extant species: *Anaphothrips obscurus* (after Jacobs & Renner 1988).

Plate 39: Thrips (Insecta: Thysanoptera) in Baltic amber.

a ×32, **b** ×30, **c** ×35, **d** ×35, **e** ×35, **f** ×32, **g** ×33, **h** ×34.

a

b

c

d

e

f

g

h

2.40 Water bugs and water striders – Insecta: Heteroptera

The Heteroptera, or true bugs, are classified together with psyllids, whiteflies, aphids, coccids (Sternorrhyncha) and cicadas, leafhoppers, treehoppers, fulgoroids (Auchenorrhyncha) in the taxonomic group of the Hemiptera. Their monophyletic origin is based in part on the common structure of the piercing-sucking mouthparts. Most true bugs are plant-sucking or predaceous terrestrial insects, while others live in or on the water. The water striders (Gerromorpha) live on the surface film of freshwater, searching the surface for food, while the water bugs (Nepomorpha) live under water, leaving it only to disperse and mate.

Expectedly, only a few specimens of water bugs (Nepomorpha) have been found. In addition to a Nepidae mentioned in the early literature (ANDER 1942, BACHOFEN-ECHT 1949) and a Notonectidae mentioned by JORDAN (1953), BACHOFEN-ECHT (1949) also reported three nymphs of a Corixidae. Further Corixidae nymphs have recently been identified. It is rather astounding that nymphs of the aquatic Nepomorpha occur in amber, considering that these insects first leave the water as adults. Extant Corixidae inhabit shallow, standing or slowly running waters. The water may have dried up at that time, forcing the nymphs to search for nearby alternatives. The corixid nymphs trapped in amber (Plate 40 f) appear as if they were using their extended swimming legs to actually swim in the liquid resin (WICHARD & WEITSCHAT 1996).

Water striders (Gerromorpha) have also been recorded (GERMAR & BERENDT 1856, BACHOFEN-ECHT 1949, LARSSON 1978, WEITSCHAT & WICHARD 1996, POPOV 1996, ANDERSEN 1982, 1998, 2000). Although the wingless adults and the nymphs of the water striders normally live on the water surface on ponds or streams, some specimen leave the semiaquatic habitats or must leave a pond or stream that dry up during a period of drought. Close to the shore they aggregate in protected terrestrial areas (ANDERSEN 2000). The resin-producing trees must have grown close to the ponds and streams so that the water striders may become trapped by flows of resin. ANDERSEN (2000) reported of the fossil semiaquatic bugs of the eocene Baltic amber and presented the fossil species of three gerromorphean families:

Gerridae
- *Electrogerris kotashevichi* ANDERSEN, 2000
- *Succineogerris larssoni* ANDERSEN, 2000
- *Gerris* FABRICIUS, 1794 (nymphs)

Veliidae
- *Electrovelia baltica* ANDERSEN, 1998
- *Balticovelia weitschati* ANDERSEN, 2000

Hydrometridae
- *Metrocephala anderseni* POPOV, 1996
- *Limnacis succini* GERMAR & BERENDT, 1856
- *Limnacis hoffeinsi* POPOV, 1996

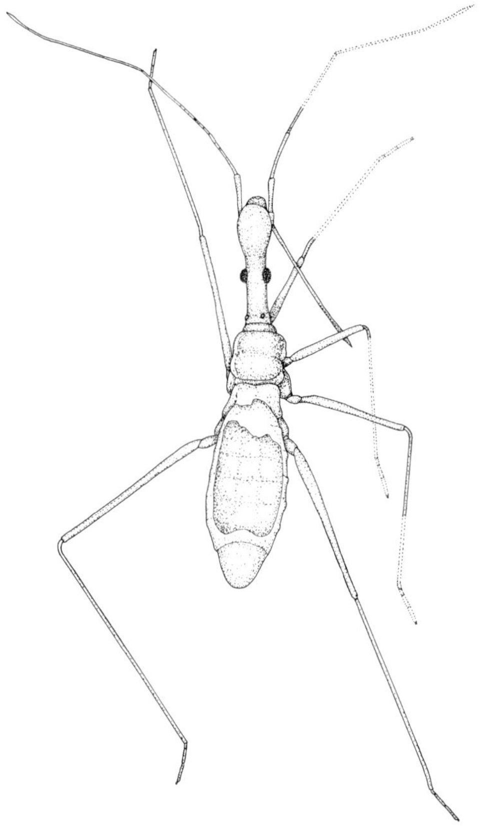

Fig. 55: Water strider *Metrocephala anderseni* POPOV, 1996 (Gerromorpha: Hydrometridae).

Plate 40: True bugs (Insecta: Heteroptera) in Baltic amber I (Nepomorpha, Gerromorpha).

a *Limnacis hoffeinsi*, holotype, × 10.
b *Limnacis* sp. (Hydrometridae), × 11.
c *Metrocephala anderseni*, holotype, × 12.
d *Metrocephala anderseni*, head, × 25.
e *Electrogerris kotashevichi,* holotype, × 9.
f Water boatman (Corixidae), larva, × 10.
g Water boatman (Corixidae), dorsal, × 11.
h Water boatman (Corixidae), lateral, × 10.

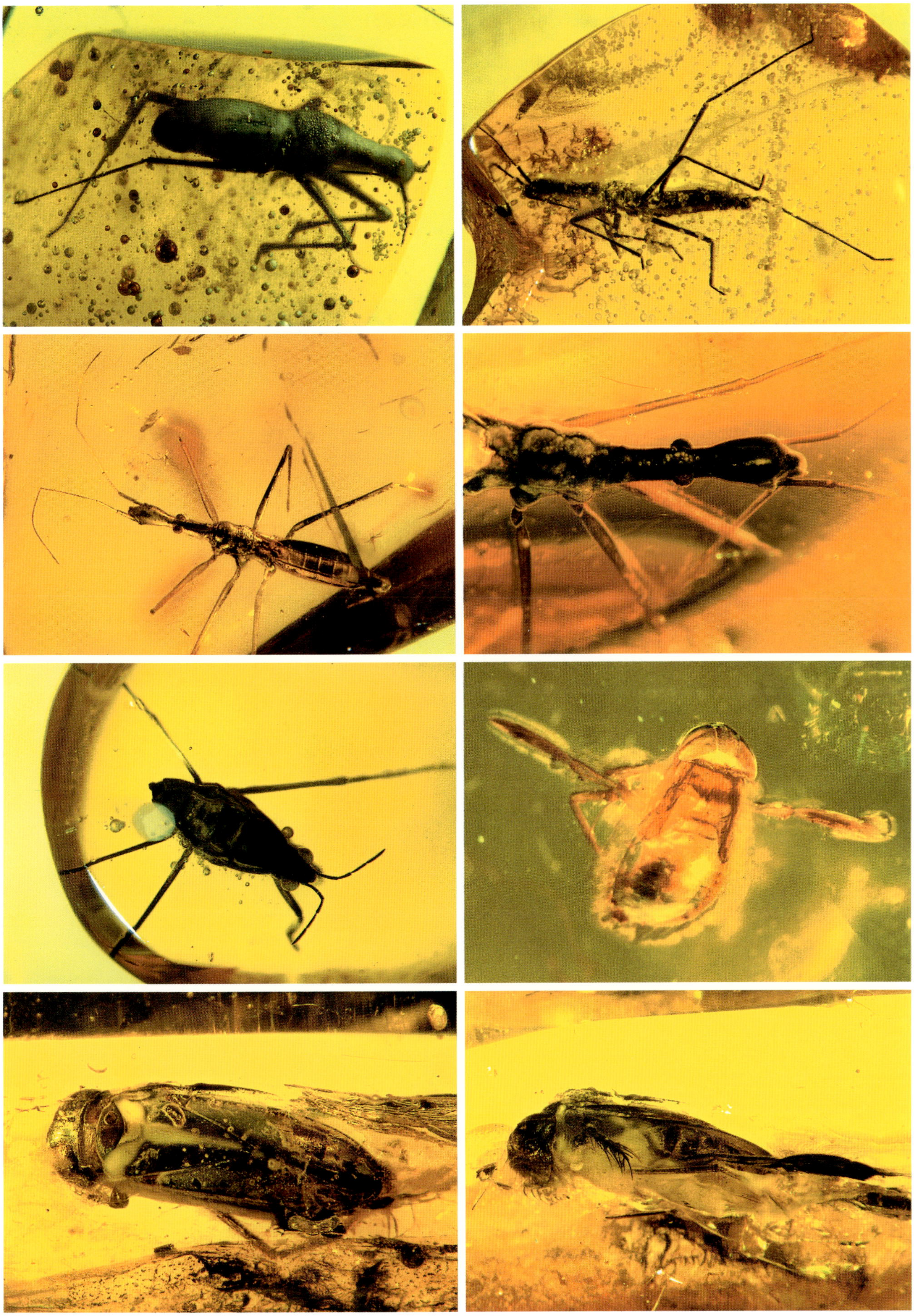
a
b
c
d
e
f
g
h

2.41 Plant bugs – Insecta: Heteroptera (Miridae)

The Miridae, or plant bugs, make up a very species-diverse family of Heteroptera. They are polyphagous plant-suckers that feed on a wide range of plants. Some species are restricted to pine trees and several are even more specialized on their host plants. As plant bugs are relatively common in amber compared to other true bugs, many polyphagous species may also be expected to occur in amber in addition to the pine and conifer specialists. GERMAR & BERENDT described the first 13 Miridae species in 1856, placing them in the genus *Phytocoris*.

According to POPOV & HERCZEK (1993 a), these plant bugs belong to the subfamily Cylapinae, which is primarily distributed in the subtropics. The subfamilies Deraeocorinae (*Deraeocoris balticus* HERCZEK & GORCYCA 1991), Mirinae and probably Orthotylinae, Phylinae or Bryocorinae, are also represented. Eocene isometopine species of the genera *Archemyiomma, Clavimyiomma, Electromyiomma, Electroisops* and *Metoisops* from Baltic amber have contributed to clarifying the phylogenetic relationship between the Psallopinae (*Isometopsallops schuhi* POPOV & HERCZEK, 1993 b) and the tropical Isometopinae. As is the case with many other families, the revision of the Miridae from Baltic amber is still only in the initial stages (JORDAN 1944, CARVALHO 1966, CARVALHO & POPOV 1984, HERCZEK 1991 a, b, 2000, HERCZEK & POPOV 1992, 1997 a, b, 1999, 2000, POPOV & HERCZEK 1992, 1993 a, b).

A list of terrestrial bugs requiring further classification was presented by SPAHR (1988) and supplemented by POPOV & HERCZEK (1993 a, POPOV 2001). They are divided among the following families:

Anthocoridae
Aradidae (Plate 42)
Ceratocombidae
Enicocephalidae (Plate 43 e, f)
Lygaeidae
Miridae (Plate 41)
Nabidae
Pentatomidae
Piesmatidae
Reduviidae (Plate 43 a-d, g, h)
Saldidae
Schizopteridae
Thaumastocoridae
Tingidae (Plate 44)

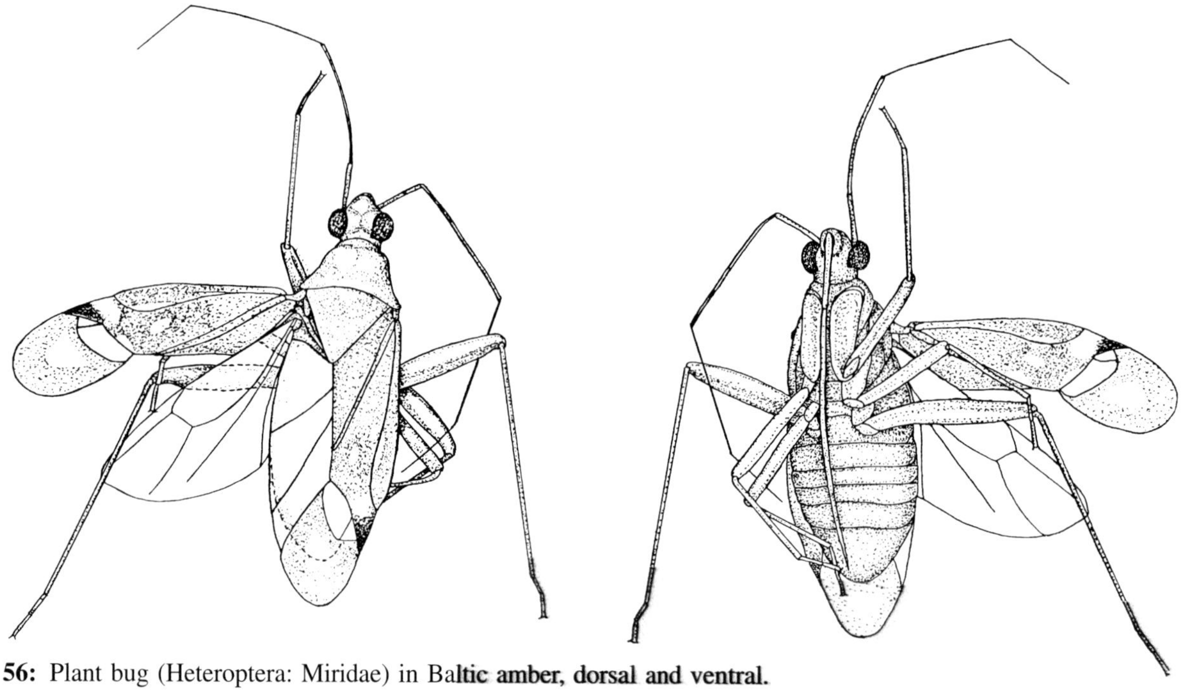

Fig. 56: Plant bug (Heteroptera: Miridae) in Baltic amber, dorsal and ventral.

Plate 41: True bugs (Insecta: Heteroptera) in Baltic amber II (Miridae).

a-h Plant bugs (Miridae) with (**b**) *Electromyiomma weitschati* POPOV & HERCZEK, 1992, holotype, **a** ×10, **b** ×20, **c** ×14, **d** ×13, **e** ×12, **f** ×16, **g** ×13, **h** ×16.

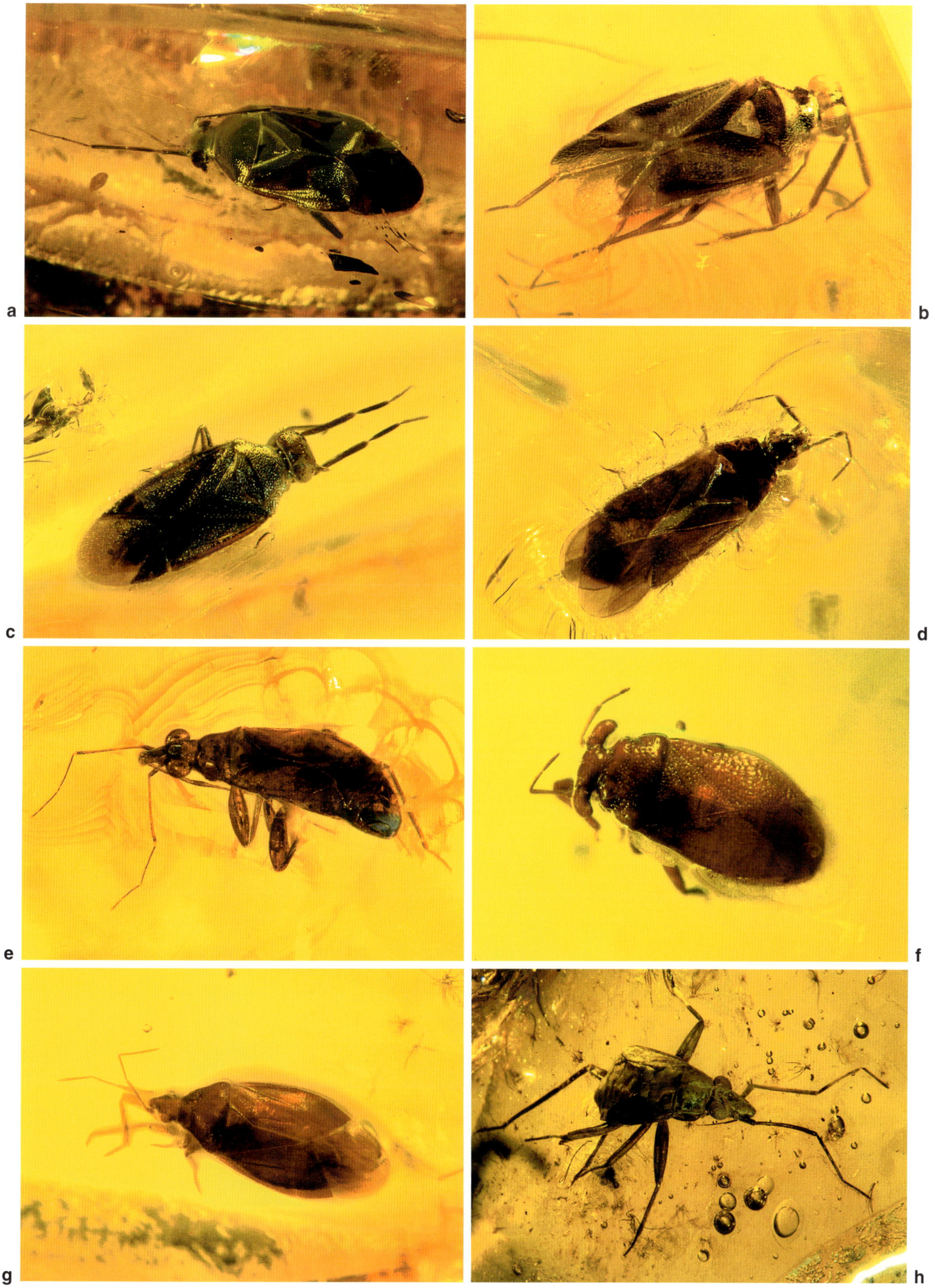
a
b
c
d
e
f
g
h

2.42 Flat bugs – Insecta: Heteroptera (Aradidae)

Publications on the Aradidae, or flat bugs, report nine described species and one nymph. HEISS (1997, 1998, 2000)) suggests that the number of species identified to date is surprisingly small in view of the fact that the Aradidae occur on fungus-infested deciduous and coniferous trees, and that the "Baltic amber forest" must have been a suitable biotope. They are divided among the following four subfamilies:

Aneurinae
Aneurus ancestralis HEISS, 1997
Aradinae
Aradus assimilis GERMAR & BERENDT, 1856
Aradus consimilis GERMAR & BERENDT, 1856
Aradus frater POPOV, 1978
Aradus frateroides HEISS,1998
Aradus popovi HEISS, 1998
Aradus superstes GERMAR & BERENDT, 1856
Calisiinae
Calisius balticus USINGER, 1941
Calisius rietscheli HEISS, 2000
Calisius vonholti HEISS, 2000
Calisius weitschati HEISS, 2000
Mezirinae
Mezira succinica USINGER, 1941

Most extant Aradidae live gregariously under the bark of deciduous and coniferous trees. They can be found in bark crevices and on the surface of tree fungi *(Polyporus, Trametes)*, where they use their long, piercing-sucking mouthparts to feed on mycelium. These piercing stylets grow to many times the body length and are rolled up in the head capsule when not in use. Only a few species are known to suck directly on plant juices, e.g. the Central European, pine-dwelling *Aradus cinnamomeus*.

With their flattened bodies, usually brownish coloring and rough surface structure, the Aradidae are well-adapted to their habitat and difficult to distinguish. This camouflage certainly protects them from falling prey to birds, for example.

Ninety percent of the roughly 2,000 species described to date inhabit tropical and subtropical rainforests. Many of them are completely wingless, while others display bizarre shapes and have developed a highly sculptured surface.

The similarity of several species described from Baltic amber to extant taxa on conifers suggests that the fossil forms lived on the resin-producing pine trees and other conifers of the "amber forest". As is the case with many other insects, it is remarkable that the body type of the Aradidae has remained unchanged, despite their great age of 40 to 50 million years, thus permitting classification of the amber specimens in contemporary genera.

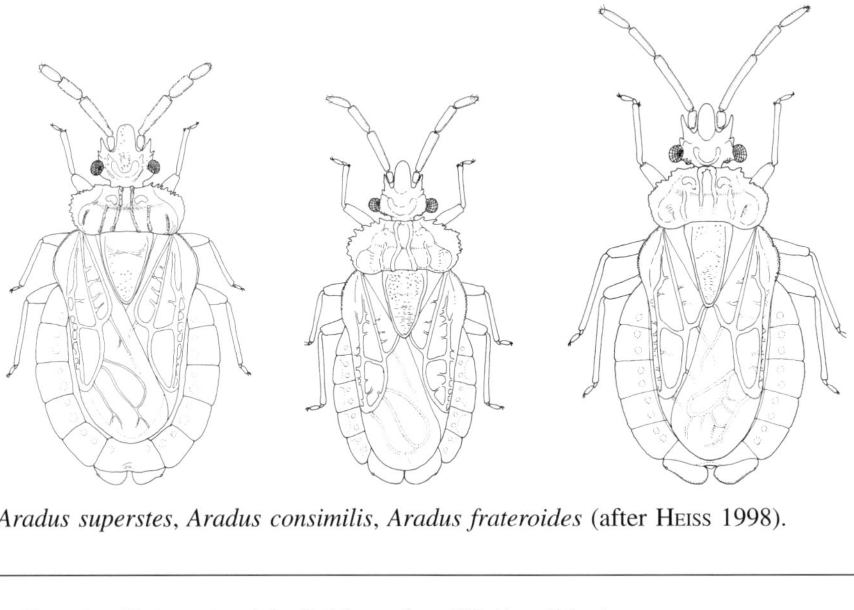

Fig. 57: Flat bugs: *Aradus superstes*, *Aradus consimilis*, *Aradus frateroides* (after HEISS 1998).

Plate 42: True bugs (Insecta: Heteroptera) in Baltic amber III (Aradidae).

a *Aradus* sp., male, × 10.
b *Aradus* sp. head and thorax, ventral, × 22.
c *Aradus* sp., female, × 11.
d *Aradus* sp., × 12.
e *Aradus* sp., × 11.
f *Aradus* sp., × 11.
g *Aneurus ancestralis* HEISS, 1997, holotype, × 12.
h *Calisius weitschati* HEISS, 2000, holotype, × 22.

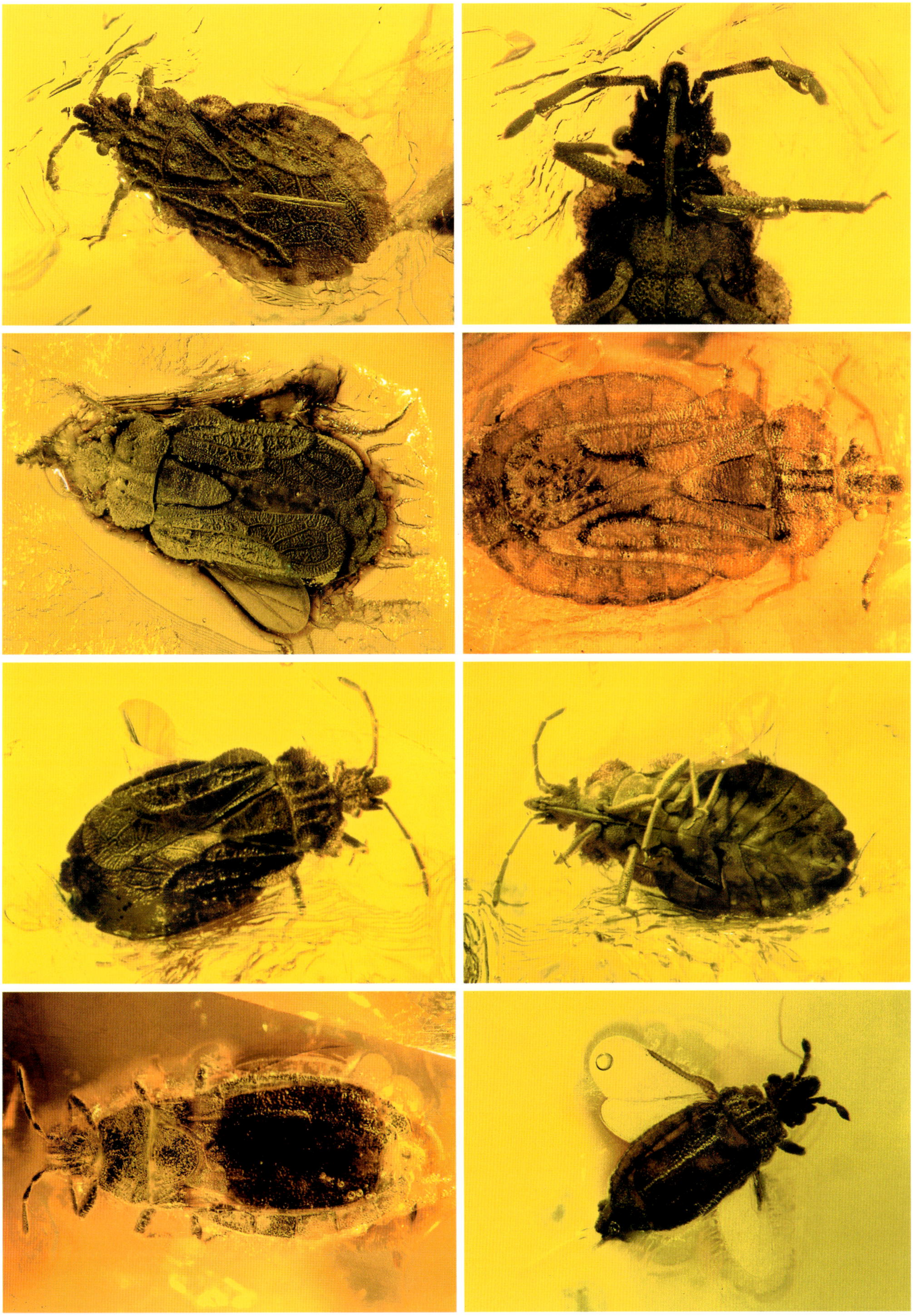

a b c d e f g h

2.43 Assassin bugs – Insecta: Heteroptera (Reduviidae)

While most assassin bugs have sturdy bodies and dark coloring, some are very delicate and slender with a midge-like shape and long antennae. The antennae are always longer than the head and pronotum, and thread-like towards the tip. The first antenna segment is distinctly angled. The three-segment proboscis is sickle-shaped and folded under the protruding head when not in use. The forelegs have developed into highly differentiated raptorial legs. Assassin bugs prefer to prey on other insects, which they pierce, paralyze and subsequently feed on by sucking. Assassin bugs are rare in Baltic amber. Only a few species have been described:

Collarhamphus mixtus Putshkov & Popov, 1995
Platymeris insignis Germar & Berendt, 1856
Proptilocerus dolosus Wasmann, 1933
Redubitus centrocnemarius Putshkov & Popov, 1993.
Redubinotus liedtkei Popov & Putshkov, 1998.

Wasmann (1933) described one particularly interesting assassin bug known as *Proptilocerus dolosus*. It is an "ant-killing, myrmecophilous bug", whose closest relative, *Ptilocerus ochraceus*, is currently native to Java. This assassin bug feeds on an ant species *(Dolichoderus bituberculatus)* that is very common in Southeast Asia. The assassin bug produces a secretion from the ventral side of the abdomen that is attractive, but toxic to ants. As soon as a worker approaches the bug, it presents the gland-covered surface of the abdomen by rising up on its middle and hind legs. As soon as the ant starts feeding on the secretions, the bug carefully wraps its forelegs around the prey and positions its proboscis on the ant's neck. Once the poison takes effect after a few minutes and the ant shows signs of paralysis, the assassin bug stabs its victim with the proboscis and sucks out the hemolymph. It quietly and inconspicuously feeds on one passing ant after the other in this way (Hölldobler & Wilson 1995). The empty ant integuments found lying next to a fossil assassin bug in amber were described by Wasmann as *Hypoclinea (Dolichoderus) tertiaria* Mayr, 1868.

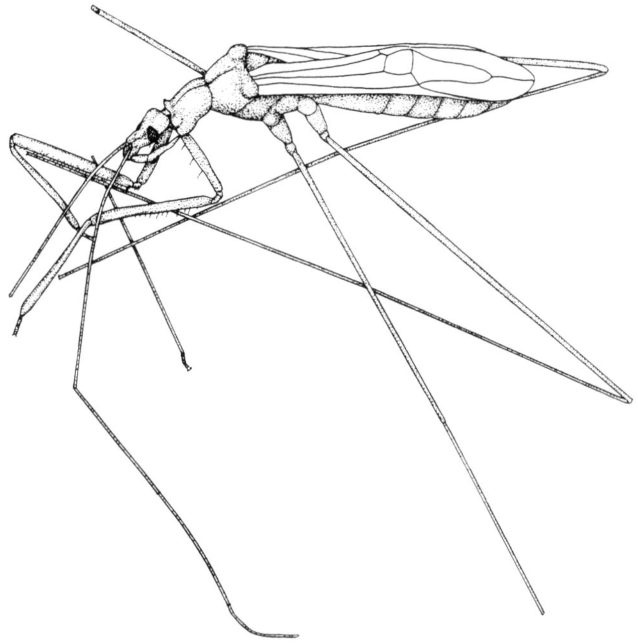

Fig. 58: Assassin bug: *Collarhamphus mixtus* Putshkov & Popov, 1995 (Reduviidae, Emesinae).

Plate 43: True bugs (Insecta: Heteroptera) in Baltic amber IV (Reduviidae, Enicocephalidae).

a *Proptilocerus dolosus* (holotype) with ant integuments, ×9
b *Proptilocerus dolosus,* head, antennae, ×17.
c *Collarhamphus mixtus* (holotype) (Emesinae), ×10.
d Assassin bug (Saicinae), ×7.
e Gnat bug (Enicocephalidae), ×22.
f Gnat bug (Enicocephalidae), ×20.
g Assassin bug, exuvia (dorsal), ×10.
h Assissan bug, exuvia (ventral), ×10.

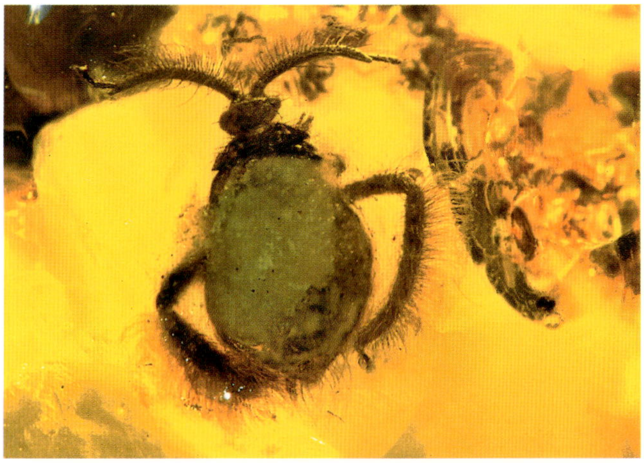
a

b

c

d

e

f

g

h

2.44 Lace bugs – Insecta: Heteroptera (Tingidae)

Lace bugs are one of the aesthetic peculiarities among Baltic amber inclusions. They are flattened bugs characterized by areolate reticulations of the pronotum and hemelytra. The variety of attractive species is reflected by this dense network of longitudinal veins and cross veins. The intermediate spaces are of varying size, more or less round or square, and generally arranged in rows. A medial keel runs down the center of the pronotum and is usually accompanied on both sides by shorter, lateral keels. The pronotum and hemelytra terminate in a lateral margin with sharp, long spines bearing setae. The head has four spines, including two jugals and two frontals (*Tingcader*, Fig. 59) bearing several curved spines. In contrast to the imagoes, the larvae are covered with spines.

Tingidae are strictly herbivorous and often restricted to specific host plants. This relationship explains the relative rareness of lace bugs in amber. As is the case with flat bugs, the lace bug species preserved in Baltic amber are those normally associated with the resin-producing trees. Other species can also occur at random and are thus very rare in amber. According to a revision by Golub & Popov (1998) and Golub (2001), the following seven Tingidae species have been reported to date:

Cantacaderinae

Paleocader avitus (Drake, 1950)
Paleocader quinquecarinatus (Germar & Berendt, 1856)
Paleocader strictus Golub & Popov, 1998
Intercader weitschati Golub & Popov, 1998
Sinalda baltica (Drake, 1950)
Sinalda froeschneri Golub & Popov, 1998
Tingicader cervus Golub & Popov, 1998

Tinginae

Archepopovia yurii Golub, 2001

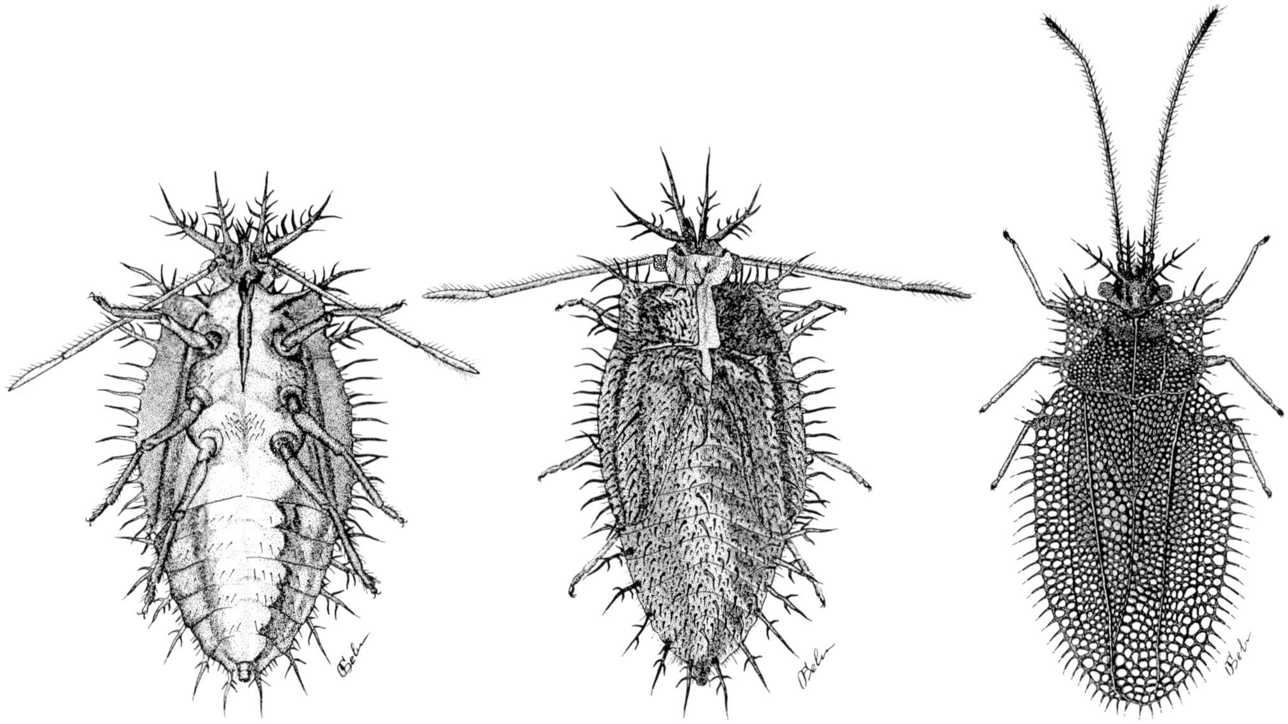

Fig. 59: Lace bug *Tingicader cervus* Golub & Popov, 1998: exuvia (dorsal), exuvia (ventral), imago (holotype).

Plate 44: True bugs (Insecta: Heteroptera) in Baltic amber V (Tingidae).

a *Sinalda baltica* (Drake, 1950), ×12.
b *Sinalda baltica* (Drake, 1950), ×18.
c *Paleocader strictus* Golub & Popov, 1998, ×13.
d *Paleocader strictus* Golub & Popov, 1998, ×12.
e *Tingicader cervus* Golub & Popov, 1998, ×22.
f *Intercader weitschati* Golub & Popov, 1998, ×30.
g *Tingicader cervus*, exuvia, dorsal view, ×30.
h *Tingicader cervus*, exuvia, ventral view, ×25.

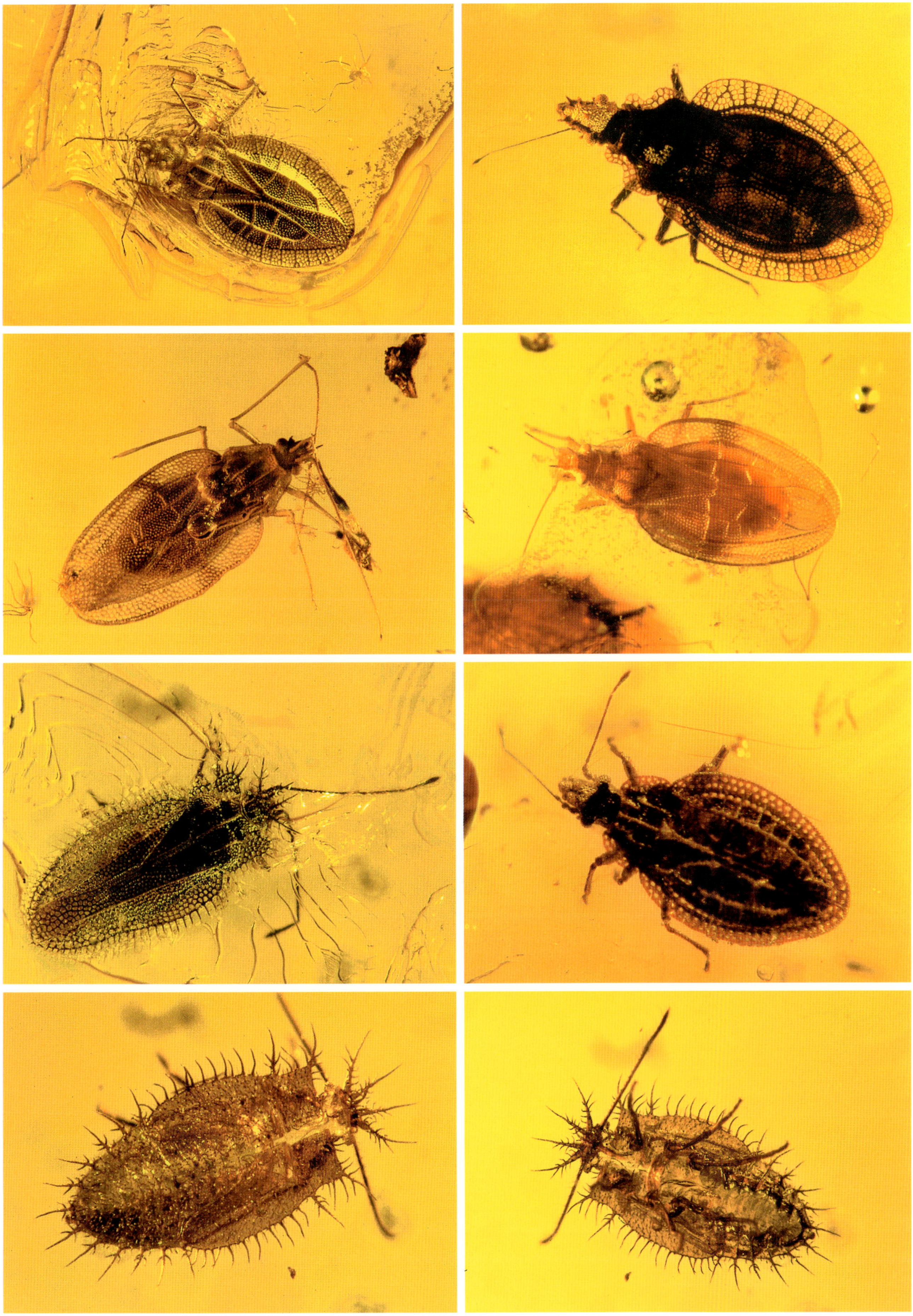
a b c d e f g h

2.45 Winged cicadas – Insecta: Auchenorrhyncha

The Auchenorrhyncha (cicadas, leafhoppers, treehoppers, fulgoroids), Heteroptera (true bugs) and Sternorrhyncha (psyllids, whiteflies, aphids, coccids) belong to the Hemiptera, all of which are characterized by piercing-sucking mouthparts. Like virtually all Hemiptera, auchenorrhynchans feed on plants by piercing them and sucking the plant juices. Most species live in the tropics and subtropics and their feeding habits make them directly dependent on plants. The females often insert their eggs into the plant tissue with the ovipositor, or deposit them in the soil in the direct vicinity of the plant. The song of the cicadas plays an important role in mating. The typical timbal organ consists of a convex cuticular plate reinforced with ribs, which is often protected by an operculum. The paired, dorsolateral timbal organs are located on the first abdominal segment. Sound is produced by the in-and-out movement of the cuticular plate caused by muscle action and inherent elasticity.

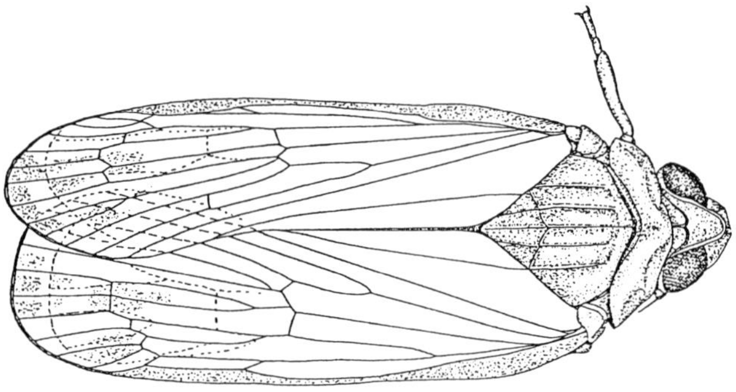

Fig. 60: Amber auchenorrhynchans (Fulgoromorpha: Cixiidae).

No comparative studies of the timbal organs of the cicadas in Baltic amber have been conducted to date, and no information is available on the paleobiology of the cicadas. The amber publications in this area concentrate on inventorying, identifying and describing cicada species on the basis of the imagoes and nymphs.

Nymphs are much more numerous in Baltic amber than adults. LARSSON (1978) linked this ratio to the assumption that the resin production of the amber trees was more active at the time of year when cicada nymphs are phenologically more numerous than imagoes. In temperate climate zones, this period occurs in spring and early summer. However, the annual developmental cycle of the cicadas becomes less distinct as the climate shifts to subtropical and tropical conditions. In addition, not all cicadas are monocyclic. They can go through two or more generations each year, or sometimes take several years to complete their developmental cycle.

These highly interesting considerations require very precise knowledge of the respective developmental cycles of the insects and, for verification purposes, should also include comparative studies of the development stages of other insects.

Plate 45: Auchenorrhynchans (Insecta: Auchenorrhyncha) in Baltic amber I.

a-d Winged fulgoromorphs (Fulgoromorpha), **a** ×8, **b** ×7, **c** ×8, **d** ×9.

e-h Winged spittlebugs (Cicadomorpha), **e** ×10, **f** ×10, **g** ×9, **h** ×10.

a
b
c
d
e
f
g
h

2.46 Cicada nymphs – Insecta: Auchenorrhyncha

Auchenorrhynchans are divided into two groups: the Fulgoromorpha and the Cicadomorpha. Both are represented in Baltic amber. The description of the fossil amber auchenorrhynchans is far from complete and primarily based on the early works of GERMAR & BERENDT (1856) and the later, less comprehensive works of BERVOETS (1910), COCKERELL (1910), JACOBI (1938) and USINGER (1939). New studies are from SZWEDO & KULICKA (1999 a, b), SZWEDO & GEBICKI (1998, 1999), and from GEBICKI & SZWEDO (2000). The classification and family grouping of the Auchenorrhyncha are based on METCALF & WADE (1966), SPAHR (1988) and CARVER et al. (1991):

FULGOROMORPHA

Cixiidae

Cixius fraternus GERMAR & BERENDT, 1856
Cixius gracilis GERMAR & BERENDT, 1856
Cixius insignis GERMAR & BERENDT, 1856
Cixius loculatus GERMAR & BERENDT, 1856
Cixius longirostris GERMAR & BERENDT, 1856
Cixius sieboldti GERMAR & BERENDT, 1856
Cixius succineus GERMAR & BERENDT, 1856
Cixius vitreus GERMAR & BERENDT, 1856
Cixius cf. *cunicularius* L.
Cixius cf. *nervosus* L.
Kulickamia jantaris GEBICKI & SZWEDO, 2000
Oliarus oligocenus COCKERELL, 1910

Dictyopharidae

Dictyophara reticulata (GERMAR & BERENDT, 1856)

Flatidae

Flata cf. *cunicularia* BURMEISTER
Flata cf. *nervosa* GRAVENHORST

Fulgoridae

Poiocera nassata GERMAR & BERENDT, 1856
Poiocera pristina GERMAR & BERENDT, 1856

Issidae

Issus reticulatus (BERVOETS, 1910)

Ricaniidae

Tritophania patruelis JACOBI, 1938

CICADOMORPHA

Aphrophoridae

Aphrophora electrina GERMAR & BERENDT, 1856
Aphrophora vetusta GERMAR & BERENDT, 1856
Ptyelus carbonarius (GERMAR & BERENDT, 1856)

Cercopidae

Cercopis melaena GERMAR & BERENDT, 1856

Cicadellidae

Ambericarda skalskii SZWEDO & GEBICKI, 1998
Cicadella minuta (BERVOETS, 1910)
Typhlocyba encaustica GERMAR & BERENDT, 1856
Typhlocyba resinosa GERMAR & BERENDT, 1856

Cicadidae

Coelidiidae

Coelidia immersa (GERMAR & BERENDT, 1856)
Coelidia spinicornis (GERMAR & BERENDT, 1856)

Iassidae

Iassus homousius (GERMAR & BERENDT, 1856)
Iassus punctatus (BERVOETS, 1910)

Macropsidae

Macropsis homousia (GERMAR & BERENDT, 1856)
Macropsis minuta (BERVOETS, 1910)

Tettigellidae

Tettigella proavia (GERMAR & BERENDT, 1856)
Tettigella terebrans (GERMAR & BERENDT, 1856)

The family Tettigometridae, from LARSSON (1978), and the genera *Cicadula*, *Deltocephalus* and *Thamnotettix* of the family Euscelidae, from BACHOFEN-ECHT (1949), are also included.

Plate 46: Auchenorrhynchans (Insecta: Auchenorrhyncha) in Baltic amber II.

a-d Fulgoromorph nymph (Fulgoromorpha), **a** ×9, **b** ×10, **c** ×9, **d** ×12.
e-h Cicadomorph nymph (Cicadomorpha), **e** ×12, **f** ×11, **g** ×13, **h** ×12.

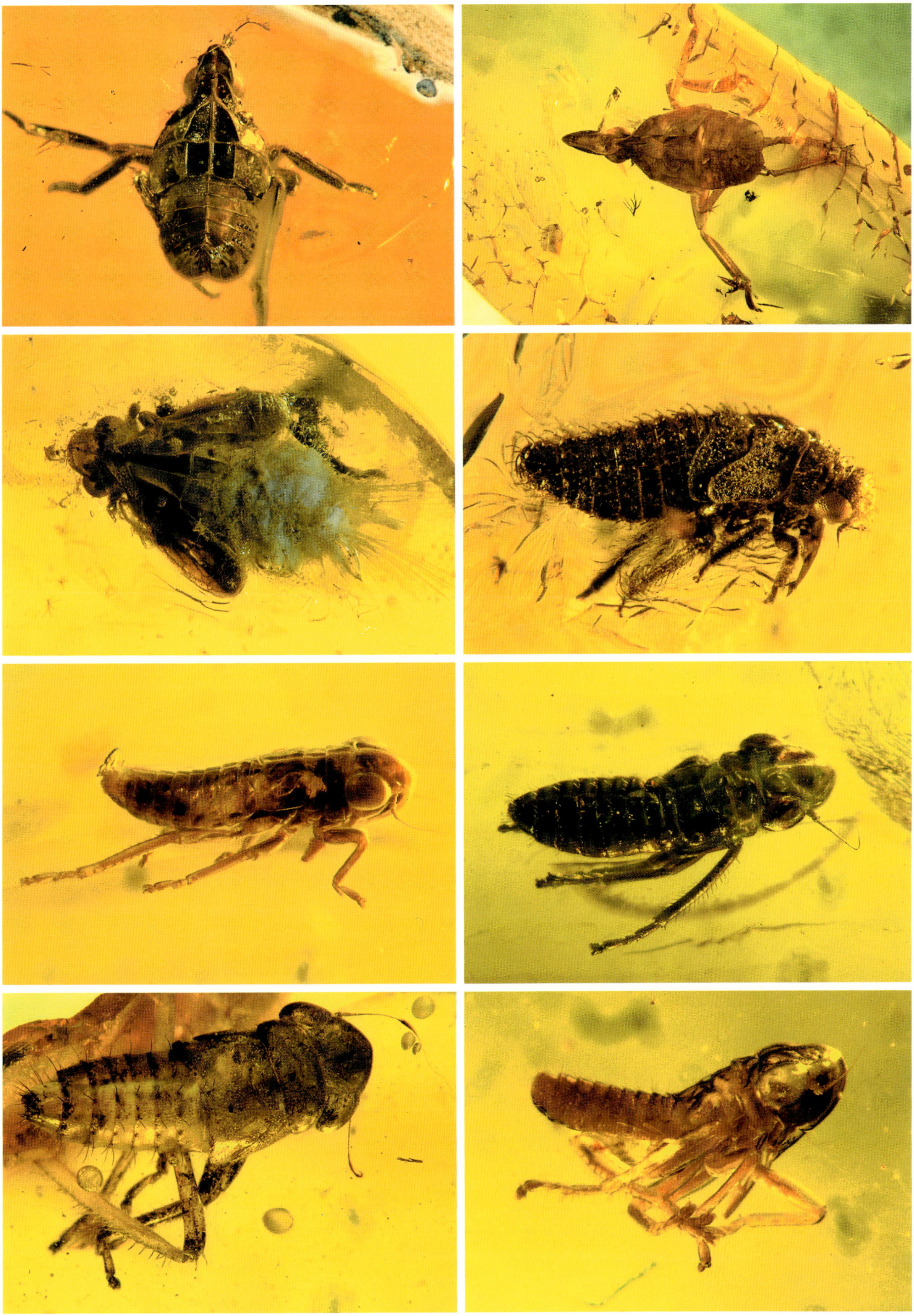

a b c d e f g h

2.47 Aphids – Insecta: Sternorrhyncha (Aphidoidea)

Aphids suck the juices of leaves and tender shoots and can therefore frequently transmit viral plant diseases. This risk is increased by the enormous reproduction rate of many species. Aphids reproduce parthenogenetically and sexually. Host plant alternation is species-specific. The females are winged or wingless, viviparous or oviparous. The males are usually winged (Aphididae) and, if wingless, are comparatively small.

The aphid reproduction rate is controlled by the food supply. They take advantage of favorable feeding conditions to rapidly expand populations by means of parthenogenetic reproduction. Viviparity helps shorten the time span of embryonic development, which begins when the mother is herself in the embryonic stage in the body of the grandmother. Female aphids in this developmental stage are wingless (Plate 47 b). This rapid, parthenogenetic development continues for several, endogenically defined generations, and is followed by sexual reproduction between a winged female and males. The fertilized eggs develop into parthenogenetic, wingless females. The goal of the aphid reproduction cycle is to achieve a high reproduction rate with the shortest possible development stage. The details of this process among the various species are much more complicated than described here.

Not surprisingly, aphids are common in Baltic amber as winged and wingless adults and as wingless nymphs. Of the aphids embedded in amber, the species represented by nymphs and females must have inhabited and utilized the resin-producing tree as a host plant, while the winged aphids may have been visitors from other plants. HEIE (1967, 1968, 1969, 1970, 1971, 1972, 1976, 1981 and 1985) compiled a list of the aphids in Baltic amber containing over 60 species, to which six more species were added by WEGIEREK (1990, 1996 a, b). The genus *Germaraphis*, comprising at least 15 species (Pemphigidae), is dominant in amber and represented by nymphs and wingless adults. Several of these fossil species are characterized by a rostrum that is longer than the entire body (Plate 47 e, f), this being interpreted as an adaptation to the bark of the resin-producing amber tree (HEIE 1967, LARSSON 1978). The amber species are distributed among the following families (HEIE 1985, WEGIEREK 1996 a, b):

Anoeciidae	**Aphididae**
Drepanosiphidae	**Electraphididae**
Hormaphididae	**Mindaridae**
Pemphigidae	**Thelaxidae**

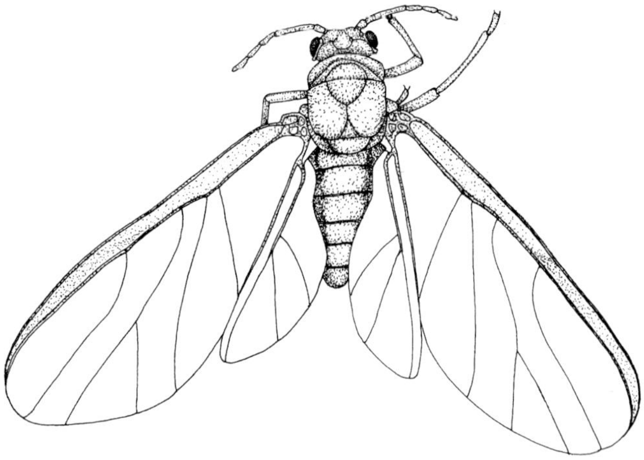

Fig. 61: Winged aphid *Mindarus magnus* (Mindaridae).

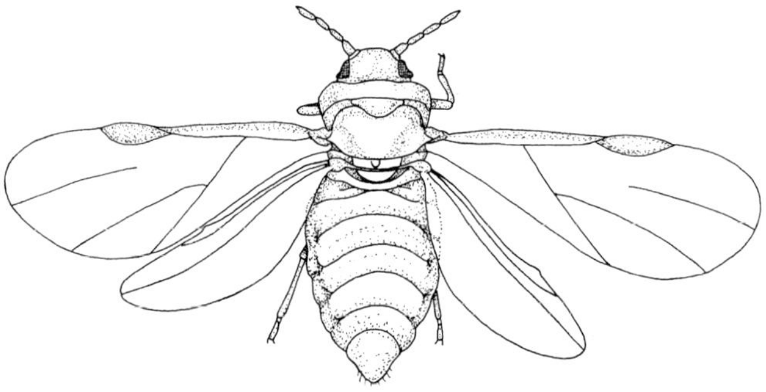

Fig. 62: Winged aphid *Schizoneurites* sp. (Electraphididae).

Plate 47: Sternorrhynchans (Insecta: Sternorrhyncha) in Baltic amber I (Aphidoidea).

a *Schizoneurites* sp. (Electraphididae), × 20.
b *Palaeosiphon hirsutus* (Drepanosiphidae), female, × 22.
c *Mindarus magnus* (Mindaridae), × 11.
d *Electromyzus acutirostris* (Drepanosiphidae), × 15.
e *Germaraphis* cf. *oblonga* (Pemphigidae), larva, × 20.
f *Germaraphis* cf. *dryoides* (Pemphigidae), larva, × 21.
g *Megapodaphis monstrabilis* (Drepanosiphidae), × 20.
h *Megapodaphis* sp. (Drepanosiphidae), larva, × 22.

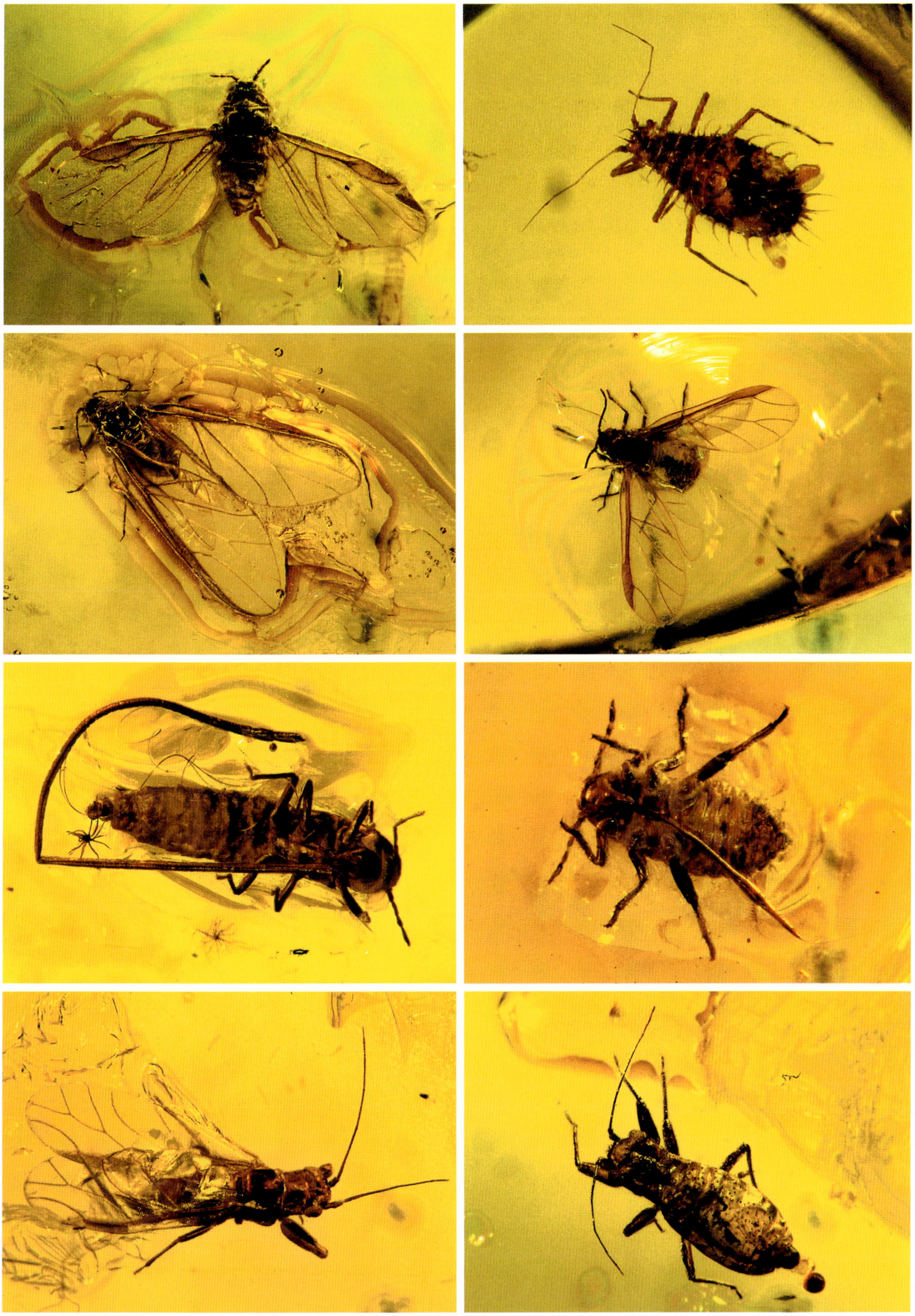

a b c d e f g h

2.48 Coccids – Insecta: Sternorrhyncha (Coccoidea)

Six-thousand extant species of scale insects occur all over the world, although most are subtropical or tropical. The Coccoidea display distinct sexual dimorphism, meaning that the males and females have very different appearances.

In accordance with the general body characteristics of adult insects, the body of the males comprises the head, thorax and abdomen. They have forewings and hind wings, the latter possibly being reduced to short rudiments or completely absent. The mouthparts are also reduced, meaning that the males do not feed. Males usually have compound eyes, although even these are reduced in some species.

All females have a larval structure, with the body segmentation being hardly distinguishable at all. They are wingless and the antennae and legs are reduced or absent. As a result, the females of many species are entirely sessile.

Fossil scale insects are rare. The oldest specimens originate from the Lower Cretaceous. Although amber inclusions are known from Lebanese, Taimyr, New Jersey and Dominican amber, nearly 90% stem from Baltic amber, including the Bitterfeld deposits. The first scale insect *(Acreagris crenata)* reported from amber is a wingless female or larva described by Koch & Berendt (1854), but mistakenly classified with the Apterygota. Ferris (1941) published a comprehensive study of the scale insects in Baltic amber. Fossil scale insects, particularly those from Baltic amber, have been the subject of intense study by Koteja since 1981 (Koteja 2000). Although the males survive for only one to three days, they are more frequent in Baltic amber than females or nymphs. Roughly 50% of all scale insects occurring in Baltic amber belong to the genus *Matsucoccus* and one of its five species (Koteja 1984). Extant matsucoccids are confined to the Holarctic and to the conifers genus *Pinus*. The fossil *Pinus succinifera* is discussed as the resin-producing conifer.

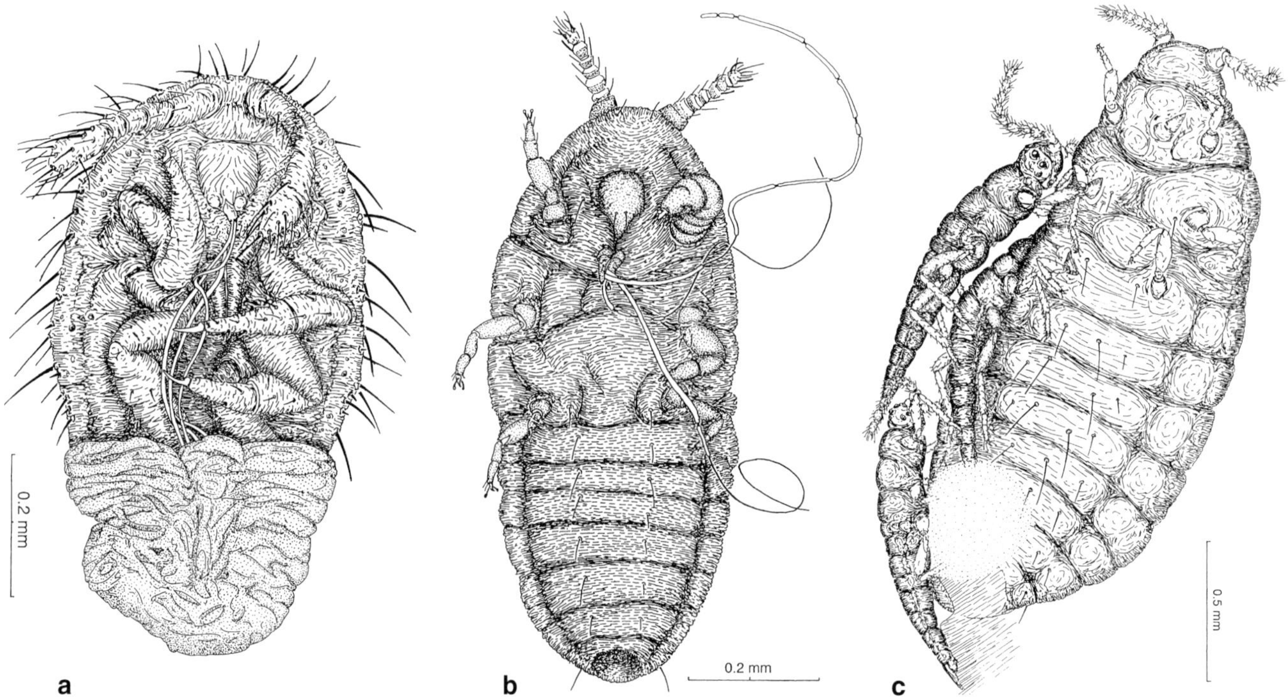

Fig. 63: **a** A hatching archeococcid larva from venter, **b** A pityococcid larva with exposed piercing stylets, **c** Three males and a female of pityococcids in copula (after Koteja 1998).

Plate 48: Sternorrhynchans (Insecta: Sternorrhyncha) in Baltic amber II (Coccoidea).

a Scale insect, (Margarodidae), male, ×12.
b *Matsucoccus pinnatus* (Matsucoccidae), ×20.
c Scale insect (Putoidae), male, ×14.
d *Matsucoccus pinnatus* (Matsucoccidae), ×20.
e Scale insect (Margarodidae), male, ×12.
f Scale insect, wingless male, ×15.
g *Arctorthezia antiqua* (Ortheziidae), female, ×12.
h Scale insect, (?Margarodidae) female, ×12.

a
b
c
d
e
f
g
h

2.49 Whiteflies, psyllids – Insecta: Sternorrhyncha (Aleyrodoidea, Psylloidea)

Whiteflies are known as such because their body and wings are often covered with a waxy coating secreted from ventral wax glands on the abdomen. They use their legs to spread the wax over their bodies, making them appear white. The remains of this wax coating are occasionally also visible on whiteflies preserved in amber, giving the wings a patchy appearance. The males and females are winged and hold the wings in a rooflike position over the abdomen when not in use. While masses of whiteflies often occur on plants today and occasionally also spread viral diseases, the few amber inclusions do not support any statements concerning the biology and ecology of their Eocene ancestors. As whiteflies have also been found in Burmese and Lebanese amber, they are preferred objects for the phylogenetic study of fossil and extant species (SCHLEE 1970). The first and currently only whitefly from Baltic amber was described nearly 150 years ago:

Aleyrodidae

Aleyrodes aculeatus MENGE, 1856

Similarly, only one psyllid species was known from Baltic amber for many years: *Strophingia oligocaenica* ENDERLEIN, 1915. This species has since been assigned to a new genus and further species belonging to the families Aphalaridae and Paleoaphalaridae have been described (KLIMASZEWSKI 1993, 1997):

Aphalaridae

Paleopsylloides oligocaenica (ENDERLEIN, 1915)
Eogyropsylla eocenica KLIMASZEWSKI, 1993
Eogyropsylla jantaria KLIMASZEWSKI, 1993
Eogyropsylla magna KLIMASZEWSKI, 1997
Eogyropsylla parva KLIMASZEWSKI, 1997
Parascenia weitschati KLIMASZEWSKI, 1997

Paleoaphalaridae

Protoscena baltica KLIMASZEWSKI, 1997

Over 1,200 extant psyllid species are distributed all over the world. They occur in the temperate climate zone, but are more diverse in subtropical and tropical regions. *Parascenia weitschati* is a typical example of the amber psyllids. This species is most closely related to the genus *Colopscenia*, whose extant species are primarily distributed from central to western Asia, but also occur in the Mediterranean region.

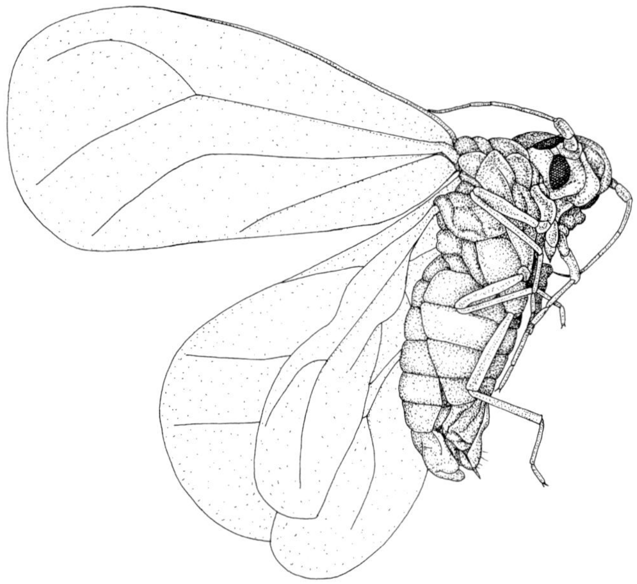

Fig. 64: Whitefly (Aleyrodoidea: Aleyrodidae).

Plate 49: Sternorrhynchans (Insecta: Sternorrhyncha) in Baltic amber III (Aleyrodoidea, Psylloidea).

a Whitefly (Aleyrodidae), ×40.
b Whitefly (Aleyrodidae), ×35.
c–h Psyllids:
c *Eogyropsylla* sp. (Aphalaridae), ×18.
d *Eogyropsylla* sp. (Aphalaridae), ×20.
e *Eogyropsylla jantaria* (Aphalaridae), holotype, ×23.
f *Eogyropsylla* sp. (Aphalaridae), ×18.
g *Parascenia weitschati* (Aphalaridae), holotype, ×18.
h *Eogyropsylla* sp. (Aphalaridae), larva, ×37.

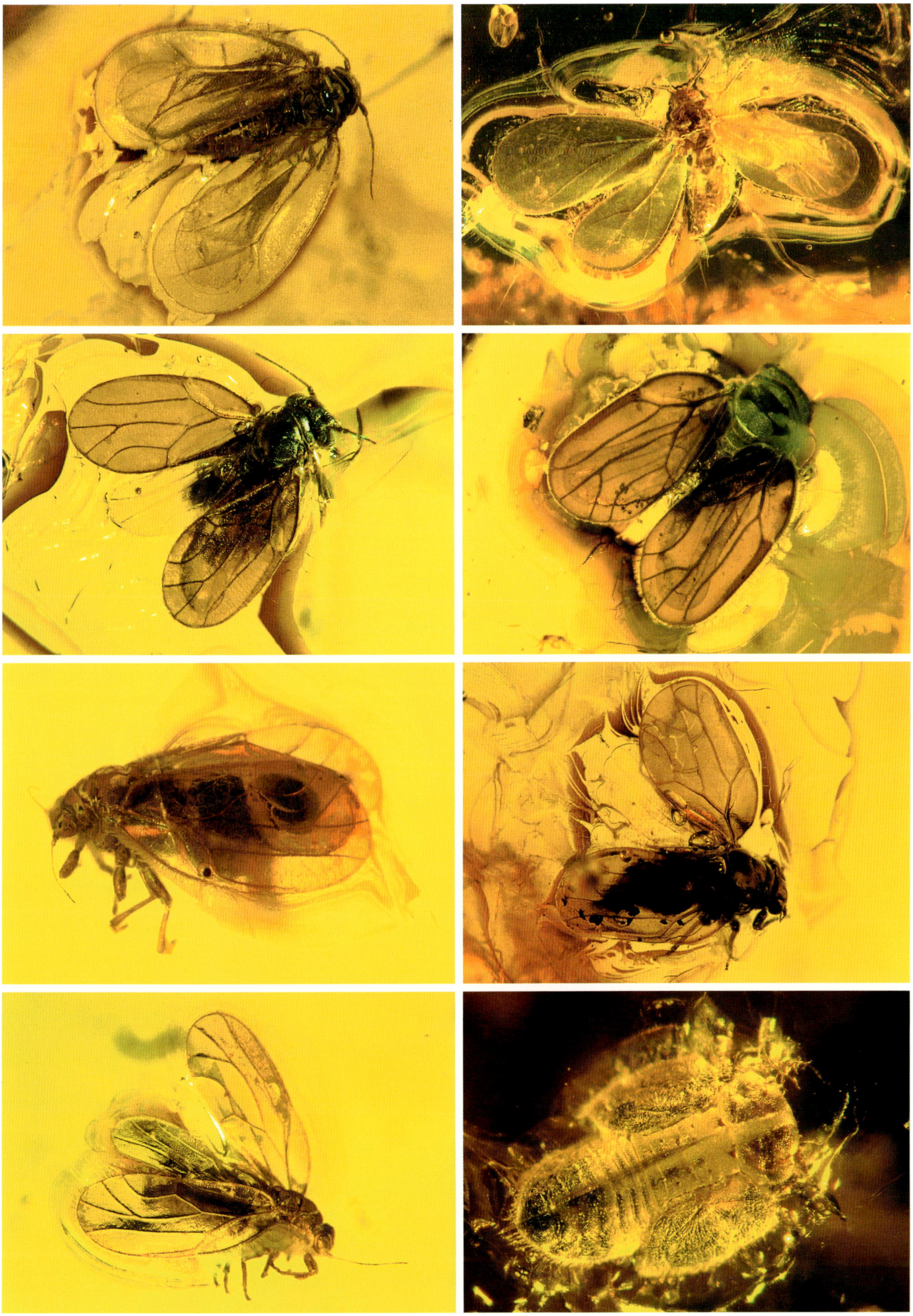

a b c d e f g h

2.50 Fishflies – Insecta: Megaloptera (Corydalidae)

The order Megaloptera belongs to the group of holometabolous insects (Holometabola, Endopterygota) that undergo complete metamorphosis. Between the last larval stage and the imago, there is a pupal stage, a brief, quiescent phase during which the larvae transform into winged and sexually mature imagoes. The Holometabola include the following insect orders: Megaloptera, Raphidioptera, Planipennia (= Neuroptera), Coleoptera, Strepsiptera, Hymenoptera, Mecoptera, Trichoptera, Lepidoptera, Diptera and Siphonoptera.

Pupae of these orders are found only rarely in amber, because they have no locomotive organs and generally lead a reclusive lifestyle during their brief developmental phase. The pupae found in amber are among the rarest inclusions. Planipennia pupae (Plate 56 g), ant pupae (Hymenoptera: Formicidae; Plate 71 d, f) and Diptera pupae of the Tipulidae (Plate 80 g, h) and Anisopodidae (Plate 82 c, d) were probably entombed in amber because, although hidden, they were not sufficiently protected from the flowing resin.

Only about 300 extant Megaloptera species are known today worldwide, these primarily occurring in the subtropics and tropics (Aspöck et al. 1980, New & Theischinger 1993). They are divided into two families: Corydalidae and Sialidae. Their larvae are aquatic and only leave the water in the last stage to pupate in the soil near the water's edge. The winged imagoes keep close to water and usually lay their eggs on overhanging plants, so that the primary larvae can fall safely into the water.

Megaloptera are primitive Holometabola with a fossil record dating back to the early Paleozoic Era. The oldest representatives originate from the Upper Permian (Riek 1976, Ponomarenko 1977): *Permosialis* (Permosialidae), *Parasialis, Sojanasialis* (Parasialidae) and *Tychtodelopterum* (Tychtodelopteridae). Both imagoes and larvae have been described. The nine pairs of appendages on both sides of the larval abdomen are indicative of an aquatic lifestyle. Larvae of extant species have seven (Sialidae) or eight (Corydalidae) pairs of abdominal gills.

Other larvae and imagoes are known from the Mesozoic. *Cretochaulus lacustris* Ponomarenko, 1976 was described from a larva and an imago from the Lower Cretaceous and is the first fossil record of the family Corydalidae. The description of *Chauliosialis sukatshevae* Ponomarenko, 1976 is based on a larva from Upper Cretaceous Taimyr amber. The familial assignment is uncertain.

While the family Corydalidae (*Chauliodes prisca* Pictet, 1854) was also found to occur later on in Eocene Baltic amber of the Cenozoic, the family Sialidae was reported for the very first time from this epoch and era on the basis of larvae (Weidner 1958) and imagoes (Wichard 1997). These discoveries support the phylogenetic theory (Ponomarenko 1977) that the family Sialidae did not appear as a young branch in the phylogenetic tree of the Megaloptera until the beginning of the Cenozoic, while the Corydalidae were already present in the Mesozoic.

Plate 50: Fishflies (Insecta: Megaloptera) in Baltic amber I (Corydalidae).

Eye region with pectinate antennae of *Chauliodes* sp. (Corydalidae), male, × 38.

2.51 Alderflies – Insecta: Megaloptera (Sialidae)

Alderflies live as larvae in the sandy bottom of slowly running waters or hidden in the muddy banks of standing waters. They feed on worms and small aquatic insects. The mature, roughly two centimeter-long larvae leave the water to pupate in the moist soil of the littoral zone. The hatched imagoes keep close to water, inhabiting the nearby vegetation. They rarely fly and usually sit on reeds and other aquatic plants, keeping their large, darkly veined wings in a rooflike position over the abdomen when not in flight. After mating and fertilization, the females immediately lay individual egg masses, each containing several hundred eggs. The primary larvae, which hatch after just a few days, must reach water in order to survive and continue the developmental process.

This highly water-dependent life cycle makes alderflies rare insects in amber. As the imagoes do not feed extensively and seldom fly, only the mature males take to the air if they are lured by the attractants secreted by the females or possibly misled by the scent of the resin. Only individual males have been reported to date from Baltic amber. PICTET (1845) was the first to describe an exotic megalopteran of the family Corydalidae, with its impressive, unilateral, pectinate antennae (Plate 50). Two additional males of the genus *Sialis* were found much later (Plate 51 a, b, Fig. 65):

Corydalidae
Chauliodes prisca PICTET, 1854
Sialidae
Sialis groehni WICHARD, 1997
Sialis (Protosialis) baltica WICHARD, 1997

Alderfly larvae are also uncommon in amber. When present-day larvae leave the water to pupate, they do not go further than five meters from the water's edge to dig a hole for pupation in the loose earth. Two larvae are known from Baltic amber, which are indicative of high, local biotope diversity. These discoveries provide clear evidence that resin-dripping trees must have stood on the banks of these waters, and that the amber forests were replete with standing and running waters.

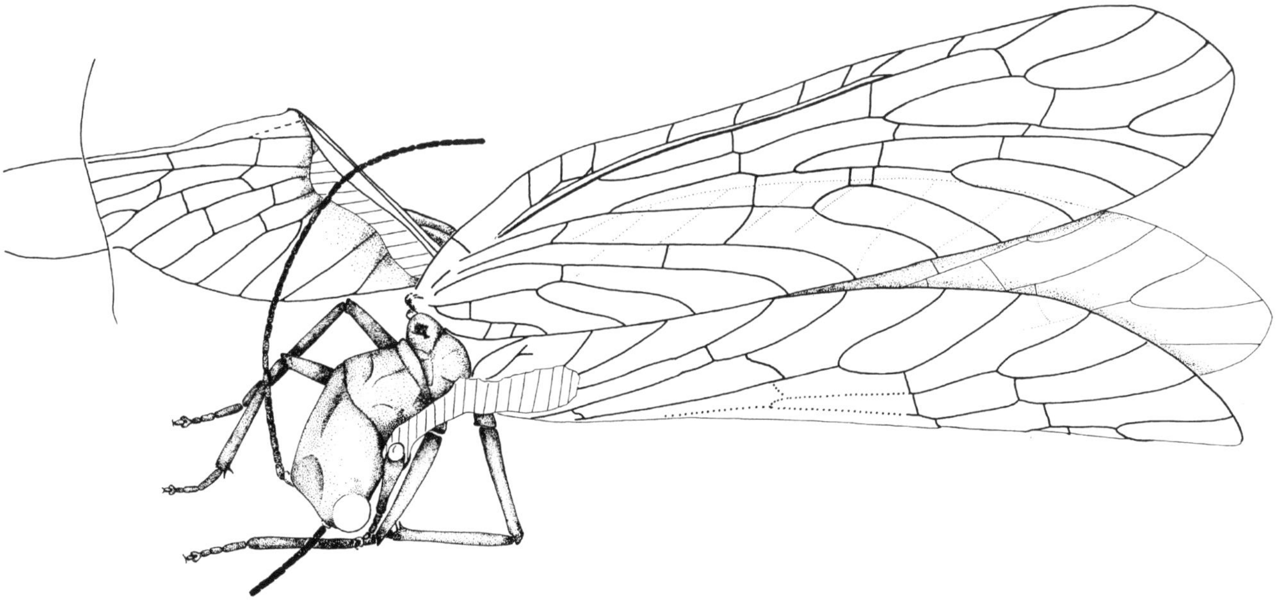

Fig. 65: *Sialis baltica* WICHARD, 1997 (Megaloptera: Sialidae), holotype, dorsolateral view.

Plate 51: Alderflies (Insecta: Megaloptera) in Baltic amber II (Sialidae).

a *Sialis groehni* WICHARD, 1997, holotype, male, dorsal, with left hind wing, ×7.
b *Sialis baltica* WICHARD, 1997, holotype, male, dorsolateral view, ×8.
c *Sialis* sp., larva with lateral tracheal gills and abdominal appendage, ×8.

a

b

c

2.52 Snakeflies – Insecta: Raphidioptera

Snakeflies are one of the rare, but highly interesting inclusions in Baltic amber. Only a few larvae have been reported to date. HAGEN (1856) described two larvae, only one of which was intact, and placed them in the genus *Raphidia*. WEIDNER (1958) had access to a further specimen from the SCHEELE collection. This larva in extended position is roughly ten millimeters long. The basic features agree with HAGEN's description of the *Raphidia* larva. One relevant feature that distinguishes larvae of the Raphidiidae from the Inocelliidae is the number of ocelli (ASPÖCK et al. 1980): the Inocelliidae have four ocelli, while the Raphidiidae have six or seven. Accordingly, the larva examined by WEIDNER belongs to the family Inocelliidae. The description by HAGEN does not take this feature into account.

Present-day snakefly larvae are terrestrial predators under the bark of trees or in the upper, epedaphic soil layers under tree and shrub vegetation. The larvae and imagoes of most species prefer warm habitats. The nearly 200 snakefly species existing worldwide today are primarily distributed in Central Asia, the Mediterranean and western North America.

Snakefly imagoes have distinct head, thorax and wing features. The depressed head bears short, usually threadlike antennae, strong, chewing mouthparts, large compound eyes, and ocelli (Raphidiidae), which may be absent (Inocelliidae). The elongate, cylindrical pronotum is what gave this insect its common name: snakefly. At rest, the wings are folded in rooflike fashion over the abdomen. The wing venation is characterized by a regular pattern of cells formed by cross-veins and the branching of the longitudinal veins. Branching increases towards the wing margin.

The following species have been described from imagoes:

Inocelliidae

Electrinocella peculiaris (CARPENTER, 1956)
Fibla carpenteri ENGEL, 1995
Fibla erigena (HAGEN, 1856)

Raphidiidae

Raphidia baltica CARPENTER, 1956

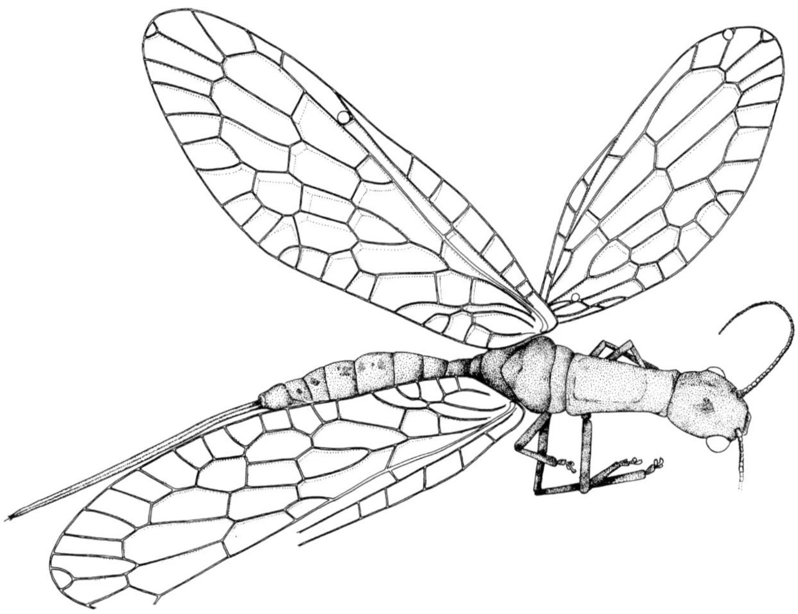

Fig. 66: Snakefly (Raphidioptera: Inocelliidae) female.

Plate 52: Snakeflies (Insecta: Raphidioptera) in Baltic amber.

a Inocelliidae (Snakeflies), female, with ovipositor, ×5.
b Inocelliidae (Snakeflies), male, ×4.
c Snakefly larva, ×13.

a

b

c

2.53 Dustywings – Insecta: Planipennia (Coniopterygidae)

MEINANDER (1990) lists 423 species in his checklist of the Coniopterygidae of the world, although tens of living species have been described during the last decade. According to OHM (pers. comm.), these extant species can currently be compared to nine fossil and two subfossil species. The oldest dustywings originate from Upper Cretaceous deposits in Kazakhstan (*Juraconiopteryx zherichini* MEINANDER, 1975). The species *Archiconiopteryx liasina* ENDERLEIN, 1909 from Upper Lias deposits in Mecklenburg was recently discovered to be a specimen of Hemiptera (ANSORGE 1996). Both species possibly belong to the subfamily Aleuropteryginae. Seven other species persevered in Cretaceous amber belong to the Aleuropteryginae: *Glaesoconis fadiacra* WHALLEY, 1980 (recently assigned to a new genus: ENGEL in press) from Lebanese amber, *Glaesoconis cretica* MEINANDER, 1975 from Taimyr amber, *Glaesoconis nearctica* GRIMALDI, 2000 in New Jersey amber and four species of *Apoglaesoconis* from New Jersey amber (GRIMALDI 2000, ENGEL in press). In addition, a single species of the subfamily Coniopteryginae is known from Lebanese amber (AZAR et al. 2000).

Several Eocene dustywings from the following six taxa have been preserved in Baltic amber, including its Bitterfeld deposits:

Archiconiocompsa prisca ENDERLEIN, 1910
Archiconis electrica ENDERLEIN, 1930
Coniopteryx (Coniortes) timidus (HAGEN, 1856)
Heminiphetia fritschi ENDERLEIN, 1930
Hemisemidalis sharovi MEINANDER, 1975
Hemisemidalis kulickae DOBOSZ & KRZEMINSKI, 2000

Archiconis electrica, which is closely related to *Glaesoconis* species from Cretaceous amber, is assigned to the tribe Fontenelleini (Aleuropteryginae) (cf. Plate 53 a, Fig. 67). *Archiconiocompsa* (Aleuropteryginae) and the extant genus *Coniocompsa* are assigned to the tribe Coniocompsini, whose species are distributed in the tropics and subtropics. Although it is inadequate according to modern taxonomic criteria, the description of *Coniortes timidus* can be linked to the genus *Coniopteryx* and thus to the subfamily Coniopteryginae (ENDERLEIN 1930, MEINANDER 1972). Other Coniopteryginae from Baltic amber include *Heminiphetia fritschi*, which is closely related to the extant genus *Neosemidalis* from the Australian region (MEINANDER 1972), and *Hemisemidalis sharovi*, whose genus is today represented by four species in the Mediterranean, the Near East, Mongolia and South Africa (ASPÖCK et al. 1980).

In order to present a comprehensive view of the dustywings, two additional species described from copal are also included here: *Coniopteryx enderleini* MEUNIER, 1910 in Togo copal and *Semidalis copalina* MEUNIER, 1910 in Madagascar copal. *Semidalis* is a species-diverse genus with worldwide distribution (excluding Australia).

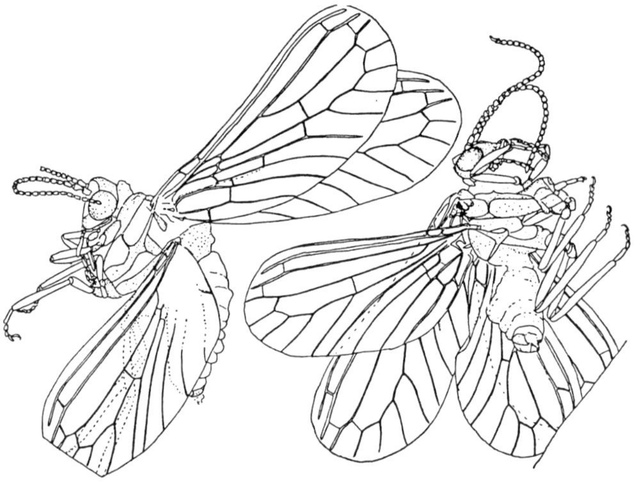

Fig. 67: Dustywing *Glaesoconis cretica* MEINANDER, 1975 (Coniopterygidae) from Taimyr amber.

Plate 53: Lacewings (Insecta: Planipennia) in Baltic amber I (Coniopterygidae).

a *Archiconis electrica* ENDERLEIN, 1930, × 12.
b *Archiconiocompsa prisca* ENDERLEIN, 1910, × 13.
c *Archiconis electrica* ENDERLEIN, 1930, × 12.
d *Archiconiocompsa prisca* ENDERLEIN, 1910, × 13.
e *Archiconis electrica* ENDERLEIN, 1930, × 12.
f *Archiconiocompsa* sp., × 12.
g dustywing (Coniopterygidae), larva, × 20.
h *Heminiphetia fritschi* ENDERLEIN, 1930, × 13.

2.54 Aquatic lacewings – Insecta: Planipennia (Sisyridae, Neurorthidae)

The Osmylidae (Chap. 2.55), Sisyridae and Neurorthidae represent families of Planipennia species whose larvae live temporarily or permanently in water. The larvae of the Osmylidae inhabit the banks of calm streams, entering the water only occasionally to hunt soft-bodied insects and other small, aquatic animals. The larvae of the Sisyridae and Neurorthidae develop entirely in water. The pupae of several extant Neurorthidae also remain in their aquatic habitat, protected from the water by a plastron enclosed in a loosely woven cocoon. The imagoes lead a reclusive lifestyle on littoral vegetation. Hagen (1856) described two species of Sisyridae from Baltic amber. But Nel & Jarzembrowski (1997) show that it is not a sisyrid but a neurorthid:

Rophalis amissa (Hagen, 1856)
Rophalis relicta (Hagen, 1856)

Some larvae displaying typical features of the Neurorthidae have also been found in Baltic amber. The elongate larvae show distinct head-thorax-abdomen segmentation. The white coloring of the slender and posteriorly tapered abdomen contrasts strongly with the brown head capsule. The thorax bears brown to light-brown sclerites on a light background. The highly sclerotized prothorax is divided into two sections, the front being narrower than the back. This "neck segment" is connected to the head by a "rolling joint" (Zwick 1967, Wichard et al. 1995). While the "neck" appears to be nearly enclosed by two sclerites, the notum to the rear of the prothorax is entirely sclerotized. The mesothorax and metathorax also display distinct sclerite pairs that are oval in shape and light-brown in color, as is characteristic for the genus *Neurorthus*. The larvae have thread-like antennae, moderately long labial palpi and piercing-sucking mouthparts. The mandibles and maxillae form a functional unit of elongate suctorial tubes. The pointed ends angle inwards in the manner of pincers.

The occurrence of *Neurorthus* larvae in amber is a curiosity, considering that the pupae develop in water and the larvae are not forced to leave it unless it dries up. Malicky (1984) and Wichard et al. (1995) pointed out that *Neurorthus fallax* can remain in its aquatic habitat for pupation. However, Aspöck et al. (1980) suspected that the aquatic larvae of *Neurorthus* species leave the water to pupate on land. This scenario would easily explain the occurrence of *Neurorthus* larvae in amber.

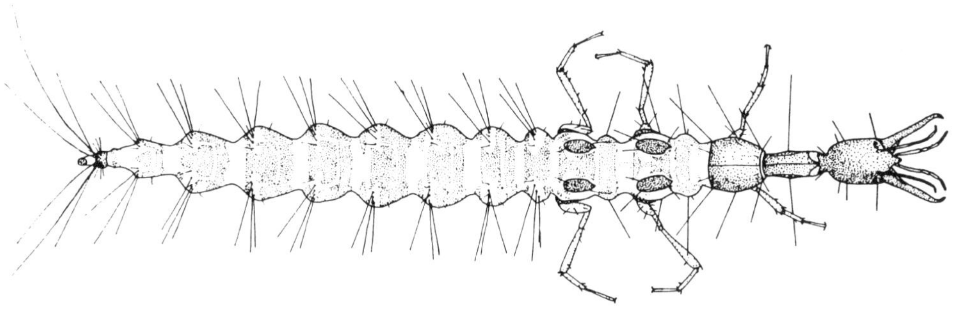

Fig. 68: Extant larva of *Neurorthus fallax* (Planipennia: Neurorthidae) (after Zwick 1967).

Plate 54: Lacewings (Insecta: Planipennia) in Baltic amber II (Sisyridae, Neurorthidae).

a Sisyridae: spongilla fly, female with left forewing in front of the genitalia, ×28.
b Sisyridae: spongilla fly, ×9.
c Sisyridae: spongilla fly, ×20.
d Neurorthidae, larva, ×10.
e Neurorthidae, larva, anterior, ×22.

a b c d e

2.55 Lacewings – Insecta: Planipennia (Berothidae, Osmylidae, Psychopsidae)

Although lacewings (Planipennia) are not frequent insects in Baltic amber, they are represented by a wide range of species. The reported families and described species include:

Ascalaphidae
Neadelphus protae MacLeod, 1970
(Plate 56h, Fig. 69)
Berothidae
Proberotha prisca Krüger, 1923
(Plate 55a, b)
Chrysopidae
(Plate 56d)
Coniopterygidae
Archiconiocompsa prisca Enderlein, 1910
Archiconis electrica Enderlein, 1930
Coniopteryx timidus (Hagen, 1856)
Heminiphetia fritschi Enderlein, 1930
Hemisemidalis sharovi Meinander, 1975
(Plate 53a-h)
Dilaridae
Cascadilar eocenicus Engel, 1999
(Plate 56a, b)
Hemerobiidae
Prophlebonema resinata (Krüger, 1922)
Prospadobius moestus (Hagen, 1856)
(Plate 56c, e, f)
Mantispidae
Fera venatrix Whalley, 1980
(British amber)
Neurorthidae
(Plate 54d, e)
Nymphidae
Pronymphes mengeanus (Hagen, 1856)
Osmylidae
Protosmylus pictus (Hagen, 1856)
(Plate 55d, f, h)
Psychopsidae
Propsychopsis hageni MacLeod, 1970
Propsychopsis helmi Krüger, 1923
Propsychopsis lapicidae MacLeod, 1970
(Plate 55c, e, g)
Sisyridae
(Plate 54a-c)

Roughly 60 extant species of the Berothidae (Neuroptera) are distributed in the warm temperate zones of all continents, but primarily in the subtropics and tropics (Aspöck et al. 1980). Both extant and amber species have a delicate wing structure consisting of parallel, longitudinal veins that branch evenly towards the margin. The offset cross-veins and branches have a light-brown coloring at the base. The typical pubescence on the wings is located along the longitudinal veins (Plate 55a, b).

Osmylidae larvae are predaceous, occasionally entering small bodies of water to hunt larvae and worms. They head the systematic list of Planipennia families arranged according to the increasingly aquatic lifestyle of their larvae: Osmylidae, Sisyridae, Neurorthidae (cf. Chap. 2.54). The Osmylidae are divided into various subfamilies, including roughly six species of Protosmylinae distributed in Japan and the Oriental region. *Protosmylus pictus* (Hagen, 1856), the only Osmylidae species reported to date from Baltic amber, purportedly belongs to this subfamily. Together with many other amber inclusions, this species illustrates the fact that the extant fauna of Southeast Asia is comparable to that of Eocene amber.

Of the Psychopsidae found in Baltic amber (Plate 55c, e, g), 21 extant species also occur in the Oriental and Ethiopian regions, as well as in Australia (New 1989). Psychopsidae can be distinguished from all other Planipennia by their large wings. The rounded front and hind wings are extremely broad and densely covered with parallel veins.

Plate 55: Lacewings (Insecta: Planipennia) in Baltic amber III (Berothidae, Osmylidae, Psychopsidae).

a Berothidae, beaded lacewing, ×5.
b Berothidae, branched veins at the wing margin, ×10.
c Psychopsidae (*Propsychopsis* sp.), ×9.
d Osmylidae (*Protosmylus* sp.), head region, ×20.
e Psychopsidae, larva, ×18.
f Osmylidae (*Protosmylus* sp.), wing region, ×15.
g Psychopsidae, larva head, ×25.
h Osmylidae (*Protosmylus* sp.), genital region, ×25.

a b c d e f g h

2.56 Lacewings – Insecta: Planipennia (Dilaridae, Chrysopidae, Ascalaphidae, Hemerobiidae)

Two new Planipennia families can be added to those reported by Larsson (1978) and Spahr (1992) from Baltic amber: Dilaridae and Chrysopidae.

Over 65 extant species of the family Dilaridae are included in the subfamilies Dilarinae, which has a Palearctic distribution, and Nallachiinae, which occurs in the Nearctic and Neotropical regions (Aspöck et al. 1980, New 1989, Oswald 1998). A female from the subfamily Dilarinae was the first to be found in Eocene Baltic amber (Plate 56 a, b) and a male was described shortly thereafter as *Cascadilar eocenicus* Engel, 1999. In contrast to the pectinate antennae of the males, the females have simple, filiform antennae. Another distinct feature of the females is the long ovipositor, which is roughly as long as the body in this amber specimen. The long ovipositor was probably used to lay eggs in the deep cracks and crevices of tree bark, possibly on the amber tree in this case.

The Chrysopidae are distributed worldwide and, with nearly 2,000 species, make up the largest extant family of Planipennia. Despite this fact, they have only recently been found in Baltic amber (Plate 56 d). Their dependence on specific prey (aphids, scale insects, insect larvae) can lead to a preference for specific plants, thus providing an explanation for the rareness of these unskilled flyers in Baltic amber.

Owlflies (Ascalaphidae), have only been found to date as larvae. Berendt (1830), Burmeister (1832) and Hope (1834) believed one specimen, which has long since disappeared, to be an antlion larva. Handlirsch (1907, 1925) and Bachofen-Echt (1949) supported this opinion. Klebs (1910) mentions another lost larva described by Ander (1942) and Andrée (1951) to be a member of the Myrmeleontidae or Ascalaphidae. Weidner (1958) definitively confirmed the occurrence of Ascalaphidae in Baltic amber on the basis of a larva from the Scheele collection (Hamburg). MacLeod (1970) described the Ascalaphidae species *Neadelphus protae* MacLeod, 1970 from yet another larva. New larvae have been found since then.

According to Aspöck et al. (1980), Ascalaphidae larvae live in the leaf litter, under stones and on tree stumps. Unlike antlions (Myrmeleontidae), they do not build funnels in loose, sandy soil to capture prey, but rather roam freely. For this reason alone, the probability of getting captured in resin is much greater for owlfly larvae than for antlion larvae, which lie concealed under their funnels in the sandy soil.

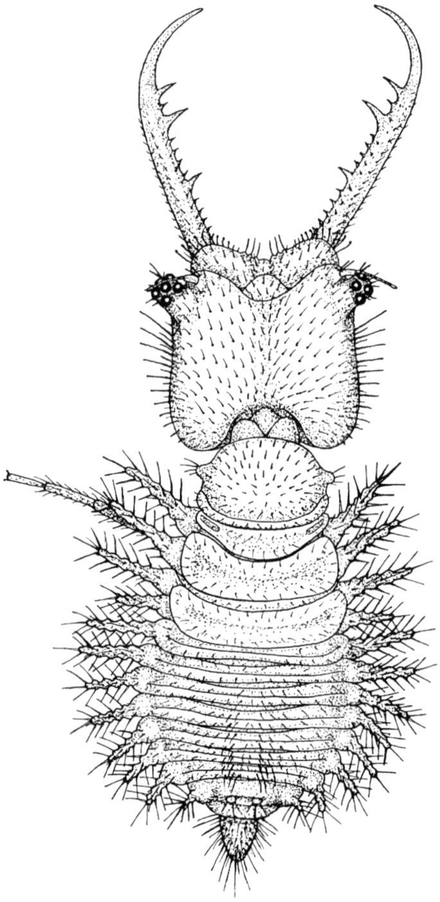

Fig. 69: Ascalaphidae larva (owlfly).

Plate 56: Lacewings (Insecta: Planipennia) in Baltic amber IV (Dilaridae, Hemerobiidae, Chrysopidae).

a Dilaridae, pleasing lacewing, female *(Cascadilar)*, × 12.
b Dilaridae, long ovipositor *(Cascadilar)*, × 24.
c Hemerobiidae, brown lacewing, × 8.
d Chrysopidae, green lacewing, × 11.
e Hemerobiidae, brown lacewing, × 12.
f Hemerobiidae, brown lacewing, × 11.
g Planipennia, pupa, × 10.
h Ascalaphidae, *Neadelphus protae*, larval head, × 17.

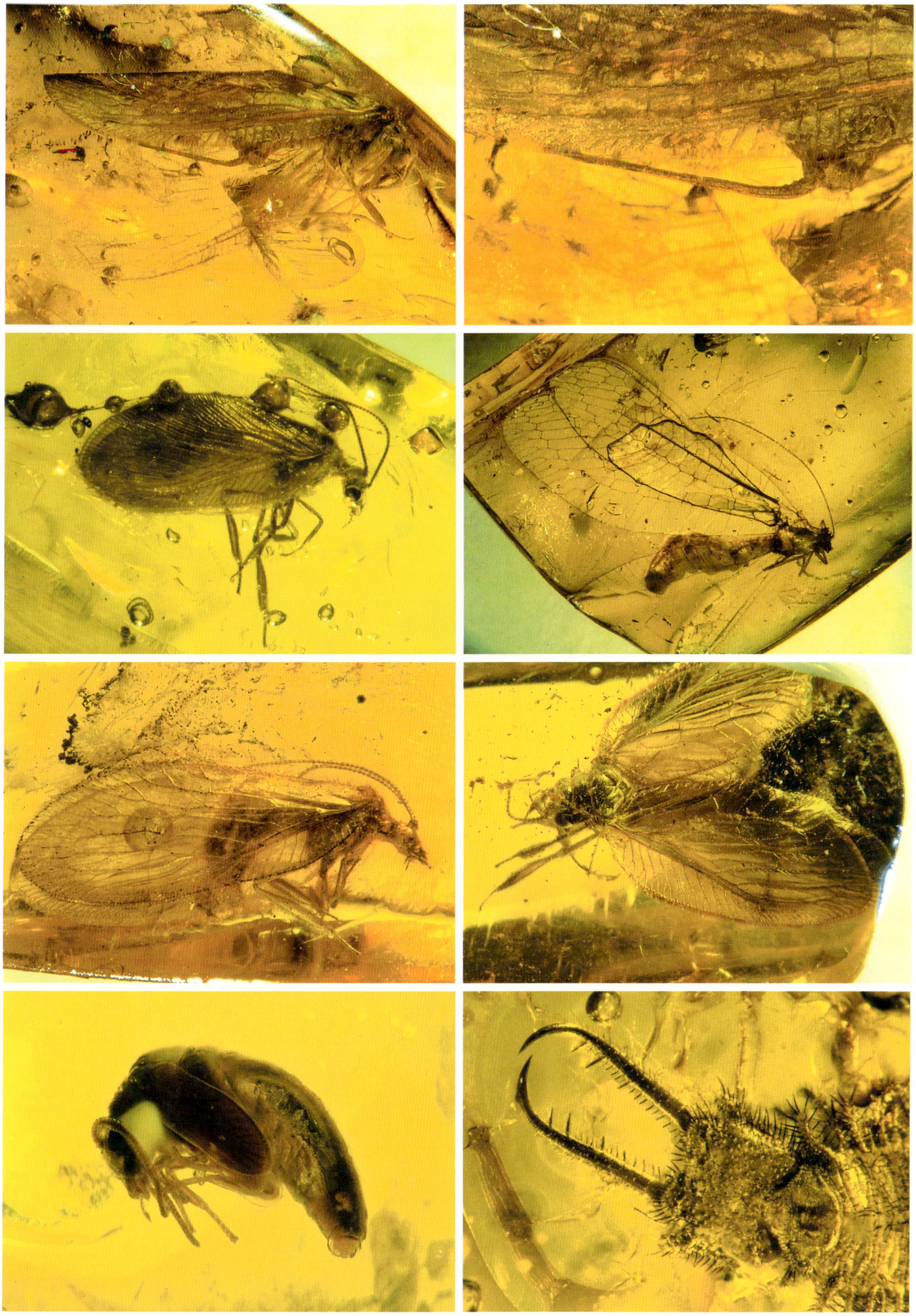

a b c d e f g h

2.57 Beetles – Insecta: Coleoptera (Cupedidae)

Beetles inhabit all continents of the world except Antarctica, displaying the greatest diversity in the tropics. The size of the adults varies from 0.3 mm (Ptiliidae) all the way to 15 cm (Scarabaeidae). The earliest fossil beetles originate from the Lower Permian (approx. 265 million years ago) (PONOMARENKO 1995). The order Coleoptera began to diverge into numerous families in the Late Paleozoic and was already very diverse by the Early Cenozoic, as illustrated by more than 72 families and subfamilies recorded to date from Baltic amber.

The order Coleoptera comprises four suborders, whose phylogenetic relationships have yet to be clarified definitively: Adephaga, Myxophaga, Polyphaga and Archostemata. One question that remains unresolved is the identity of the most primitive phylogenetic group of the Coleoptera. Either the Archostemata or the Polyphaga are the sister group of all other Coleoptera (KLAUSNITZER 1975, KUKALOVÁ-PECK & LAWRENCE 1993).

The Archostemata, which usually live on wood as larvae and adults, include primitive forms that display a close relationship to Late Paleozoic beetles. Several fossil beetles are classified with the Cupedidae (Archostemata) on the basis of their primitive elytron structure. Fossil wood fragments believed to show feeding marks of the Cupedidae have been found from the Triassic. The earliest confirmed records of this family originate from the Jurassic. Cupedidae of the genera *Cupes* and *Priacma* also occur in Eocene Baltic amber (Plate 57). They belong to taxa that currently have a very limited and discontinuous distribution. According to ANDER (1942), roughly 20 species of Cupedidae occur in East Asia, Australia, South Africa, the Tanganyikan region, Madagascar, North America, Brazil and Chile. While they today represent relict groups, they previously inhabited vast, continuous areas in the Tertiary, when the tropical and subtropical climate zone extended far to the north and south.

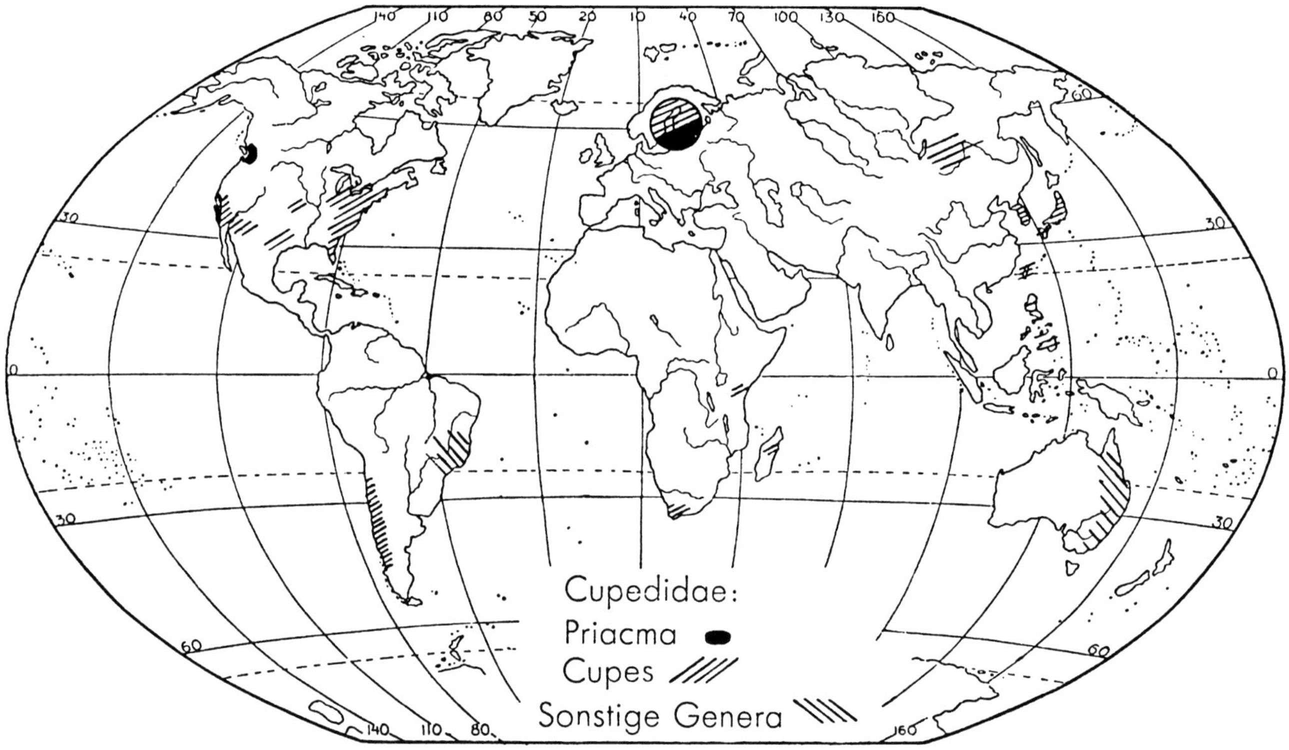

Fig. 70: Distribution of the Cupedidae (genera *Priacma* and *Cupes*) today and in Baltic amber (after: ANDER 1942).

Plate 57: Beetles (Insecta: Coleoptera) in Baltic amber I (Cupedidae), ×13.

2.58 Beetles – Insecta: Coleoptera (Carabidae)

In addition to water beetles, the Adephaga also include ground beetles of the family Carabidae, which is further divided into the subfamilies Carabinae (ground beetles), Cicindelinae (tiger beetles) and Paussinae (paussids). All three taxa are frequently treated as separate families. The Carabidae have a worldwide distribution. They lead a predaceous lifestyle, primarily inhabiting the ground surface of a wide variety of biotopes.

The Carabinae in amber are dominated by forest-dwelling species. According to a survey of the amber beetles in the collection of the Berlin Museum of Natural History (Hieke & Pietrzeniuk 1984), the Carabinae are represented in descending order by the Lebiini (18), Agonini (15), Harpalini (3), Pterostichini (2), Clivinini (2), Stenolophinini (1), Trechini (1) and Zuphiini (1). Of the Lebiini, the relatives of *Dromius* prefer to hunt under bark and on dead wood. Extant Agonini colonize the soil layers of mixed and deciduous forests. Characteristic ground beetles from arid biotopes do not appear to occur at all in Baltic amber.

One, well-known and frequently cited (Larsson 1978) species of the tiger beetles (Cicindelinae) is *Tetracha carolina* L, which has been reported to occur in amber and, according to Horn (1906), is identical to an extant species. This 17 mm-long tiger beetle is currently indigenous to the southern USA, the West Indies and Central America. It prefers open terrain and, like its larvae concealed in the sandy soil, is predaceous on other soil insects. Consequently, it is surprising that this tiger beetle occurs in Baltic amber. Klausnitzer (1982) correctly remarked that individual inclusions cannot necessarily be classified as extant species merely on the basis of external similarities. This statement addresses a general problem in the study of amber inclusions. Furthermore, in the case of *T. carolina*, recent studies have demonstrated specific differences between this species and the fossil, indicating that they are indeed likely not conspecific (Röschmann, 1999).

The paussids (Paussinae) are nocturnal predators of the tropics and subtopics. Most Paussinae are characterized by complex, noticeably enlarged antennae. These beetles inhabit ant nests. Glands on the antennae and the body produce a secretion the ants feed on. Paussids suck on the larvae and pupae of their hosts (Nagel 1980, 1987). Wasmann (1926 a, b, 1927, 1929) was the first to study the paussids in Baltic amber. According to a revision of the taxa (Nagel 1987), 20 confirmed species are divided among the following genera:

Succinarthropterus
- *Succinarthropterus helmi* (Schaufuss, 1896)
- *Succinarthropterus kühnlii* (Stein, 1877)
- *Succinarthropterus kolbei* (Wasmann, 1926)

Pleurarthropterus (Pleurarthropterus)
- *Pleurarthropterus hermenaui* (Wasmann, 1926)

Pleurarthropterus (Balticarthropterus)
- *Pleurarthropterus andreei* (Wasmann, 1928)
- *Pleurarthropterus skwarrae* (Wasmann, 1929)
- *Pleurarthropterus balticus* (Wasmann, 1926)
- *Pleurarthropterus subtilis* (Wasmann, 1926)
- *Pleurarthropterus hagedorni* (Wasmann, 1926)
- *Pleurarthropterus antiquus* (Wasmann, 1925)
- *Pleurarthropterus aterrimus* (Wasmann, 1929)
- *Pleurarthropterus simoni* (Wasmann, 1926)
- *Pleurarthropterus schaufussi* (Wasmann, 1926)
- *Pleurarthropterus fritschi* (Wasmann, 1929)

Pleurarthropterus (Acmarthropterus)
- *Pleurarthropterus kuntzeni* (Wasmann, 1927)

Cerapterites
- *Cerapterites primaevus* Wasmann, 1925

Protocerapterus
- *Protocerapterus primigenius* Wasmann, 1926
- *Protocerapterus incola* Wasmann, 1927

Arthropterites
- *Arthropterites klebsi* Wasmann, 1925

Eopaussus
- *Eopaussus balticus* Wasmann, 1926

Plate 58: Beetles (Insecta: Coleoptera) in Baltic amber II (Carabidae).

a-f Carabinae (ground beetles), **a, b, c, d** × 10, **e, f** × 11.
g-h Paussinae (paussid beetles), **g** × 10, **h** × 18.

a b c d e f g h

2.59 Beetles – Insecta: Coleoptera (Dytiscidae, Gyrinidae)

Of the adephagous water beetles, the predaceous diving beetles (Dytiscidae), whirligig beetles (Gyrinidae) and their larvae have been recorded from Baltic amber. While the Scirtidae are very frequent (Chap. 2.62), true water beetles of the families Gyrinidae and Dytiscidae are extremely rare.

One whirligig beetle from the KÜNOW collection at the Berlin Museum of Natural History has been reliably identified to date. The specimen is in very good condition, although it is covered by a reddish patina caused by the aging of the amber (Plate 59 g). This beetle probably belongs to the genus *Gyrinus*. Only one Gyrinidae species has been described: *Gyrinoides limbatus* MOTSCHULSKY, 1856. In the list of beetles in his amber collection, KLEBS (1910) mentions another gyrinid genus, *Orectochilus*, without any further commentary. If this genus is correct, it is indicative of the existence of running waters in the Baltic amber forest. A recently reported larva of a whirligig beetle (WICHARD & WEITSCHAT 1996) may have been captured by dripping resin as it left the water to pupate on land (Plate 59 h).

HIEKE & PIETRZENIUK (1984) pointed out that the predaceous diving beetle at the Berlin Museum of Natural History (Plate 59 a) is the only record of an imago from the family Dytiscidae. Another predaceous diving beetle has been found (WICHARD & WEITSCHAT 1996) since then (Plate 59 b). The Dytiscidae larvae preserved in amber are of particular interest in this context. They currently represent three subfamilies: Hydroporinae (Plate 59 f), Colymbetinae (Plate 59 c, d) and Laccophilinae (Plate 59 e). It is highly probable that one specimen belongs to the Laccophilinae, as evidenced by the temporal spines and the structure of the antennae and maxillary palpi. WEIDNER (1958) provided the first record of Colymbetinae larvae by describing a *Rhantus* (?) sp. larva belonging to the tribe Colymbetini. The subfamily Hydroporinae is represented by a larva already mentioned by BERENDT (1845), which was initially mistaken to be a Thysanura until HANDLIRSCH (1907) placed it among the Dytiscidae larvae related to *Hyphydrus*. *Hyphydrus* is a primarily tropical genus with only two species occurring in Europe. The head of the larvae extends in spoon-like fashion towards the front and, in the larvae preserved in Baltic amber, displays unusual, antler-like structures (Fig. 71).

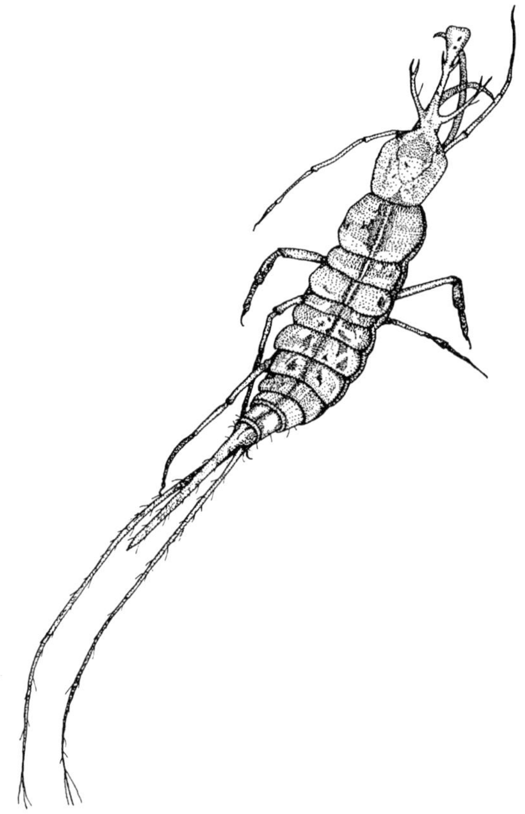

Fig. 71: Larva of *Hyphydrus* sp. (Dytiscidae).

Plate 59: Beetles (Insecta: Coleoptera) in Baltic amber III (Dytiscidae, Gyrinidae).

a Dytiscidae (predaceous diving beetles), × 11.
b Dytiscidae (predaceous diving beetles), × 10.
c Dytiscidae, Colymbetinae larva (?*Rhantus* sp.), × 7.
d Dytiscidae, Colymbetinae larva, × 9.
e Dytiscidae, Laccophilinae larva, × 11.
f Dytiscidae, Hydroporinae larva, × 28.
g Gyrinidae (whirligig beetles), × 10.
h Gyrinidae, larva, × 8.

a
b
c
d
e
f
g
h

2.60 Beetles – Insecta: Coleoptera (Staphylinidae, Pselaphidae, Scydmaenidae)

The family Staphylinidae (rove beetles) is distributed worldwide. It comprises nearly 50,000 extant species, making it just as successful as the Curculionidae (Chap. 2.65) in terms of species diversity. Their habitats are accordingly varied, although most species are epedaphic predators. The Staphylinidae can be found in and near the soil surface, in decaying plant material, under bark and on fungi. The majority prefer moist biotopes.

Hieke & Pietrzeniuk (1984) provided a helpful overview of the complex rove beetle family in Baltic amber. According to this overview, the collection of amber beetles at the Berlin Museum of Natural History is dominated by the subfamilies Tachyporinae (30), Paederinae (>20), Aleocharinae (30) and Staphylininae (>20), while the diverse subfamily Oxytelinae (1-2) is underrepresented. The Omalinae (5-6) and Proteininae (1-2) are also represented, albeit by only a few specimens. The purely euedaphic Leptotyphlinae do not occur at all.

The Klebs collection includes a number of Staphylinidae, which are primarily from the genera *Tachyporus*, *Anthobium*, *Lathrobium*, *Atheta*, *Bryocharis*, *Medon*, *Philonthus* and *Scopaeus* (Klebs 1910), and basically correspond to the range of the Berlin collection.

Species of Scydmaenidae and Pselaphidae are mostly surface-dwelling beetles. The Scydmaenidae are predaceous on mites, living under leaves and bark, in decaying tree material and moss.

The family of minute Pselaphidae (1 to 2.5 mm) comprises numerous, mostly tropical species. The short-winged mold beetles of the subfamily Pselaphinae are also predaceous on mites in moss, leaves and decaying plant material in the leaf litter. The ant-loving beetles of the subfamily Clavigerinae live exclusively in ant nests. The imagoes stroke the ants with their antennae in order to be fed. In return, the imagoes release secretions from under a tuft of hair, which the ants ravenously consume.

Numerous species of both families are represented in Baltic amber. The first descriptions of fossil species were provided by Schaufuss (1890 a, b), followed later on by Franz (1976).

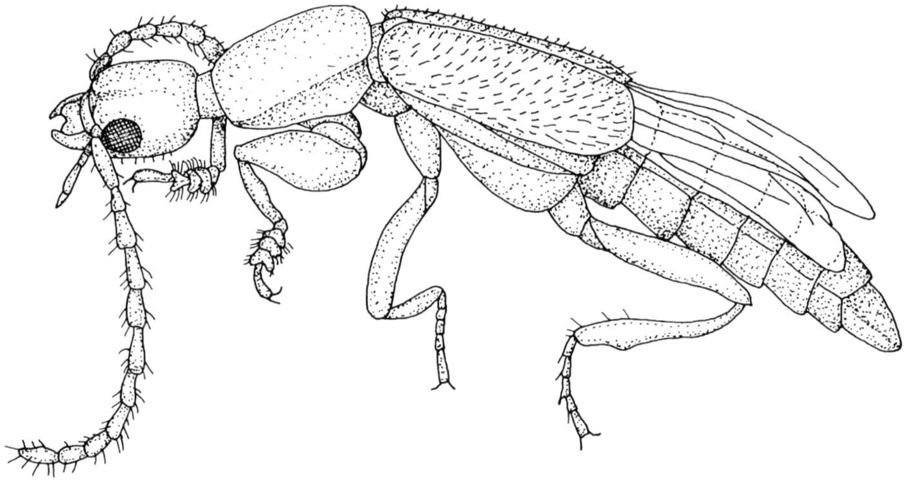

Fig. 72: Rove beetle (Coleoptera: Staphylinidae).

Plate 60: Beetles (Insecta: Coleoptera) in Baltic amber IV.

a-d Staphylinidae (rove beetles), **a** × 18, **b** × 11, **c** × 12, **d** × 10.
e-f Pselaphidae (short-winged mold beetles), **e** × 12, **f** × 15.
g-h Scydmaenidae (antlike stone beetles), **g** × 10, **h** × 21.

a b c d e f g h

2.61 Overview of amber beetles – Insecta: Coleoptera

The systematic and taxonomic classification of the beetles in Baltic amber is still only in the initial stages for most groups, even though (or precisely because) beetles are among the most frequent amber inclusions. Hieke & Pietrzeniuk (1984) compiled a very informative overview of the amber beetle families, taking into consideration their biology and paleoecology. The preliminary list of families presented here incorporates the work of Keilbach (1982), Spahr (1981 a, b), Hieke & Pietrzeniuk (1984) and Wunderlich (1996). The beetle families belonging to three suborders are listed according to the classification by Lawrence & Britton (1991). Some of the taxa previously classified as separate families are treated as subfamilies:

Archostemata
- Cupedidae – Reticulated beetles

Adephaga
- Carabidae – Ground beetles
- Dytiscidae – Predaceous diving beetles
- Gyrinidae – Whirligig beetles

Polyphaya

Staphyliniformia
- Hydrophilidae – Water scavenger beetles
- Histeridae – Hister beetles
- Ptiliidae – Feather-winged beetles
- Leiodidae – Round fungus beetles
- Scydmaenidae – Antlike stone beetles
- Silphidae – Carrion beetles
- Staphylinidae – Rove beetles
- Pselaphidae – Short-winged mold beetles

Eucinetiformia
- Scirtidae – Marsh beetles
- Eucinetidae – Plate-thigh beetles
- Clambidae – Fringe-winged beetles

Scarabaeiformia
- Lucanidae – Stag beetles
- Scarabaeidae – Scarab beetles

Elateriformia
- Dascillidae – Soft-bodied plant beetles
- Buprestidae – Metallic wood-boring beetles
- Byrrhidae – Pill beetles
- Dryopidae – Long-toed water beetles
- Elmidae – Riffle beetles
- Limnichidae – Minute marsh-loving beetles
- Heteroceridae – Variegated mud-loving beetles
- Ptilodactylidae
- Artematopidae
- Cerophytidae
- Eucnemidae – False click beetles
- Throscidae
- Elateridae – Click beetles
- Lycidae – Net-winged beetles
- Lampyridae – Fireflies or lightning bugs
- Cantharidae – Soldier beetles

Bostrichiformia
- Dermestidae – Dermestid beetles
- Bostrichidae – Branch and twig borers
- Anobiidae – Deathwatch beetles

Cucujiformia
- Lymexylidae – Ship-timber beetles
- Trogossitidae – Bark-gnawing beetles
- Cleridae – Checkered beetles
- Melyridae – Soft-winged flower beetles
- Sphindidae – Dry-fungus beetles
- Nitidulidae – Sap beetles
- Rhizophagidae – Root-eating beetles
- Cucujidae – Flat bark beetles
- Phalacridae – Shining flower beetles
- Cryptophagidae – Silken fungus beetles
- Erotylidae – Pleasing fungus beetles
- Byturidae – Fruitworm beetles
- Endomychidae – Handsome fungus beetles
- Coccinellidae – Ladybird beetles
- Corylophidae – Minute fungus beetles
- Lathridiidae – Brown scavenger beetles
- Mycetophagidae – Hairy fungus beetles
- Ciidae – Minute tree fungus beetles
- Melandryidae – False darkling beetles
- Mordellidae – Tumbling flower beetles
- Rhipiphoridae – Wedge-shaped beetles
- Colydiidae – Cylindrical bark beetles
- Tenebrionidae – Darkling beetles
- Oedemeridae – False blister beetles
- Meloidae – Blister beetles
- Pythidae
- Pyrochroidae – Fire-colored beetles
- Anthicidae – Antlike flower beetles
- Aderidae – Antlike leaf beetles
- Scraptiidae
- Cerambycidae – Long-horned beetles
- Chrysomelidae – Leaf beetles
- Anthribidae – Fungus weevils
- Urodontidae
- Curculionidae – Snout beetles and weevils

Plate 61: Beetles (Insecta: Coleoptera) in Baltic amber V (Staphylinidae, Scarabaeidae), × 22.

2.62 Beetles – Insecta: Coleoptera (Scirtidae)

The Scirtidae (Helodidae, Elodidae), or marsh beetles, are the most frequent beetle family in Baltic amber. HIEKE & PIETRZENIUK (1984) compared beetles in three large amber collections and found that the Scirtidae make up ten to twenty percent of all beetle inclusions. Their frequency provides information on the characteristics of the biotopes in the "Baltic amber forest". For example, the Scirtidae indicate a wealth of standing waters, because their larvae are exclusively aquatic, mostly inhabiting still waters. They prefer bogs, puddles littered phytotelma. Small water holes between tree roots are often sufficient for their survival. Small bodies of water of this kind exist in swampy or boggy terrain and, in the presence of (resin-producing) trees, are reminiscent of the biotopes in a swamp forest. Marsh beetle imagoes live on herbaceous plants in the immediate vicinity of these waters. As many as fifteen fossil species of the genera *Cyphon, Elodes*, *Microcara* and *Plagiocyphon* have been described (KLAUSNITZER 1976, YABLOKOV-KHNZORIAN 1961). The genus *Cyphon* encompasses marsh beetles in half-bog, standing waters. Based on the habitats of extant species, fossil species of the genus *Elodes* may have been rheophilous stream-dwellers (KLAUSNITZER 1976).

Only one species each of the Ptilodactylidae and Dryopidae, the closest relatives of these marsh beetles, has been recorded to date.

Scirtidae

Brachelodes motschulskyi YABLOKOV-KHNZ., 1961
Cyphon pallasi YABLOKOV-KHNZORIAN, 1961
Cyphon krynyckyi YABLOKOV-KHNZ., 1961
Cyphon shevyrevi YABLOKOV-KHNZ., 1961
Cyphonogenius zakhvatkini YABLOKOV-KHNZ., 1961
Helodes modesta KLAUSNITZER, 1976
Helodes transversa KLAUSNITZER, 1976
Helodes egregia KLAUSNITZER, 1976
Helodes setosa KLAUSNITZER, 1976
Helodes minax KLAUSNITZER, 1976
Helodopsis solskyi YABLOKOV-KHNZOR., 1961
Microcara dokhturovi YABLOKOV-KHNZ., 1960
Microcara kuznezovi YABLOKOV-KHNZ., 1960
Microcara znoijkoi YABLOKOV-KHNZ., 1960
Microcara zubkovi YABLOKOV-KHNZ., 1960
Plagiocyphon plavilschikovi YABLOKOV-KHNZORIAN, 1960

Ptilodactylidae

Ptilodactyloides stipulicornis MOTSCHULSKY, 1856

Dryopidae

Palaeoriohelmis samlandica BOLLOW, 1940

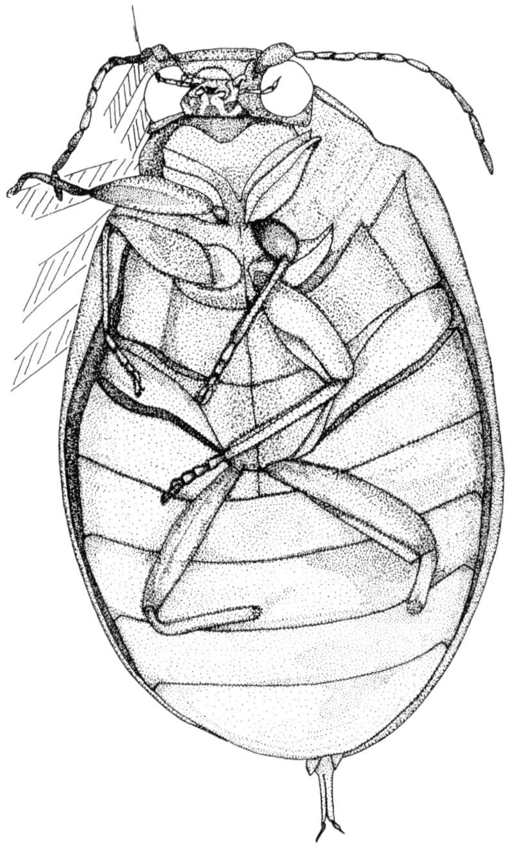

Fig. 73: *Cyphon* sp. (Coleoptera: Scirtidae).

Plate 62: Beetles (Insecta: Coleoptera) in Baltic amber VI.

a Scirtidae (marsh beetles), larva, × 18.
b Scirtidae (marsh beetles), × 10.
c Scirtidae (marsh beetles), × 12.
d Scirtidae (marsh beetles), × 11.
e Elateridae (click beetles), × 9.
f Cantharidae (soldier beetles), × 9.
g Cleridae (checkered beetle), × 9.
h Cleridae (checkered beetle), larva, × 12.

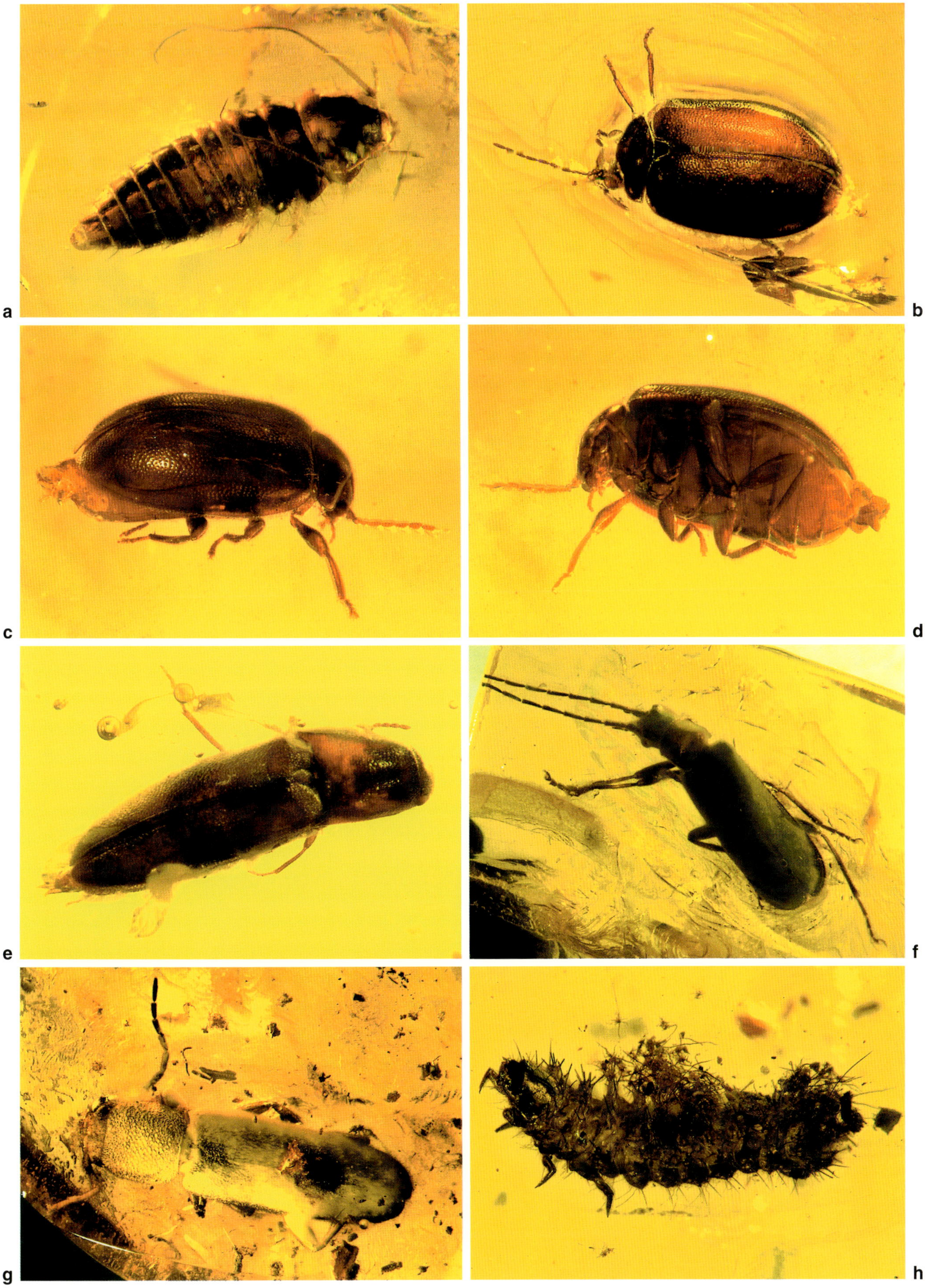
a
b
c
d
e
f
g
h

2.63 Preference of the beetles in the "amber forest" – Insecta: Coleoptera

The beetle fauna in Baltic amber is extremely diverse, as illustrated by the great number of families. The frequency of occurrence of these families varies in terms of both the number of species and the abundance of individuals. A breakdown of the families according to their frequency in amber also indicates the preference of the beetles for the biotope of the "Baltic amber forest".

The Scirtidae (marsh beetles) head the list. Their high frequency in amber provides key information on the nature of the amber forest (Chap. 2.62). They are followed in order of decreasing preference by the Elateridae which, like the similarly numerous Mordellidae and Scraptiidae, are typical inhabitants of moist, shady fields in the immediate vicinity of a forest. Their larvae develop in decaying trees, live under loose bark and feed on moldy wood. Our interpretation is based on the work of Hieke & Pietrzeniuk (1984), who compared three, large amber collections (Berlin Museum of Natural History, Copenhagen Museum of Natural History, Klebs collection) and listed the beetle fauna according to the frequency of the families.

The Scirtidae and Elateridae are followed by the Anobiidae on the frequency scale. They point to a third component of the "Baltic amber forest". Together with the frequent Aderidae, Eucnemidae, Scolytidae, Melandryidae, Anthribidae, Anthicidae, Cerambycidae and Mycetophagidae, the Anobiidae are indicative of microhabitats characterized by decaying trees, rotting trunks, moldy wood and numerous sporophores of the Polyporaceae. The beetles bring to mind scenarios of a jungle.

Fourth place is occupied by the Staphylinidae, which inhabit moist leaf litter and decaying plant matter. They are joined there by other frequent families in amber, such as the Scydmaenidae, Pselaphidae, Lathridiidae and Cryptophagidae.

According to Hieke & Pietrzeniuk (1984), the 20 most frequent families listed below indicate that the amber forest was a jungle with sparse tree stands, heavy undergrowth, rotting and moldy wood, patches of herbaceous vegetation and numerous standing waters in swampy terrain (particularly as implied by the number of Scirtidae (Chap. 2.62)):

1.	Scirtidae	999
2.	Elateridae	893
3.	Anobiidae	532
4.	Staphylinidae	350
5.	Mordellidae	253
6.	Aderidae	202
7.	Scydmaenidae	200
8.	Scraptiidae	196
9.	Curculionidae	155
10.	Pselaphidae	150
11.	Curculionidae (only Scolytinae)	148
12.	Eucnemidae	140
13.	Mycetophagidae	136
14.	Melandryidae (= Serropalpidae)	129
15.	Carabidae	126
16.	Anthicidae	114
17.	Lathridiidae	112
18.	Cantharidae	110
19.	Cerambycidae	100
20.	Cryptophagidae	89

Preference of the beetle families with the number of inclusions recorded from three collections (after Hieke & Pietrzeniuk 1984).

Plate 63: Beetles (Insecta: Coleoptera) in Baltic amber VII.

a Mycetophagidae (hairy fungus beetles), × 11.
b Mycetophagidae (hairy fungus beetles), × 10.
c Colydiidae (cylindrical bark beetles), × 15.
d Lathridiidae (brown scavenger beetles), × 18.
e Scraptiidae, larva, × 10.
f Scraptiidae, × 14.
g Rhipiphoridae (wedge-shaped beetles), × 10.
h Aderidae, antlike leaf beetle, × 12.

a

b

c

d

e

f

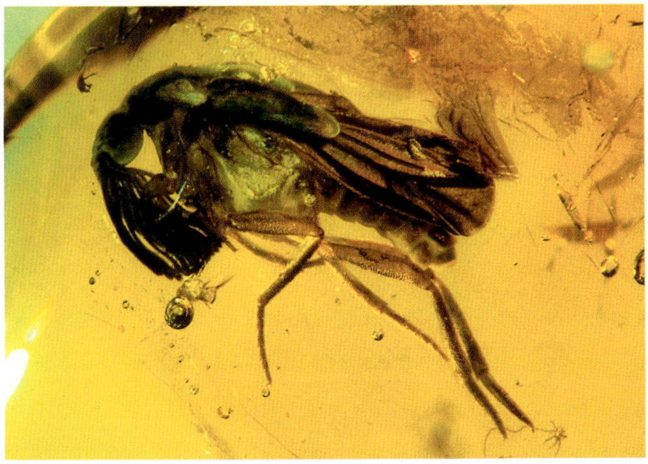

g

h

2.64 Beetles as forest-dwellers – Insecta: Coleoptera

The beetles can help narrow down the kind of (resin-producing) trees that existed in the "Baltic amber forest". Forest-dwelling beetles have adapted to the forest environment in a variety of ways. They colonize all biotopes and inhabit all strata, from the soil, to the herbaceous, shrub and tree layers. Their life cycles are closely dependent on their host plants, in order to allow undisturbed development in the microhabitats of the forest and its flora – from oviposition, to the feeding and growth of the larvae, the metamorphosis of the pupae and the emergence of the adult beetles. The interaction between beetles and their host plants is often so close, that knowledge of the beetles can, in turn, allow conclusions to be made about the forest environment and tree flora. This retrospective view is extensive enough to support a cautious assessment of the Eocene amber forest.

For example, it is interesting to note that, of the 100 Cerambycidae (long-horned beetles) surveyed (Hieke & Pietrzeniuk 1984), 38 inclusions can reliably be classified in the Mediterranean genus *Notorrhina*. The Asemini, with the genus *Notorrhina*, and the numerous Lepturini are a good indication of conifers, their exclusive habitat. If conifers dominated the "amber forest", then the relatively high number of these long-horned beetles is not surprising. The Anobiidae (deathwatch beetles) are frequent (260 inclusions), with the subfamily Anobiinae primarily inhabiting dead wood from conifers. A fairly small percentage of the Anobiidae inclusions belongs to the subfamily Dorcatominae, which inhabited fungus on trees. The numerous bark beetles (Curculionidae: Scolytinae) colonize coniferous more often than deciduous trees, living with their larvae in the bark.

The majority of beetle families in Baltic amber were forest-dwellers that lived on dead wood, fungi, the soil, or as generalists on deciduous and coniferous trees. According to Hieke & Pietrzeniuk (1984), some families are only represented by a few individuals, because they were apparently restricted to deciduous trees. These include the Cerophytidae (2 specimens), Lucanidae (1) and the mostly tropical Bostrichidae (2). The Scarabaeidae, which are represented by six specimens in the list compiled by Hieke & Pietrzeniuk (1984), also prefer deciduous trees.

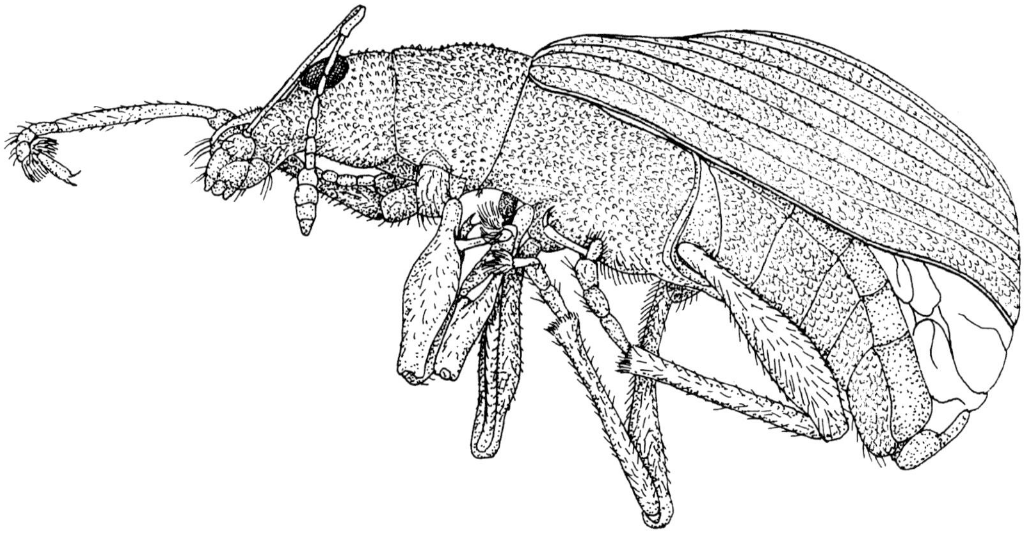

Fig. 74: Weevil (Coleoptera: Curculionidae).

Plate 64: Beetles (Insecta: Coleoptera) in Baltic amber VIII.

a Anobiidae (deathwatch beetles), × 12.
b Anobiidae (deathwatch beetles), × 10.
c Anthribidae, *Glaesotropis weitschati*, holotype, × 10.
d Anthribidae, *Glaesotropis minor*, holotype, × 12.
e Curculionidae, Scolytinae, × 16.
f Curculionidae, Scolytinae, × 16.
g Cleridae (checkered beetles), × 10.
h Chrysomelidae (leaf beetles), × 10.

a

b

c

d

e

f

g

h

2.65 Beetles – Insecta: Coleoptera (Curculionidae)

According to the classification by LAWRENCE & BRITTON (1991), the weevil family (Curculionidae) includes the subfamilies of pin-hole borers (Platypodinae) and bark beetles (Scolytinae). Even without these two subfamilies, the Curculionidae still make up the largest beetle family with more than 50,000 extant species (and thereby the most diverse group of all animal life!). They are exclusively phytophagous insects and utilize the entire range of food offered by higher plants, including seeds, fruits, flowers, buds, leaves, branches, bark, bast and roots. Consequently, it is no surprise that weevils are relatively frequent in Baltic amber.

Current knowledge of the weevils in amber has been compiled from the older work of BURMEISTER (1832), KEFERSTEIN (1834), BERENDT (1845) and HELM (1896 a, b, 1899), and from the 20th-century work of KLEBS (1910), WAGNER (1924) and VOSS (1953, 1972). The dominant subfamilies of the Curculionidae in Baltic amber are the Brachyderinae, Otiorrhynchinae, Trachodinae, Cossoninae and Cryptorhynchinae. Most species of these subfamilies inhabit deciduous trees. The subfamilies Apioninae, Anthonominae and Nanophyinae recorded from amber comprise bud and flower-piercing weevils.

Roughly 800 species of pin-hole borers (Platypodinae) are known worldwide today, the majority of which live in relatively warm regions. Although they are occasionally listed among the beetles of Baltic amber, reliable evidence of their occurrence has yet to be provided. The Platypodinae occur in copals from various warm regions and frequently in Dominican amber.

Bark beetles (Scolytinae) are typical inclusions in Baltic amber. According to HIEKE & PIETRZENIUK (1984), they are one of the 20 most frequent taxa (cf. 2.63). Virtually without exception, the amber beetles are classified among the Hylesinini, and not the Scolytini or Ipini, which today inhabit coniferous forests in the temperate zone. The authors suspect that the Eocene amber forest was dominated by other conifers preferred by the Hylesinini. Bark beetles are often strictly monophagous insects, rarely colonizing closely related tree species. Only a few extant bark beetles are polyphagous on both coniferous and deciduous trees (BRAUNS 1964). In view of the fact that bark beetles are frequent in Baltic amber, a detailed comparison of the taxa with extant relatives may provide valuable information on the composition of tree species in the amber forest.

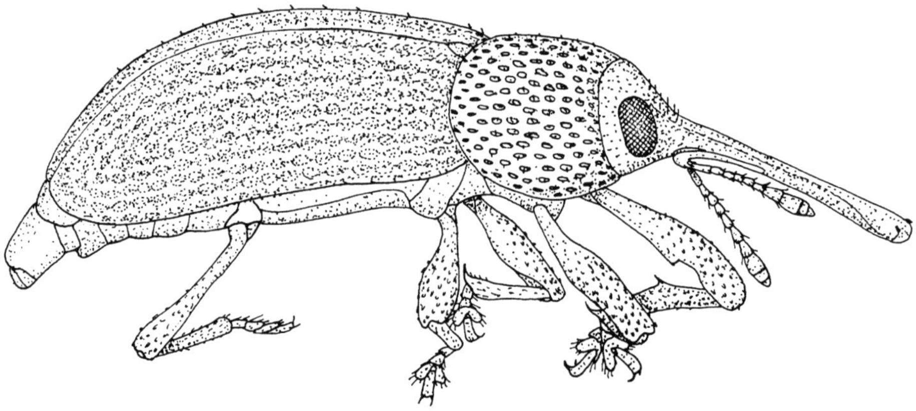

Fig. 75 a-h: Weevil (Curculionidae).

Plate 65: Beetles (Insecta: Coleoptera) in Baltic amber IX (Curculionidae).
a ×9, **b** ×10, **c** ×11, **d** ×10, **e** ×10, **f** ×12, **g** ×13, **h** ×12.

a b c d e f g h

2.66 Twisted-winged parasites – Insecta: Strepsiptera

The twisted-winged parasites (Strepsiptera) make up an order of holometabolous insects comprising roughly 550 extant species in nine families. Their systematic classification has not been clarified conclusively. They have a sister group relationship with either the Coleoptera or probably the Diptera (KINZELBACH 1990, KUKALOVA-PECK & LAWRENCE 1993, AFZELIUS & DALLAI 1994, WHITING & WHEELER 1994, WHITING et al. 1997). The biology of the Strepsiptera is characterized by the endoparasitic lifestyle of most developmental stages, which also explains their extreme sexual dimorphism and the polymetabolism of the larvae. The hosts are always insects, including the Zygentoma (Lepismatidae), Blattodea, Mantodea, Orthoptera, Heteroptera, Auchenorrhyncha, Hymenoptera (Apoidea, Vespoidea) and Diptera (KATHIRITHAMBY 1989, KINZELBACH 1990, KINZELBACH & POHL 1994).

The oldest specimen of a fossil strepsipteran was found in Eocene brown coal in Halle, Germany (KINZELBACH & LUTZ 1985). This fossil is a primary larva of the Myrmecolacidae, *Strichotrema eocaenicum* (HAUPT, 1950). As the Myrmecolacidae make up the phylogenetically youngest family of the Strepsiptera (KINZELBACH 1971), POHL & KINZELBACH (1995) concluded that all nine families of the Strepsiptera or their stem-groups already existed by the Eocene. At present, the three strepsipteran species recorded from Baltic amber represent two families:

Mengeidae
Mengea tertiaria (MENGE, 1866)
Myrmecolacidae
Stichotrema triangulum POHL & KINZELBACH, 1995.
Stichotrema weitschati KINZELBACH & POHL, 1994.

Only free-flying males have been found in amber to date. There have been no reports of the free-living, wingless, or permanently endoparasitic females. The males probably flew into the sticky resin accidentally upon locating a female. The family Myrmecolacidae, which is most frequent in amber, is currently distributed in the tropical, Australian and western Palearctic (one species) regions. It displays the greatest density in the Neotropics.

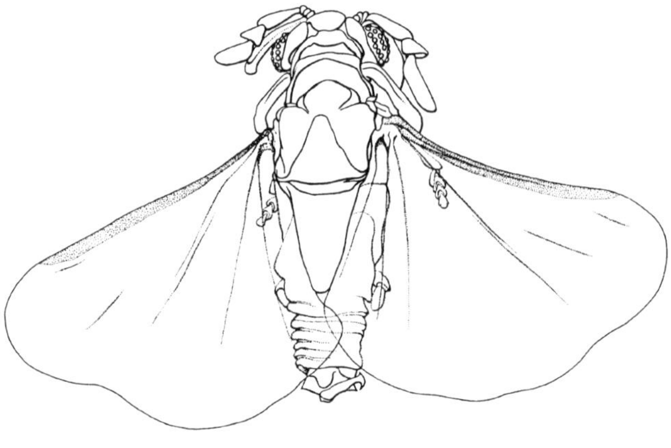

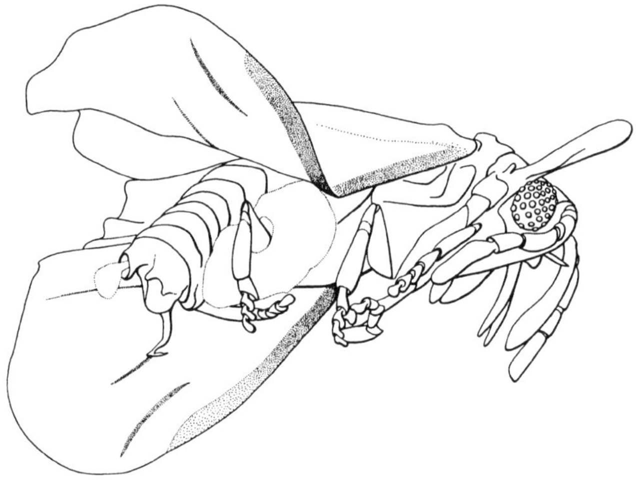

Fig. 76 a: *Stichotrema triangulum* (Myrmecolacidae), **b** *Stichotrema weitschati* (Myrmecolacidae), holotype.

Plate 66: Twisted-winged parasites (Insecta: Strepsiptera) in Baltic amber.

- **a** *Stichotrema weitschati* (Myrmecolacidae), head, ×40.
- **b** *Mengea tertiaria* (Mengeidae), head, ×60.
- **c** *Stichotrema weitschati* (Myrmecolacidae), dorsal, ×20.
- **d** *Mengea tertiaria* (Mengeidae), dorsal, ×28.
- **e** *Stichotrema weitschati* (Myrmecolacidae), lateral, ×20.
- **f** *Mengea tertiaria* (Mengeidae), lateral, ×28.
- **g** *Stichotrema* sp., ×20.
- **h** *Stichotrema* sp., ×20.

a
b
c
d
e
f
g
h

2.67 Hymenopterans – Insecta: Hymenoptera

Hymenopterans belong to the group of holometabolous insects (Endopterygota) that undergo complete metamorphosis (eggs, larvae, pupae, imagoes). The order encompasses many forms with a body length of up to five centimeters, as well as an enormous number of small to very small insects reaching a maximum of two millimeters in length. The latter group includes the Baltic amber fairyflies (Mymaridae, Plate 69 b). Many hymenopteran species practice brood care and form social communities with highly developed modes of behavior. The order Hymenoptera comprises at least 160,000 species. It is divided into two suborders: the Symphyta (sawflies and horntails with caterpillar-like larvae), of which only the Cephidae, Siricidae and Tenthredinidae have been reliably recorded from Baltic amber, and the Apocrita, which constitute by far the more species-diverse suborder both today and in Eocene amber.

Brues (1933 b) made a major contribution to our knowledge of Hymenoptera in Baltic amber with his comprehensive work entitled "The Parasitic Hymenoptera of the Baltic Amber". Although further studies have been published since then, the taxonomy and systematics are based on the information originally provided by Brues (1933b).

The Megalyridae (Plate 67 a) are represented by two species from the genus *Prodinapsis* and can easily be mistaken for members of the Braconidae, as was the case in Bachofen-Echt (1949). Only a few dozen species of Megalyridae exist today in Southeast Asia and Australia. They parasitize beetle larvae living under tree bark.

The Megaspilidae are characterized by an unusually large pterostigma combined with reduced wing venation. They belong to the superfamily Ceraphronoidea and are represented in Baltic amber by the two genera *Conostigmus* and *Lagynodes*. The genus *Lagynodes* displays definite sexual dimorphism, with wingless females and winged males.

The Evaniidae have a distinct appearance characterized by a slender abdominal stalk attached high above the hind coxae (propodeum). With over 400 species occurring in tropical regions, the Evaniidae are very diverse today and thus frequently found in copal. Brues (1933 b) described three species, the largest being *Evania producta* (Plate 67 c). The females of all extant species oviposit in the egg cases of cockroaches. For this reason, several species can also occasionally be observed in supermarket cockroach populations.

The Braconidae and Ichneumonidae make up the superfamily Ichneumonoidea, which is currently the most diverse family of Hymenoptera with an estimated diversity over 100,000 species. Brues (1933 b) studied the Braconidae in great detail, covering 126 species in 42 genera, including *Microtypus* and *Electrohelcon* (Plate 67 e, f). Although the Ichneumonidae are just as frequent in Baltic amber, only a few species have been described to date (e.g. *Astiphromma brischkei* Brues, 1923) and the family has yet to be subjected to detailed study.

The Diapriidae and Proctotrupidae belong to the superfamily Proctotrupoidea. Only a few genera are represented in Baltic amber. Maneval (1938) described two Diapriidae of the genera *Cinetus* and *Pantolyta*, while Brues (1940) described six species of the genus *Cryptoserphus* and one species of the genus *Proctotrupes* Brues, (1923).

Plate 67: Hymenopterans (Insecta: Hymenoptera) in Baltic amber I.

a Megalyridae *(Prodinapsis succinalis)*, × 24.
b Megaspilidae (*Conostigmus* sp.), × 25.
c Evaniidae *(Evania producta)*, × 3.5.
d Ichneumonidae, × 8.5.
e Braconidae *(Microtypus triangulifer)*, × 12.
f Braconidae (? *Electrohelcon*), × 10.
g Diapriidae, × 25.
h Proctotrupidae (*Cryptoserphus* sp.), × 16.

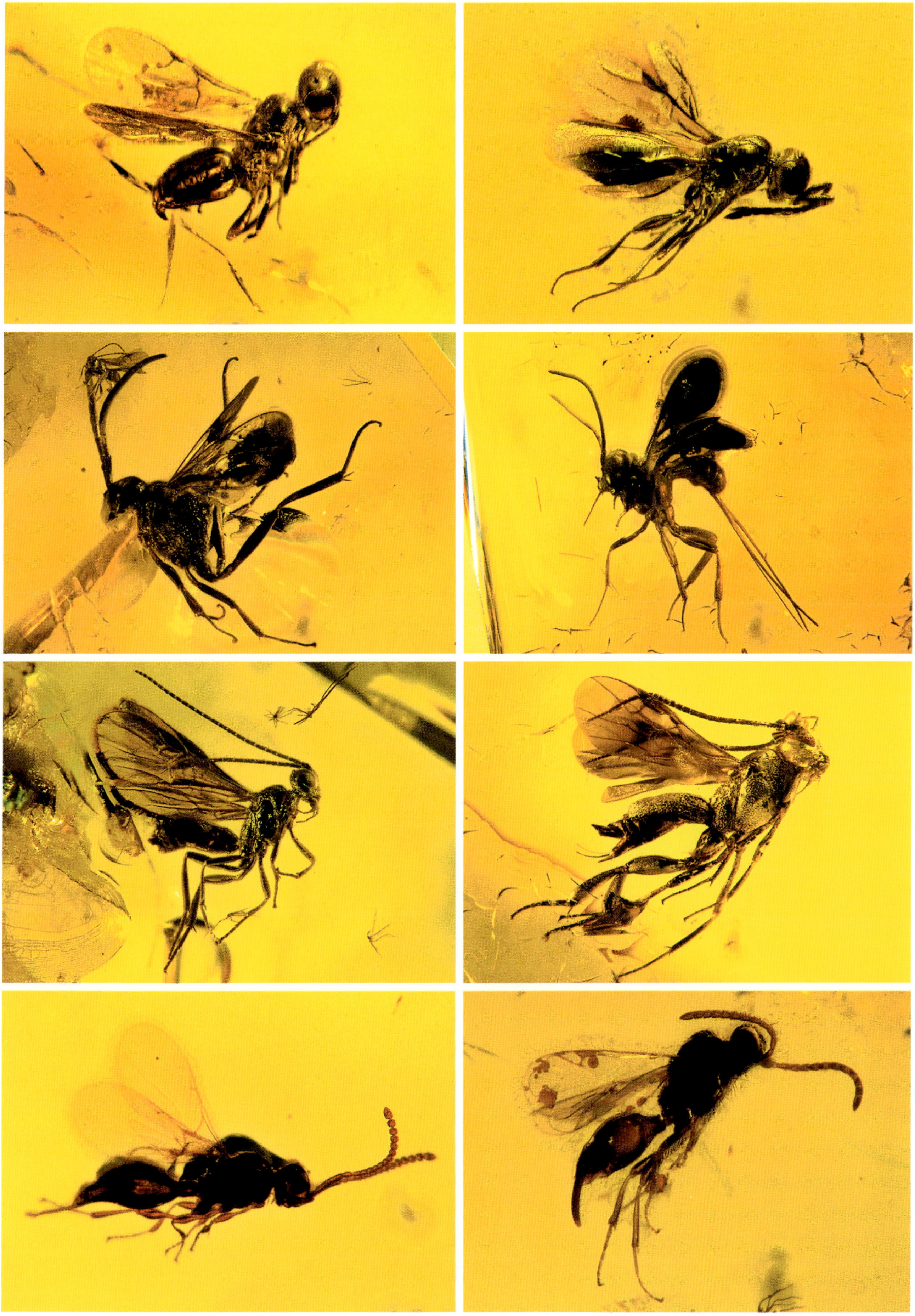

a b c d e f g h

2.68 Hymenopterans – Insecta: Hymenoptera (Chalcidoidea)

The Scelionidae (Plate 68 a, b) are among the more frequent Hymenoptera in Baltic amber. Several thousand extant species have been described, most being solitary endoparasites in insects and spider eggs. BRUES (1940 c) mentions roughly 20 genera from Baltic amber, including *Sembilanocera* and *Ceratoteleia*. Several species are very small and frequently mistaken for chalcid wasps (Chalcidoidea).

Gall wasps (Cynipoidea) are uncommon Hymenoptera in amber. Present-day phytophagous gall insects are particularly familiar representatives of this superfamily. Little known, however, is the fact that the majority of species are parasitic on other insects, such as Diptera or Hymenoptera. PRESL described the first species from Baltic amber as early as 1822. A second species description was not provided until 170 years later by KOVALEV (1995), who also studied the phylogenetics of the Cynipoidea (KOVALEV 1994).

Previously classified as chalcid wasps, the Mymarommatoidea are now treated as a separate superfamily due to their two-segmented petiole. The only genus is *Palaeomymar* (Plate 68 d), which has also been recorded from Baltic amber (STEIN 1877, RASNITZIN & KULICKA 1990).

The Chalcidoidea of Baltic amber have been little studied. In view of the fact that there are over 3,000 genera with 20,000 extant species, this reticence is probably due to the extremely difficult taxonomy of the group. On the other hand, the Chalcidoidea occur in Baltic amber with a very tempting diversity of families and species.

The Encyrtidae (Plate 69 a) currently encompass over 700 genera and over 3,800 known species, making them one of the most structurally diverse families of chalcid wasps. Because many genera of the Chalcidoidea belong to phylogenetically younger, highly specialized families, the description of chalcid wasps from Baltic amber is likely to require the establishment of new genera and possibly also new families. A comparison of extant and extinct Eocene forms indicates that the Chalcidoidea apparently continued to evolve.

MEUNIER (1905) described a few species of Mymaridae (Plate 69 b). Descriptions then followed of a Torymidae (BRUES 1923 c), an Eupelmidae (TRJAPITZIN 1963) (Plate 68 f, h) and a Tetracampidae (TRJAPITZIN & MANUKYAN 1995). The chalcid wasps of the Chalcididae and Pteromalidae shown here (Plate 68 e, g) are the first records of these families in Baltic amber.

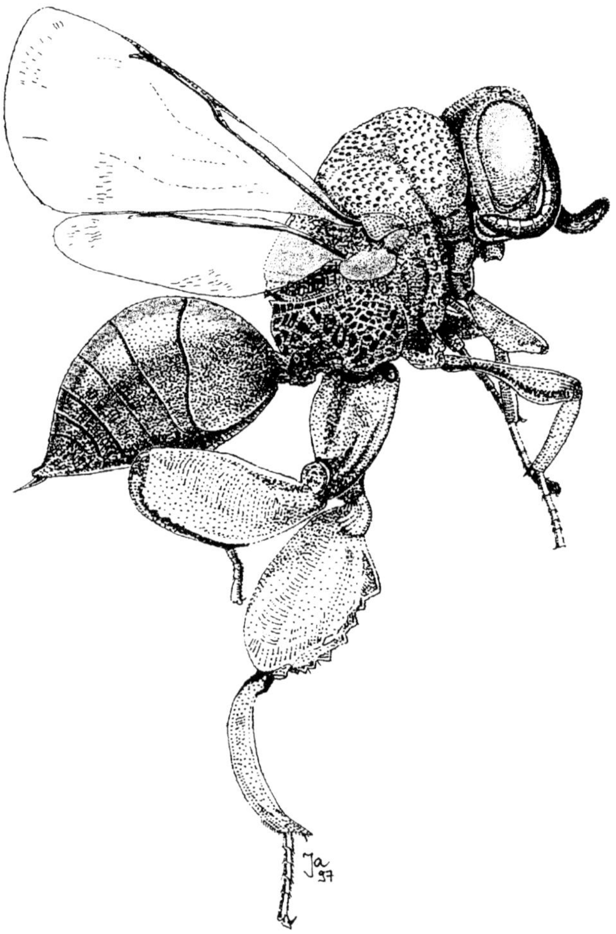

Fig. 77: Chalcid wasp (Hymenoptera: Chalcididae) (original J.-W. JANZEN).

Plate 68: Hymenopterans (Insecta: Hymenoptera) in Baltic amber II.

a Scelionidae *(Sembilanocera clavata)*, × 55.
b Scelionidae *(Ceratoteleia proleptica)*, × 16.
c Figitidae (? *Palaeofigites*), × 18.
d Mymarommatidae *(Palaeomymar)*, × 100.
e Chalcididae (? *Brachymeria*), × 10.
f Torymidae (*Monodontomerus* sp.), × 22.
g Pteromalidae, × 28.
h Eupelmidae, × 15.

a
b
c
d
e
f
g
h

2.69 Hymenopterans – Insecta: Hymenoptera (Chrysidoidea)

Brues (1933b) studied parasitic Hymenoptera of the Baltic amber and selected genera of Chrysidoidea, such as those in the families Dryinidae and Bethylidae. He described six species of the genera *Lestodryinus*, *Neodryinus* and *Thaumatodryinus* from the family Dryinidae. Haupt (1944) revised *Thaumatodryinus* and established the new genus *Harpactospecion.* The Dryinidae are one of the most specialized families of Aculeata. This differentiation is manifested by the development of the forelegs into chelate pincers, which is unique among the Hymenoptera (Plate 69c, d). Only the females have pincers, which they use to grasp their hosts (cicadas) to lay eggs. The larvae develop on the cicadas, with the anterior end of the body extending inside the host. Egg membranes and shed exuviae form sacs that remain attached to the host until it has been exhausted as a source of food.

The family Bethylidae is represented in Baltic amber by numerous genera and species. It comprises approximately 2,000 extant species and is thus the most species-diverse family among the Chrysidoidea. The Bethylidae are primarily tropical. Brues (1933b) described 16 genera of the family Bethylidae from Baltic amber, including *Isobrachium*, *Palaeobethylus* and *Lythopsenella*. Ohl (1995) later added *Lythopsenella kerneggeri* to the list of amber Bethylidae (Plate 69f, g, h).

The family Scolebythidae was first described by Evans (1963). Brothers & Janzen (1998) placed *Pristapenesia primaeva* Brues, 1933 from Baltic amber in the family Scolebythidae and revised the species description (Plate 69e, Fig 78). Little is known about the biology of the Scolebythidae. They are believed to be external parasites on long-horned beetles (Cerambycidae). The three extant species are divided into three genera occurring in Brazil, South Africa, Madagascar and Australia.

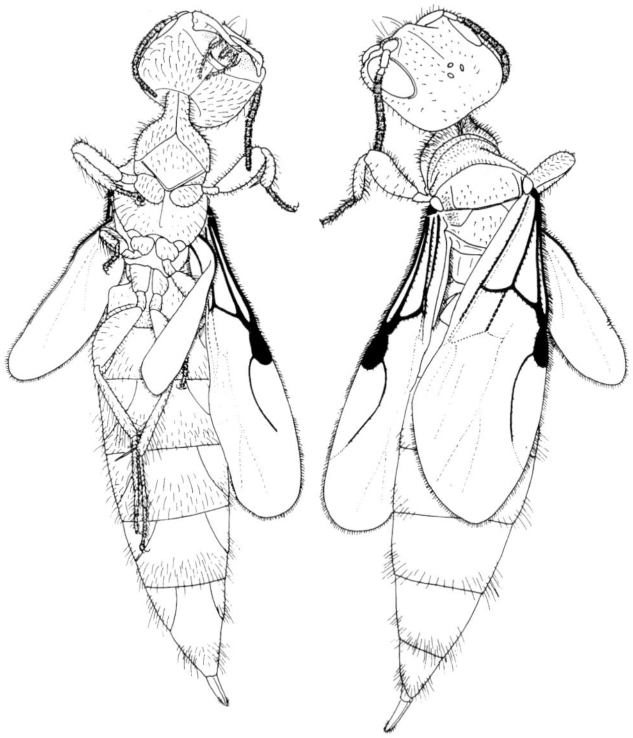

Fig. 78: Scolebythidae *Pristapenesia primaeva* Brues, 1933, female (ventral and dorsal) (after Brothers & Janzen 1998).

Plate 69: Hymenopterans (Insecta: Hymenoptera) in Baltic amber III.

a Encyrtidae, × 20.
b Mymaridae, × 30.
c Dryinidae *(Harpactospecion filicornis)*, × 12.
d Dryinidae (pincer), × 120.
e Scolebythidae *(Pristapenesia primaeva)*, × 13.
f Bethylidae (*Isobrachium* sp.), × 20.
g Bethylidae (*Palaeobethylus* sp.), × 12.
h Bethylidae (*Lythopsenella* sp.), × 25.

a b c d e f g h

2.70 Ants – Insecta: Hymenoptera (Vespoidea)

Ants make up the large family of Formicidae, which includes nearly 10,000 described species. While Brown (1954) previously divided this family into nine subfamilies, Baroni Urbani et al. (1992) believed it necessary for taxonomic reasons to distinguish between seventeen subfamilies. This discrepancy is problematic for amber ants, because the fundamental systematics and taxonomy were established by Mayr in 1868, supplemented in following decades, but never subjected to a comprehensive, modern revision. Spahr (1987) presented an overview for Baltic amber, including the following subfamilies: Dolichoderinae, Dorylinae, Formicinae, Myrmicinae, Ponerinae. A revision would likely lead to the identification of additional subfamilies.

The Dorylinae are commonly known as wandering ants. Handlirsch (1925) made the first reference to their occurrence in Baltic amber.

The European Myrmicinae are easily distinguished by their two-segment pedicel consisting of a petiole and postpetiole that form two nodes. Unfortunately, this identifying characteristic fails when it comes to the tropical forms, because the tropical Pseudomyrmecinae and Dorylinae also display this nodelike structure. Myrmicines of the following genera occur in Baltic amber: *Aphaenogaster*, *Cremastogaster*, *Monomorium*, *Myrmica*, *Stenamma*, *Leptothorax*.

In the Ponerinae the nodelike and erect pedicel comprises a single segment. The pedicel transitions into the gaster, both of which make up the abdomen. The Ponerinae have a constriction between the first and second gastral segments and a well-developed sting on the end of the abdomen. The amber ants of this subfamily include the genus *Ponera.*

The Formicinae (including the Camponotinae), or mound ants, are distinguished by the nearly upright projection on the single-segment pedicel. More significant, however, is the absence of a sting and its replacement with an acidipore (an opening at the gastral apex for the secretion of formic acid). This characteristic applies at least to the European species of the subfamily. The mound ants in Baltic amber include species of the genera *Plagiolepis, Camponotus, Lasius* and *Formica.*

The Dolichoderinae are those with a single-segment petiole, whose small, sometimes undifferentiated projection is inclined anteriorly. The sting is rudimentary. This subfamily is widely distributed and particularly diverse in the tropics. The dolichoderines are represented by amber species of the genera *Dolichoderus, Bothriomyrmex* and *Liometopum.*

A further subfamily is represented from Baltic amber by the fossil genus *Prionomyrex* which is most closely related to the recent Australian genus *Myrmecia* (Mayr 1868, Wheeler 1915) and the recent Australian genus *Nothomyrmecia* (Baroni Urbani 2000). As a consequence of the analysis of the relationships between the three related genera the subfamily Prionomyrmecinae is suggested by Baroni Urbani 2000. *Prionomyrex janzeni* Baroni Urbani, 2000 is a fossil species of the subfamily Prionomyrmecinae.

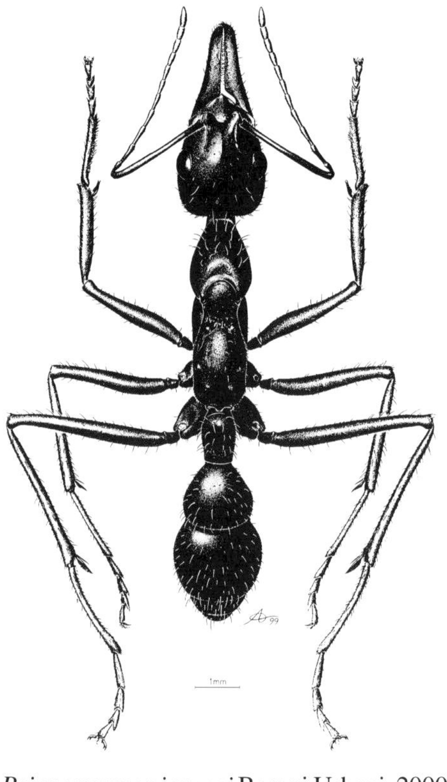

Fig. 79: *Prionomyrmex janzeni* Baroni Urbani, 2000 (Formicidae: Prionomyrmecinae).

Plate 70: Hymenopterans (Insecta: Hymenoptera) in Baltic amber IV (Vespoidea I).

a Ponerine ant (Ponerinae), × 10.
b Ponerine ant (Ponerinae), × 13.
c Myrmicine ant (Myrmicinae), × 21.
d Myrmicine ant (Myrmicinae), × 12.
e Dolichoderine ant (Dolichoderinae), × 16.
f Mound ant (Formicinae), × 14.
g Mound ant (Formicinae), × 14.
h Mound ant (Formicinae), × 15.

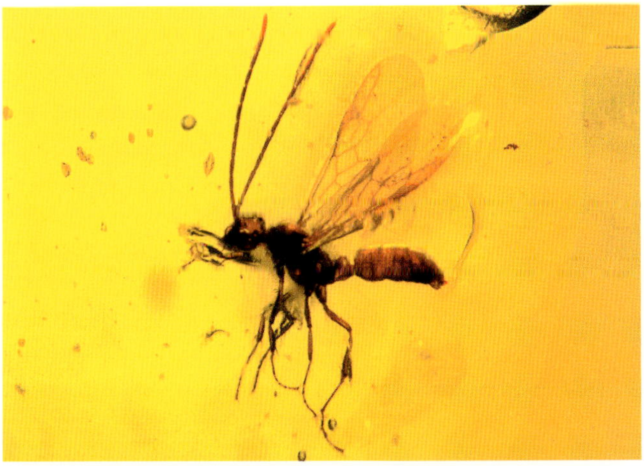
a

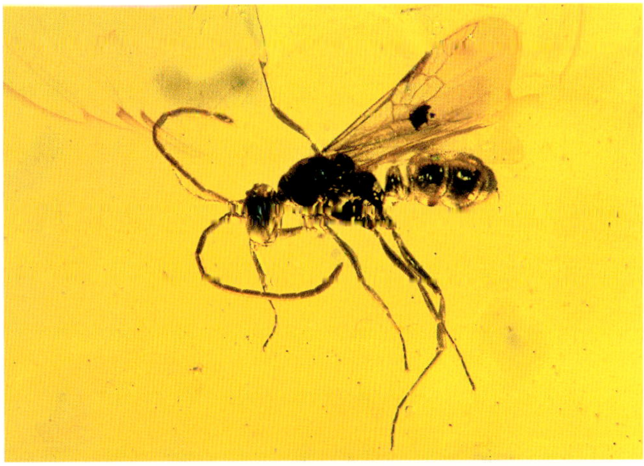
b

c

d

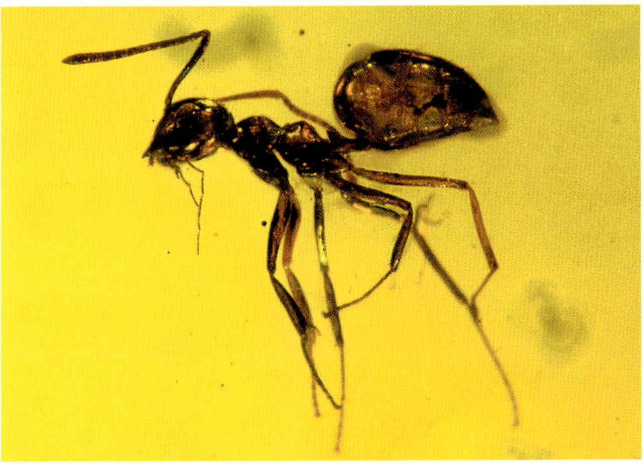
e

f

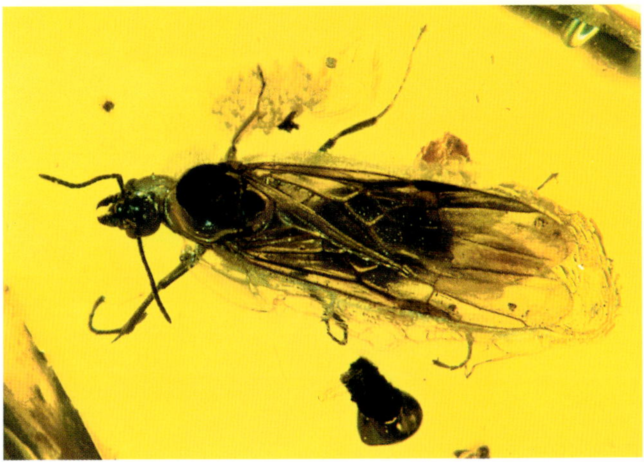
g

h

2.71 Ants – Insecta: Hymenoptera (Vespoidea)

Much has been written about the biology of ants. One of the first, classical monographs was prepared by ESCHERICH (1917). As a result of new research on the fascinating world of ants, the literature has been greatly expanded with studies of their social organization (GÖSSWALD 1985), ecology and behavior in ecosystems such as forests (GÖSSWALD 1989-1990). HÖLLDOBLER & WILSON (1990, 1995) published their many years of experience and exciting discoveries in an outstanding account of the biology of ants.

Nearly 10,000 ant species have been described to date and an equal number of unknown species are suspected to exist. The various ant species live according to fixed rules governing the organization of their colonies, in which neither males, sterile females or workers could survive on their own. The entire ant colony becomes like a single, biological organism (GÖSSWALD 1985). Be they hunters, breeders, grain or honey gatherers, or the guests, parasites, predators or slaveholders of other species, the activity of the ants is closely tied to the mechanism of the society. The colony practices an impressive form of brood care, thus ensuring its own survival.

Comparatively little is known about the biology of ants from Baltic amber. We are dependent on "snapshots" captured in amber, which together give us a general picture of their past life. Consequently, these fatal accidents have come to play an important role. Disturbing an ant hill provokes behavior that can occasionally also be observed in amber, when larvae and pupae are found scattered in the specimen: the ants apparently grabbed their brood in order to bring them to safety. These scenes are found in Dominican amber (SCHLEE 1980) and older Baltic amber (Plate 71 c-f).

Amber has captured other unfortunate scenes for ants, including spiders tying up their ant prey to feed on them later (Fig. 37), or assassin bugs luring passing ants, feeding on them and discarding the empty integuments (Plate 43 a).

The symbiosis between ants and aphids is a rather more pleasant example. HEIE (1967) reported an amber specimen from Königsberg containing fifteen ants together with a colony of aphids. When the ants stroke the aphids with their antennae, the aphids secrete drops of sugary "honeydew" greedily consumed by the ants (e.g. *Lasius* species). In return, the ants protect aphids from their enemies. Aphids and ants are indeed found together in the taphocoenoses of individual pieces of amber, thus providing an indication of this symbiotic relationship. According to HEIE (1967), however, symbiosis does not exist when ants appear in amber with aphids of the genus *Germaraphis*, which secrete wax instead of honeydew and are therefore not "milked" by the ants. Considering the relative frequency of aphids and ants, it is very possible for them to occur together in Baltic amber, be it a random coincidence or the result of a symbiotic relationship, this being determined by the identification of the taxa.

Plate 71: Hymenopterans (Insecta: Hymenoptera) in Baltic amber V (Vespoidea II).

a Mound ant (Formicinae), × 8.
b Mound ants (Formicinae), × 14.
c Disturbed ant brood, × 10.
d Ant pupa and larva, × 25.
e Worker with ant larva, × 15.
f Pupa with ant immediately prior to hatching, × 30.
g Ant larvae, × 15
h Emerged ant, × 12.

a
b
c
d
e
f
g
h

2.72 Bees and spheciform wasps – Insecta: Hymenoptera (Apoidea)

The superfamily Apoidea includes bees and their close relatives, the spheciform wasps. The two families are represented in Baltic amber by several genera and species. Bees and spheciform wasps are characterized by a broad, posterior lobe on the prothorax that extends towards the tegula but does not reach it. The main difference between the two families is the shape of the hind tarsus, which is much broader in bees than in spheciform wasps, and by the presence of plumose setae in bees.

Bees gather pollen and nectar to feed their larvae, with most genera thus possess special gathering organs located on the abdomen or hind legs. Due to their familiarity and popularity as busy honey-gatherers, bees are coveted inclusions among amber collectors and, as a result, they have been the subject of several, detailed studies. The most recent monograph of Baltic amber bees recognized 36 valid species in 18 genera and from five families (Engel 2001 b). These records include species of sweat bees (Halictidae), melittid bees (Melittidae), lithurgine bees (Megachilidae: Lithurginae), leafcutting bees (Megachilidae: Megachilinae), carpenter bees (Apidae: Xylocopinae) and "true bees" (Apidae: Apinae) (Buttel-Reppen 1906, Cockerell 1909, Kelner-Pillaut 1974, Engel 2001 b). In addition, a poorly understood extinct family of bees, intermediate between short- and long-tongued bees, is known from Baltic amber (*Paleomelitta nigripennis* Engel, 2001: Paleomelittidae). Baltic amber presently contains the most diverse fossil bee fauna and includes a remarkable array of highly social bees, many of which became extinct at the end of the Eocene (Engel 2001 b). The records of andrenine bees (Andrenidae) (Salt 1931) and bumble bees (*Bombus* spp.) have not been confirmed and are likely misidentifications (Engel 2001 b). Digger and cuckoo bees (Apidae; formerly Anthophoridae) have also been mentioned but remain unconfirmed.

Spheciform wasps clearly differ from bees in terms of their biology. Like spider wasps (Pompilidae), they dig their nests in the soil, usually placing several, paralyzed insects into each brood cell. The various families and genera have been observed to have a broad range of prey.

Relatively small, plain representatives of the families Pemphredonidae and Crabronidae have primarily been found in Baltic amber. The Pemphredonidae were covered in detail by Budrys (1993), who described several genera, most of which are exclusively fossil. Budrys (1993) estimated that the number of species of *Passaloecus*, *Eoxyloecus* and other, closely related genera in the Baltic amber fauna may be three times greater than in any extant fauna known today. Nonetheless, only two species of Crabronidae have been recorded from Baltic amber to date (Cockerell 1909). New species can certainly be expected to be found among the amber Crabronidae, possibly in new genera.

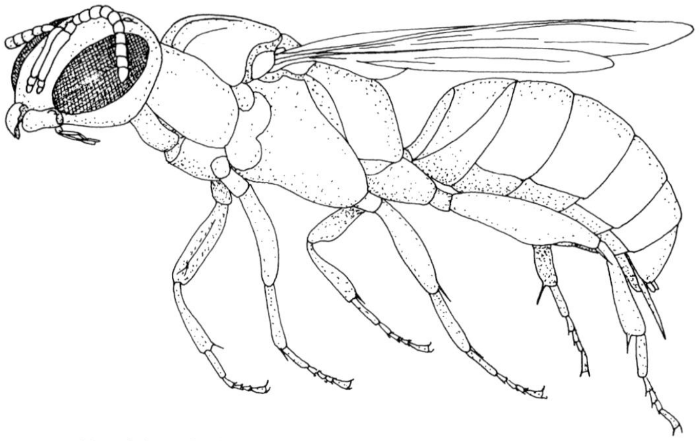

Fig. 80: Spheciform wasp (Apoidea: Spheciformes).

Plate 72: Hymenopterans (Insecta: Hymenoptera) in Baltic amber VI (Apoidea).

a Honey bee (Apidae), × 9.
b *Succinapis micheneri* (Apidae), × 9.
c *Electrapis krishnorum* (Apidae), × 8.
d Honey bee (Apidae), × 9.
e Spheciform wasp (Crabronidae), × 14.
f *Eoxyloecus* sp. (Pemphredonidae), × 15.
g *Ctenoplectrella viridiceps* (Megachilidae), × 10.
h *Ctenoplectrella viridiceps* (Megachilidae), × 10.

a

b

c

d

e

f

g

h

2.73 Overview of hymenopterans in Baltic amber – Insecta: Hymenoptera

The following list includes all hymenopteran families that have been recorded from Baltic amber on the basis of at least one, described species. The superfamilies and families are grouped according to the classification by GOULET & HUBER (1993):

Suborder: **SYMPHYTA**

CEPHOIDEA
- Cephidae

SIRICOIDEA
- Siricidae

TENTHREDINOIDEA
- Tenthredinidae

Suborder: **APOCRITA**

STEPHANOIDEA
- Stephanidae

MEGALYROIDEA
- Megalyridae

CERAPHRONOIDEA
- Megaspilidae

EVANIOIDEA
- Evaniidae
- Aulacidae

ICHNEUMONOIDEA
- Ichneumonidae
- Braconidae

PROCTOTRUPOIDEA
- Diapriidae
- Pelecinidae
- Proctotrupidae

PLATYGASTROIDEA
- Scelionidae

CYNIPOIDEA
- Figitidae
- Cynipidae

MYMAROMMATOIDEA
- Mymarommatidae

CHALCIDOIDEA
- Torymidae
- Eupelmidae
- Tetracampidae
- Mymaridae

CHRYSIDOIDEA
- Scolebythidae
- Embolemidae
- Dryinidae
- Bethylidae
- Chrysididae

VESPOIDEA
- Pompilidae
- Mutillidae
- Vespidae
- Formicidae

APOIDEA (Spheciformes)
- Ampulicidae
- Crabronidae
- Pemphredonidae

APOIDEA (Apiformes)
- Halictidae
- Melittidae
- Paleomelittidae
- Megachilidae
- Apidae

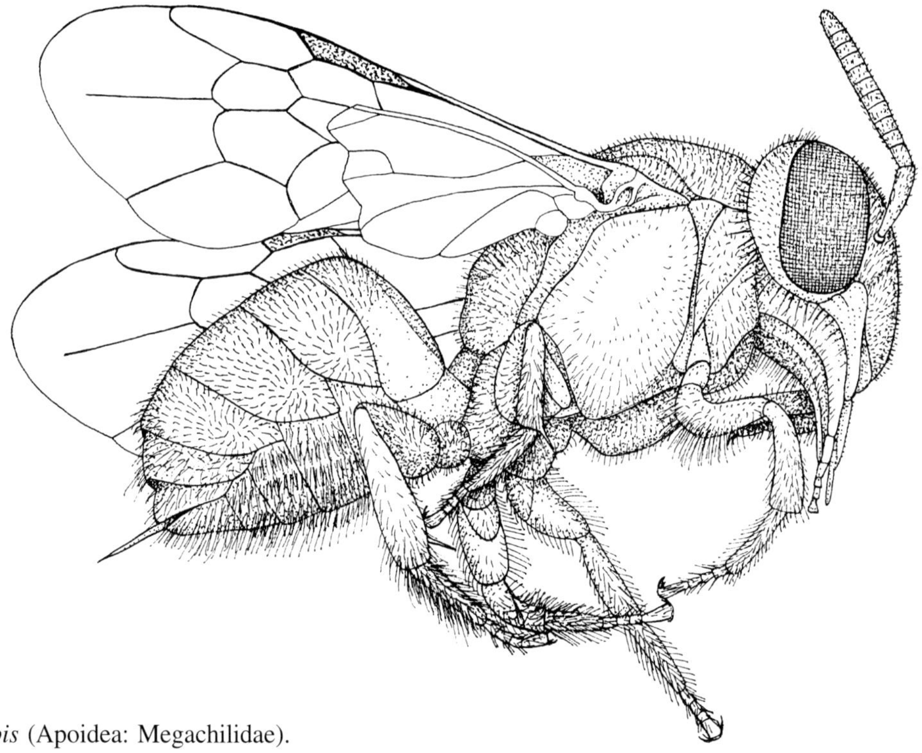

Fig. 81: Bee of the genus *Glyptapis* (Apoidea: Megachilidae).

Plate 73: Hymenopterans (Insecta: Hymenoptera) in Baltic amber VII (Apoidea).

Glyptapis fuscula COCKERELL, 1909 (Apoidea: Megachilidae), ×25.

2.74 Scorpionflies – Insecta: Mecoptera

Scorpionflies are characterized by an elongate rostrum. Over 500 species of this order of holometabolous insects have been described. The larvae are caterpillar-like and pupate in the soil. The European scorpionflies include the Boreidae (snowflies), which are wingless, the Panorpidae (common scorpionflies), the males of which have dorsally recurved genitalia superficially resembling the tail of a scorpion, and the long-legged Bittacidae (hanging-flies), which use the front legs to hang down from small twigs and the rear legs to capture flying insects.

Competing hypotheses have been proposed for the Mecoptera. Willmann (1989) considered the order to be a monophyletic group that was already represented in the Permian by the family Nannochoristidae. More extensive studies based on morphology, DNA sequences, ethology, and development have recognized the Mecoptera to be potentially paraphyletic, with the Boreidae being more closely related to the fleas, order Siphonaptera (Whiting et al. 1997). Certainly much work remains to be done on the phylogeny of this ancient group of insects and fossils will play a critical role in resolving the competing hypotheses. Pictet-Baraban & Hagen (1856) and, most recently, Carpenter (1931, 1954, 1955, 1976) have covered the fossil scorpionflies of Baltic amber. According to their revisions, three families occur in Baltic amber:

Bittacidae

Bittacus fossilis Carpenter, 1954
Bittacus minimus Carpenter, 1954
Bittacus succinus Carpenter, 1954
Electrobittacus antiquus (Pictet, 1854)

Panorpidae

Panorpa mortua Carpenter, 1954
Panorpa obsoleta Carpenter, 1954

Panorpodidae

Panorpodes brevicauda (Hagen, 1856)
Panorpodes hageni Carpenter, 1954

The wing venation of the Panorpidae is very similar to that of the Panorpodidae. The arrangement of the cross-veins in the forewing of a panorpid from Baltic amber (Plate. 74 b, Fig. 82 a) indicates a phylogenetic relationship to the Japanese genus *Panorpodes*. *Bittacus fossilis* is a frequent species among the Bittacidae. The species is characterized by the lack of a second cross-vein (r-rs) at the pterostigma of the forewing (cf. *Hylobittacus*, Byers 1979).

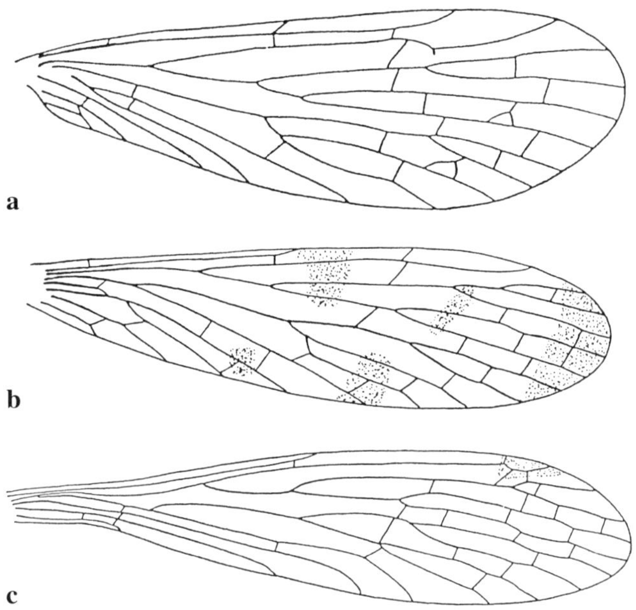

Fig. 82: Forewings of fossil Mecoptera genera.
a *Panorpodes brevicauda* (Hagen, 1856).
b *Panorpa obsoleta* Carpenter, 1954.
c *Bittacus succinus* Carpenter, 1954.

Plate 74: Scorpionflies (Insecta: Mecoptera) in Baltic amber.

a Panorpidae: *Panorpa* sp., × 20.
b Panorpodidae: *Panorpodes* sp., × 15.
c Bittacidae: *Bittacus* sp., × 15.

a

b

c

2.75 Caddisflies – Insecta: Trichoptera

Caddisflies (Trichoptera) are among the insects that occur with significant regularity in Baltic amber, but also worldwide in other fossil resins. While they comprise up to 5 or 6 percent of all inclusions in Baltic amber, they make up only 0.1 to 0.01 percent of all inclusions in Dominican amber, which is just as rich in inclusions overall. Their size range of 0.5 to 2 centimeters makes them conspicuous insects in amber.

Ulmer (1912) published a comprehensive monograph of the caddisflies from Baltic amber. The material on which the study was based originated from various museums and private collections, encompassing a total of 5,060 pieces that have unfortunately deteriorated almost completely. The vast majority (3,900 pieces) belonged to two Königsberg collections: 1,600 pieces from the Klebs collection and 2,300 from the university's Institute of Geology and Paleontology. Ulmer described 152 fossil species, which he placed in 56, mostly fossil genera. New caddisfly species have since been described from Baltic amber and its Bitterfeld deposits (Mey 1985, 1986, 1988, Wichard 1986, Wichard & Caspers 1991, Wichard & Sukatsheva 1992, Wichard & Weitschat 1996, Johanson & Wichard 1997). Caddisflies of Baltic amber, including its Bitterfeld deposits, are currently under revision (Wichard in prep.)

At rest, imagoes are characterized by anteriorly recurved, filiform antennae and brownish-gray wings covering the abdomen in a rooflike fashion. The compound eyes often appear to be green or blue in Baltic amber (Plates 75, 76 a, b). These colors are optical effects caused by the refraction of light (Weitschat & Wichard 1992). The key features of identification are the first antennal segments (Plate 76 g, h), ocelli and maxillary palpi (Plate 76 e, f), as well as the number of tibial spurs and the wing venation. The description of extant species is primarily based on the external male genitalia, which are also indispensable in the description of new Trichoptera from amber (Plate 76 c, d).

In addition to numerous imagoes, a few larvae also occur in Baltic amber. One of the first records was published by Larsson (1978). Further caddisfly larvae were described by Wichard (2000, 2001). These fossil larvae do not belong to the few, terrestrial trichopteran species, but rather have tracheal gills in adaptation to an aquatic life style. They all belong to the suborder Integripalpia, whose larvae always bear cases. However, none of the fossil larvae in amber have been preserved with cases, having probably left them beforehand. Extant larvae initially retreat into their cases when disturbed. If they are deprived of water for an extended period, so that the residual water in the case drains or evaporates, the caddisfly larvae change their behavioral strategy, leaving the case to crawl around in an unfamiliarly dry environment, where they are exposed to all occurrences without any protection whatsoever. The Integripalpia larvae in amber thus lead to the assumption that the bodies of water they inhabited ran dry, and that they were unable to synchronously adapt their life cycles. Today, bodies of water that run dry irregularly and periodically are common in arid, subtropical regions.

Fig. 83: *Holocentropus affinis* Ulmer, 1912 (Trichoptera, Polycentropodidae).

Plate 75: Caddisflies (Insecta: Trichoptera) in Baltic amber I.

Polycentropodidae from Baltic amber of the Bitterfeld deposits, ×42.

2.76 Systematic overview of amber caddisflies – Insecta: Trichoptera

Current list of the caddisflies from Baltic amber (after: WICHARD & WEITSCHAT 1996):

Suborder: **SPICIPALPIA**

Rhyacophilidae
- *Rhyacophila* (6 species)

Glossosomatidae
- *Electragapetus* (1)

Hydroptilidae
- *Agraylea* (3)
- *Electrotrichia* (1)
- *Palaeagapetus* (1)

Suborder: **ANNULIPALPIA**

Philopotamidae
- *Philopotamus* (1)
- *Ulmerodina* (1)
- *Wormaldia* (6)
- *Electracanthinus* (1)

Stenopsychidae
- *Stenopsyche* (1)

Ecnomidae
- *Archaeotinodes* (14)

Hydropsychidae
- *Diplectrona* (2)
- *Electrodiplectrona* (1)
- *Hydropsyche* (1)
- *Potamyia* (1)

Polycentropodidae
- *Archaeoneuroclipsis* (2)
- *Holocentropus* (21)
- *Neureclipsis* (4)
- *Nyctiophylax* (23)
- *Nyctiophylacodes* (1)
- *Plectrocnemia* (24)

Dipseudopsidae
- *Phylocentropus* (4)

Psychomyiidae
- *Lype* (5)

Suborder: **INTEGRIPALPIA**

Phryganeidae
- *Phryganea* (7)
- *Trichostegia* (1)

Brachycentridae
- *Brachycentrus* (1)

Goeridae
- *Goera* (1)
- *Lithax* (2)
- *Silo* (1)

Lepidostomatidae
- *Electraulax* (2)
- *Palaeocrunoecia* (3)
- *Palaeolepidostoma* (1)
- *Archaeocrunoecia* (3)
- *Electrocrunoecia* (1)
- *Maniconeurodes* (1)

Calamoceratidae
- *Ganonema* (1)
- *Georgium* (1)

Molannidae
- *Molanna* (1)
- *Molannodes* (2)

Leptoceridae
- *Setodes* (1)
- *Erotesis* (3)
- *Triplectides* (3)

Odontoceridae
- *Electrocerum* (1)
- *Electropsilotes* (1)
- *Marilia* (2)

Beraeidae
- *Bereodes* (1)

Helicopsychidae
- *Adelomyia* (1)
- *Electrohelicopsyche* (1)
- *Helicopsyche* (2)
- *Ogmomyia* (1)
- *Palaeohelicopsyche* (2)
- *Perissomyia* (1)

Sericostomatidae
- *Aulacomyia* (1)
- *Pseudoberaeodes* (1)
- *Sphaleropalpus* (1)
- *Stenoptilomyia* (1)

Plate 76: Caddisflies (Insecta: Trichoptera) in Baltic amber II.

a Polycentropodidae: *Holocentropus* sp., green eyes.
b Polycentropodidae: *Plectrocnemia* sp., blue eye.
c Polycentropodidae: *Holocentropus consobrinus*, male genitalia.
d Leptoceridae: *Erotesis* sp., female genitalia.
e Helicopsychidae: *Palaeohelicopsyche* sp., maxillary palpi.
f Goeridae: *Lithax herrlingi*, maxillary palpi, holotype.
g Helicopsychidae: *Electrohelicopsyche taenica*, antenna.
h Lepidostomatidae: *Palaeolepidostoma proavum*, antenna.

a
b
c
d
e
f
g
h

2.77 Caddisflies in Baltic amber – Insecta: Trichoptera

The order Trichoptera is represented in Baltic amber by 21 families. Extant caddisflies are classified in 43 families worldwide, 21 of which occur in Europe. A comparison of the families in Baltic amber to extant, European forms reveals many common faunal elements. Although there is no phylogenetic or, more importantly, geographic relationship between the two faunas, closer examination shows that they display extensive similarities in terms of their climatic and ecological characteristics (WICHARD 1988).

The Spicipalpia families from Baltic amber are identical to the extant European fauna. Neither fauna includes the family Hydrobiosidae, whose members are native to Australia and the Nearctic region. Several species also occur in the Oriental, southern Nearctic and Palearctic regions.

The Annulipalpia also display extensive similarities. For example, the two faunas differ with regard to only one, small family: the Stenopsychidae. While this family does not occur at all today in Europe, one species has been recorded from Baltic amber. Roughly 70 extant species are distributed in Southeast Asia, Australia and in the Oriental and Ethiopian regions.

Only the Intergripalpia display somewhat greater differences. Of the 13 families native to Europe, two are not found in Baltic amber: Thremmatidae and Limnephilidae. Three species of Thremmatidae are restricted to southern Europe, and one species reaches as far north as the Black Forest. In contrast, roughly 1,000 species of the Limnephilidae are today distributed in the Holarctic region, with a few species pushing south to the northern rim of the Oriental region. The approximately 300 European species mainly occur in the northern regions.

Ignoring the subtropical-tropical Stenopsychidae and the Limnephilidae from the cold climate zones, the trichopteran fauna of Baltic amber largely corresponds to the present-day European fauna and is indicative of a temperate climate. This observation is commensurate with the information on other aquatic insects from Baltic amber. The more clearly these aquatic insects belong to the forms preferring running waters, the more strongly they indicate a temperate climate. Aquatic insects are primarily characterized by the biology of their larvae, many of which are adapted to cold-stenothermic habitats in running waters. Based on the great number rheophilous caddisflies in Baltic amber, ULMER (1912) described a mountainous landscape with numerous streams and rivers. This mountainous landscape apparently had elevation-related climate zones ranging from subtropical lowlands to temperate highlands.

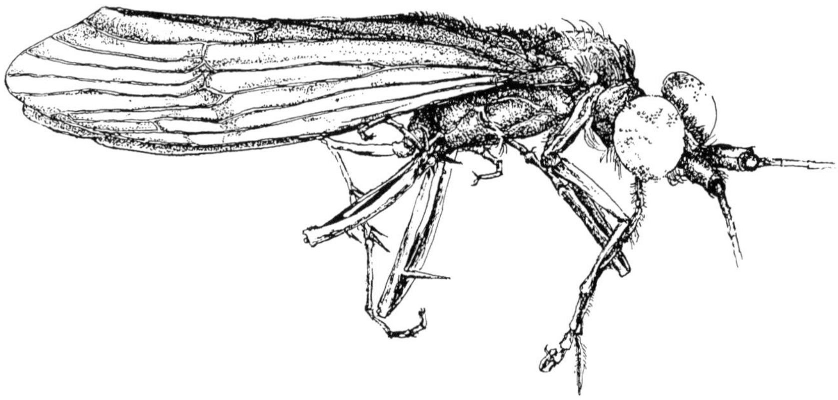

Fig. 84: *Marilia altrocki* WICHARD, 1986 (Odontoceridae), holotype.

Plate 77: Caddisflies (Insecta: Trichoptera) in Baltic amber III.

a Polycentropodidae: *Plectrocnemia* sp., × 9.
b Polycentropodidae: *Holocentropus scissus*, × 9.
c Leptoceridae: *Erotesis* sp., × 8.
d Helicopsychidae: *Palaeohelicopsyche* sp., × 10.
e Odontoceridae: *Marilia altrocki* WICHARD, 1986, holotype, in old, red amber, × 8,5.

a b c d e

2.78 Butterflies and moths in Baltic amber – Insecta: Lepidoptera

Butterflies and moths (Lepidoptera) make up one of the largest insect orders with over 150,000 species. They are surpassed only by the beetles, and perhaps the Hymenoptera, with regard to the number of species. Lepidopterans are holometabolous insects (Endopterygota), i.e. their development progresses from the eggs, caterpillars (larvae) and pupae, to the adults. The body displays distinct head-thorax-abdomen segmentation. The mesothorax and metathorax each bear one pair of wings, which are connected during flight to ensure synchronous strokes. The top and bottom sides of the wings are covered with dense, fine scales, which contain pigments or have special structures that cause optical effects and produce structural colors. These scales are what give the wings their splendid colors. The size of the lepidopterans is usually measured by the span of the unfolded wings, which ranges from 3 millimeters to 25 centimeters.

Most butterflies and moths in amber are small, rarely larger than one centimeter. Larger flying insects were able to avoid the danger of getting trapped in resin much more easily than small animals. In some instances, fairly large insects left behind wing fragments that remained stuck to the resin when they flew away. In contrast, small lepidopterans are often fully embedded, thus enabling precise determination of the families and genera, and facilitating the detailed description of new species. The butterflies and moths in Baltic amber represent the vegetation of the "amber forest". Because caterpillars are closely associated with their host plants, without which they usually cannot survive, lepidopterans are an excellent source of information for describing the biotope they inhabited in the "amber forest". Based on comparisons with present-day lepidopterans and their host plants, every new lepidopteran inclusion found helps complete the picture of the "amber forest", even though the leaves, buds, flowers and fruits of the plants themselves have only rarely been preserved. Although most caterpillars are restricted to specific plants, this comparison to related, extant species still only gives us an approximation of the situation that prevailed in the Eocene Epoch.

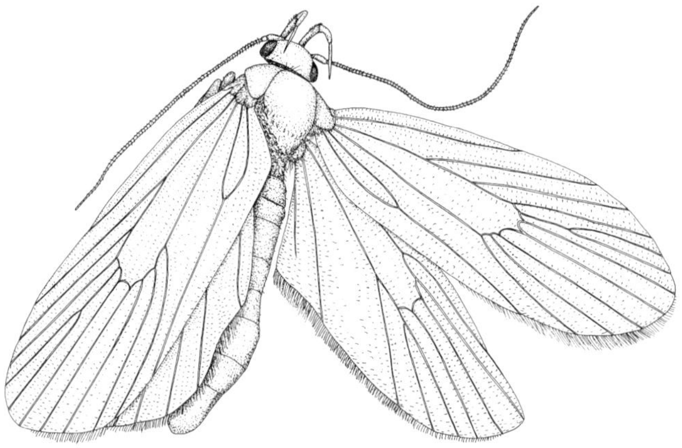

Fig. 85: Concealer moth *Borkhausenites bachofeni* REBEL, 1934 (Lepidoptera: Oecophoridae).

Plate 78: Butterflies and moths (Insecta: Lepidoptera) in Baltic amber I.

a-d Various caterpillars, **a** ×11, **b** ×10, **c** ×10, **d** ×9.
e-h Various "Microlepidoptera", **e** ×8, **f** ×6, **g** ×9, **h** ×7

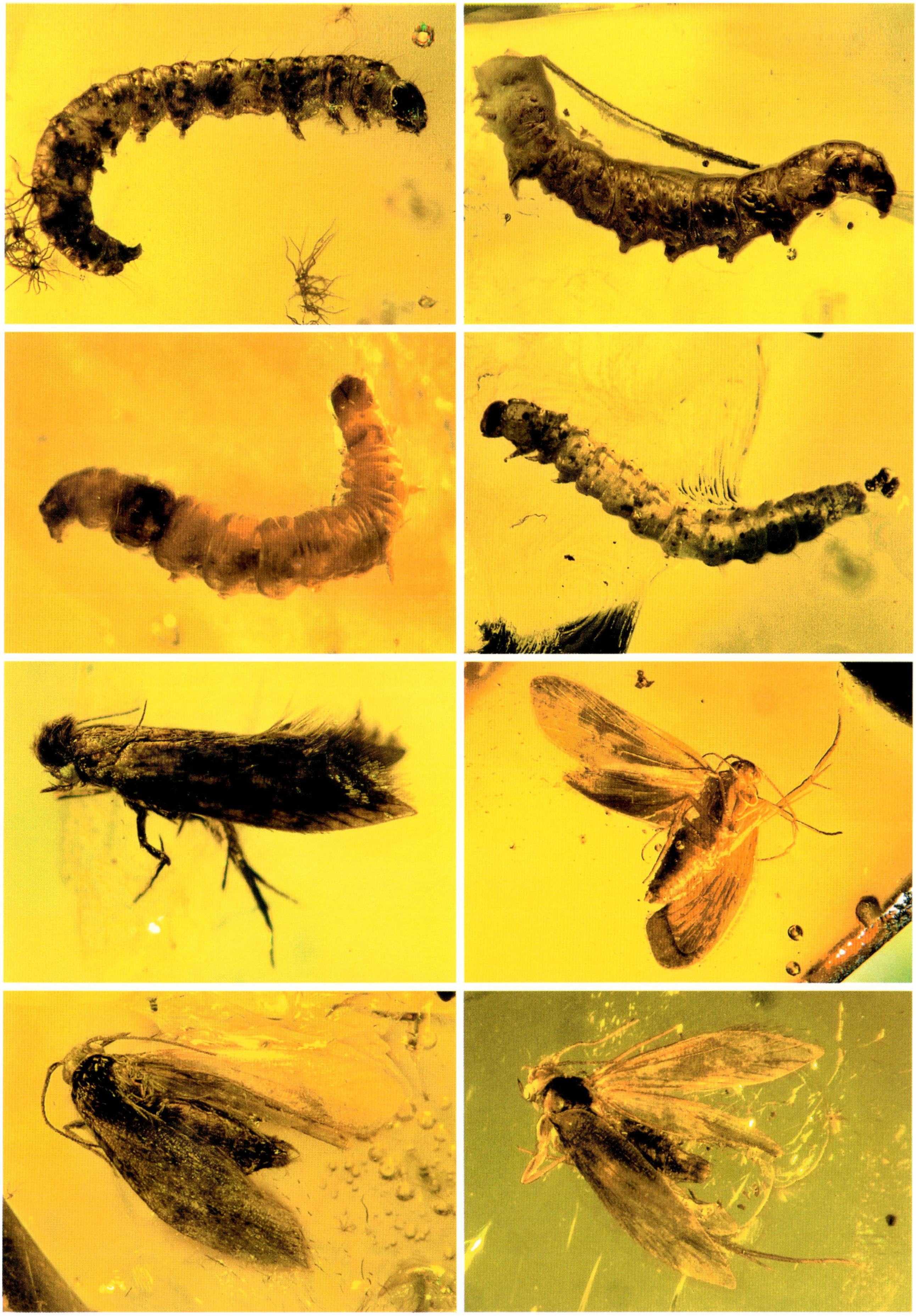

a b c d e f g h

2.79 Caterpillars – Insecta: Lepidoptera

It is no surprise that caterpillars are just as frequent in amber as imagoes. Unlike the caterpillars, the imagoes were probably lured by the odor of the resin and its illumination in the sunlight. Several caterpillar specimens must have been surprised by dripping resin that fell to the ground and enveloped them. However, the majority of inclusions involve caterpillars that fed on the resin-producing host plant and were embedded in resin flowing down the tree. This assumption would mean that numerous caterpillars and other bark-dwelling insects lived without competition on only one host plant, namely *Pinus succinifera*, to which amber is attributed. Many species-specific bark-dwellers could only have been captured in amber because their host plant was also the resin-supplier for Baltic amber.

We must abandon the assumption that only one resin-producing tree *(Pinus succinifera)* could have been the source of Baltic amber. The close association between many tree-dwellers and their host plants, as well as the competition among them, makes it much more likely that several plants, particularly conifers, contributed to resin production and the formation of Baltic amber.

The Baltic amber forest was a primeval forest that varied from one location to the next and changed over time in make-up and extent, but nevertheless persisted for over ten million years. During this period, resin-producing plants, primarily conifers, generated enormous amounts of resin in a completely natural manner.

The study of Lepidoptera in Baltic amber is mainly associated with the names of Rebel (1935, 1936, 1937) and Skalski (1973 ff.), who described new species and compiled a preliminary systematic overview (classification according to Nielsen & Common 1991):

Zeugloptera
Micropterygidae – Mandibulate moths

Heterobathmiina
Glassata
Nepticulidae – Pygmy moths
Heliozelidae – Shield bearer moths
Adelidae – Long-horned fairy moths
Incurvariidae – Leafcutter moths
Ditrysia
Psychidae – Bagworm moths
Tineidae – Clothes moths
Gracillariidae – Leafminer moths
Yponomeutidae – Ermine moths
Argyresthiidae
Plutellidae – Diamondback moths
Heliodinidae – Sun moths
Lyonetiidae – Leaf skeletonizer moths
Oecophoridae – Concealer moths
Elachistidae – Grass miner moths
Gelechiidae – Twirler moths
Symmocidae
Scythrididae – Flower moths
Tortricidae – Leafroller moths
Sesiidae – Clearwing moths
Pyralidae – Snout moths
Papilionidae – Swallowtail butterflies
Sphingidae – Hawkmoths
Arctiidae – Tiger moths
Noctuidae – Owlet moths

Plate 79: Butterflies and moths (Insecta: Lepidoptera) in Baltic amber II.

a-h Various caterpillars with their cases, **a** ×9, **b** ×18, **c** ×9, **d** ×9, **e** ×10, **f** ×8, **g** ×9, **h** ×10.

a
b
c
d
e
f
g
h

2.80 Nematocera in Baltic amber – Insecta: Diptera

Flies belong to the complex and extensive order Diptera (HENNIG 1973). They are winged, holometabolous insects with a first pair of fully developed wings and a second pair that is reduced to balancers or halteres. The numerous species and families of the order Diptera are divided into two suborders: the Nematocera and Brachycera. These suborders are further subdivided to reflect the phylogeny of the group more closely (COLLESS & MCALPINE 1991).

Flies are very common in Baltic amber. At least 70% of all inclusions are dipterans, although they are often overlooked, because those preserved in amber are usually quite small. Consequently, a great number of unknown fossil species is expected to exist. The study of the Diptera in Baltic amber is still in the early stages, even though the groundwork was laid in several, significant, early works (LOEW 1845, 1850, 1861, 1864, 1873, MEUNIER 1894-1922, ALEXANDER 1931). Modern, taxonomic analysis of the fossil Diptera in Baltic amber began with HENNIG, who applied the method of phylogenetic systematics (HENNIG 1950, 1966a) to the study of the Nematocera and Brachycera in Baltic amber (HENNIG 1938-1972).

According to the systematic classification by WOOD & BORKENT (1989), the Nematocera from Baltic amber are divided into eighteen families. Several families have been combined, e.g. the Fungivoridae and Keroplatidae with the Mycetophilidae. The Tipulidae were grouped with the Limoniidae studied by KRZEMINSKI (1985b, 1998a, b, 2000a, b) and PODENAS (1999a, c, d).

Suborder NEMATOCERA

- **Tipulidae** – Crane flies
- **Nymphomyiidae** – Nymphomyiid flies
- **Bibionidae** – March flies
- **Mycetophilidae** – Fungus gnats
- **Sciaridae** – Dark-winged fungus gnats
- **Cecidomyiidae** – Gall midges
- **Psychodidae** – Moth and sand flies
- **Trichoceridae** – Winter crane flies
- **Anisopodidae** – Wood gnats
- **Scatopsidae** – Scavenger flies
- **Tanyderidae**
- **Dixidae** – Dixid midges
- **Corethrellidae**
- **Chaoboridae** – Phantom midges
- **Culicidae** – Mosquitoes
- **Simuliidae** – Blackflies
- **Ceratopogonidae** – Biting midges
- **Chironomidae** – True midges

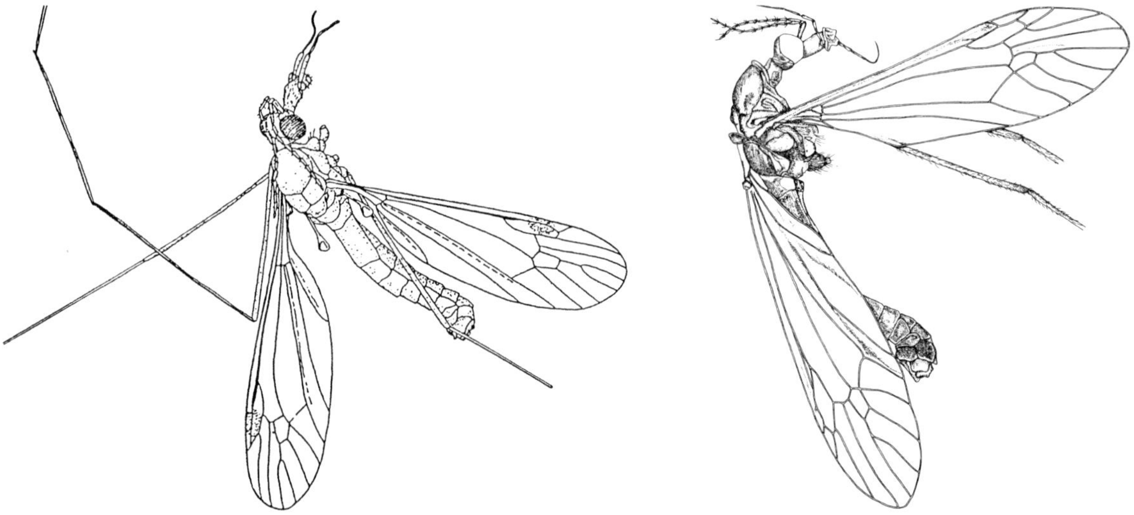

Fig. 86: Two crane flies (Diptera: Tipulidae).

Plate 80: Flies (Insecta: Diptera) in Baltic amber I (Tipulidae).

- **a** Limoniinae: Crane fly *Neolimnomyia* sp., ×10.
- **b** Limoniinae: Crane fly *Pseudolimnophila producta,* ×4.
- **c** Limoniinae: Crane fly *Palaeogonomyia* sp., ×7.
- **d** Limoniinae: Crane fly *Elephanomyia pulchella*, ×6.
- **e** Tipulinae: Crane fly, ×3.
- **f** Tipulinae: Crane fly, head region, ×10.
- **g** Tipulinae: Crane fly, pupa, dorsal, ×9.
- **h** Tipulinae: Crane fly, pupa, ventral, ×10.

a b c d e f g h

2.81 Nematocera in Baltic amber – Insecta: Diptera

A comparison of the Nematocera from three amber collections (**a** Warsaw Earth Museum, **b** Danzig collection and **c** Copenhagen Zoological Museum) provides information on the frequency of the families (Kulicka et al. 1985). Even though the content of the collections has changed and the taxonomic study of the families has progressed, the data still establishes a basis for orientation and reveals trends in their frequencies:

Families	Collections (%)		
	a	b	c
Tipulidae	2.11	1.17	3.30
Nymphomyiidae	–	–	–
Bibionidae	0.07	0.03	0.20
Mycetophilidae	13.38	8.99	18.60
Sciaridae	22.56	23.34	14.90
Cecidomyiidae	2.62	2.77	7.60
Psychodidae	4.67	6.23	6.30
Trichoceridae	0.01	0.00	0.00
Anisopodidae	0.05	0.06	0.00
Scatopsidae	0.17	0.22	1.00
Tanyderidae	0.00	0.00	0.10
Dixidae	0.05	0.00	0.00
Corethrellidae	–	–	–
Chaoboridae	0.04	0.03	0.10
Culicidae	–	–	–
Simuliidae	0.26	0.03	0.40
Ceratopogonidae	7.87	6.92	11.00
Chironomidae	46.13	49.85	36.50

The families Nymphomyiidae, Corethrellidae and Culicidae have since been reported, the systematics and taxonomy of which were established respectively by Wagner et al. (2000), Borkent & Szadziewski (1992), Szadziewski et al. (1994), Szadziewski (1998), and Podenas (1999b).

The larvae and pupae of extant Nymphomyiidae live in between mosses and stones in rapidly flowing mountain streams characterized by minimal temperature fluctuation. The apneustic larvae breathe through the body surface and feed on decaying plant material. The imagoes are up to 3 millimeters long and have reduced mouthparts, meaning that they do not feed during the brief adult phase. Nymphomyiid flies display three, distinctive morphological characteristics: 1. Long, narrow wings (which may also be absent) with reduced venation and a marginal fringe, 2. Ventrally adjoined compound eyes, i.e. forming a ventral, not dorsal, bridge, 3. Paired, threadlike appendages on the fifth and sixth abdominal segments of the males, the function of which is still a subject of speculation. Eight species of this family have been described up to now, to which the fossil species *Nymphomyia succinea* from Baltic amber and its Bitterfeld deposits can now be added (Wagner et al. 2000).

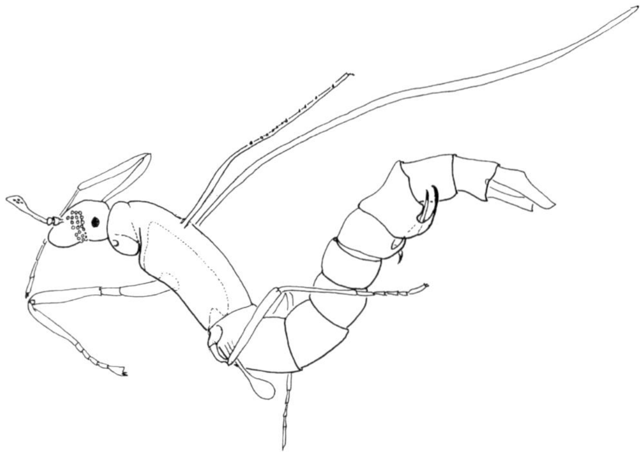

Fig. 87: Nymphomyiid fly (after Wagner et al. 2000b).

Plate 81: Flies (Insecta: Diptera) in Baltic amber II.

a Nymphomyiidae: Nymphomyiid fly, ×25.
b Bibionidae: March fly, ×7.
c Sciaridae: Dark-winged fungus gnat, ×11.
d Sciaridae: Dark-winged fungus gnat, ×12.
e Mycetophilidae (Macrocerinae): Fungus gnat, ×7.
f Mycetophilidae: Fungus gnat, ×9.
g Psychodidae: Moth fly, ×12.
h Scatopsidae: Scavenger fly, ×8.

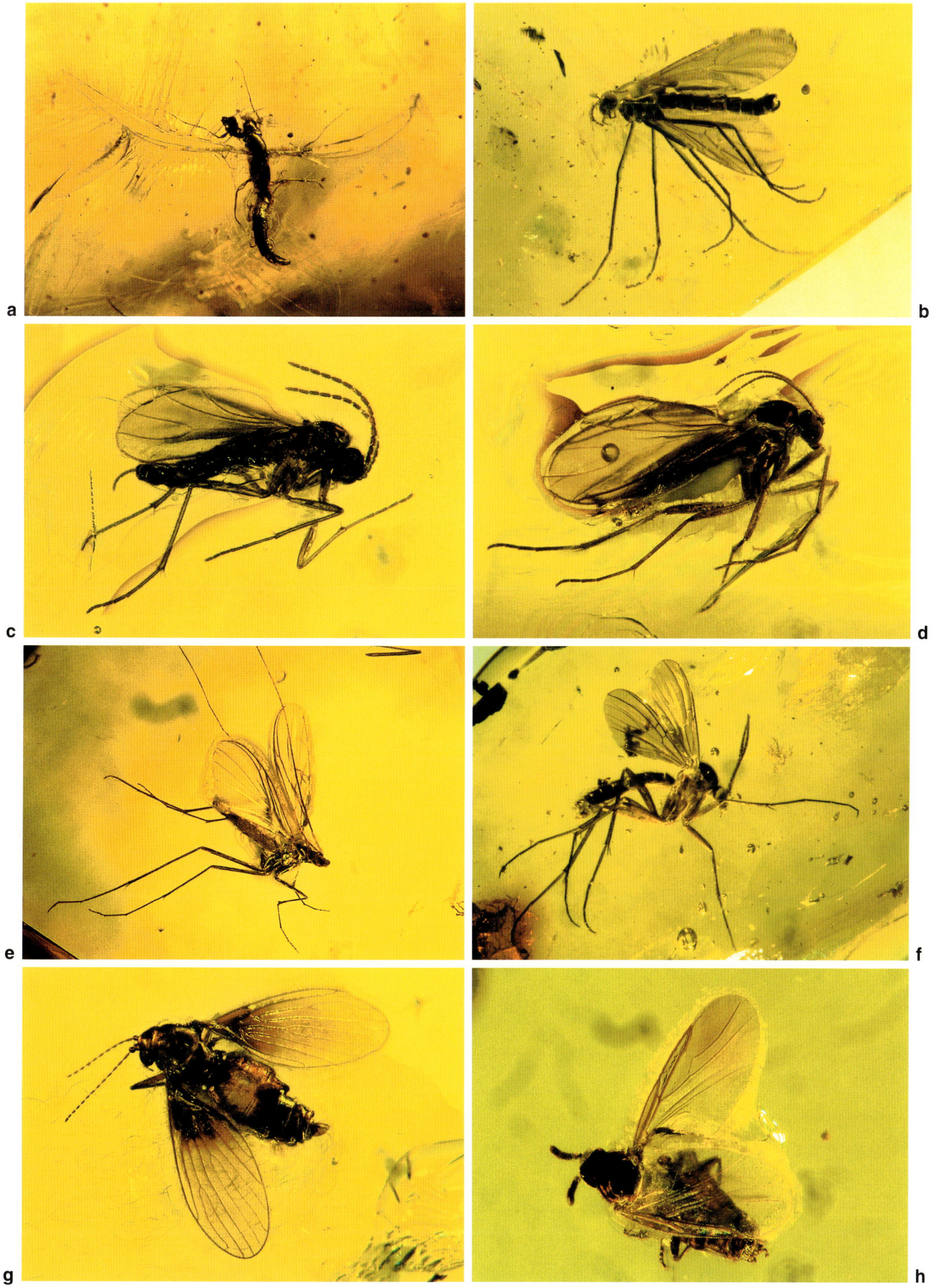
a
b
c
d
e
f
g
h

2.82 Anisopodidae and Tanyderidae in Baltic amber – Insecta: Diptera

The wood gnats (Anisopodidae) were described by Loew (1850) and Meunier (1899, 1904), and the nomenclature was later revised by Edwards (1921, 1928). According to these sources, six species are divided between the two genera *Mycetobia* and *Anisopus*:

Mycetobia connexa Meunier, 1899
Mycetobia defectiva Loew, 1850
Mycetobia longipennis Meunier, 1899
Mycetobia platyuroides Meunier, 1899
Anisopus (Sylvicola) splendida (Meunier 1904)
Anisopus (Sylvicola) thirionis (Meunier 1904)

Although wood gnats are rare in Baltic amber, they are still of special interest, because all developmental stages have been preserved in both Baltic and Dominican amber (Grimaldi 1991). Plate 82 a-d shows all of these stages: a female with discharged egg string (a), a larva that could easily be mistaken for a worm (b), a pupa with conspicuous spines (c), and an imago "just" leaving the pupal exuvia (d). The hatching process in another inclusion is shown in Fig. 88.

The family Tanyderidae is considered to be a primitive and rare group of Diptera. It is represented by 37 extant species in Australia, Asia and the Nearctic region, and by one species in South Africa. At least several of the extant species are typical aquatic insects, whose larvae and pupae live in water. The larvae of the Australian *Eutanyderus* sp. have been studied in detail. They have a closed tracheal system, breathe through the body surface and are additionally equipped with filamentous tracheal gills on the posterior end of the abdomen. The pupae have paired spiracular gills on the thorax (prothorax), with a plastron network in the cuticula of the gill surface that holds a cushion of air (plastron). This network enables a gas exchange to take place at the boundary surface between the plastron and the surrounding water (Hinton 1966, Wichard et al. 1995).

Adults of extant species live along the banks of large rivers; only a few imagoes have been described from Baltic amber:
Macrochile spektrum Loew, 1850
Macrochile baltica Podenas, 1997.

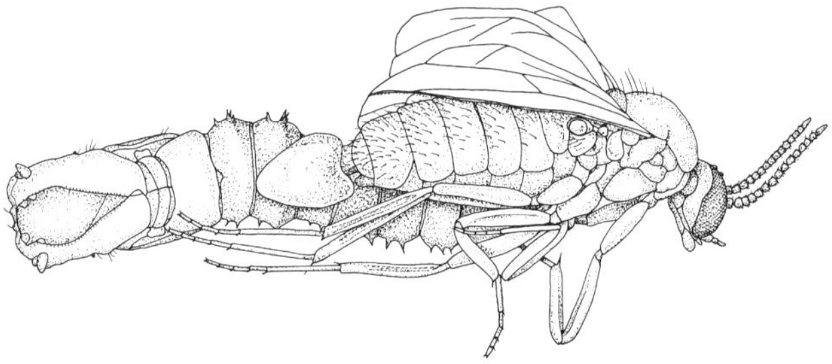

Fig. 88: Wood gnat (Anisopodidae) hatching from the pupal exuvia.

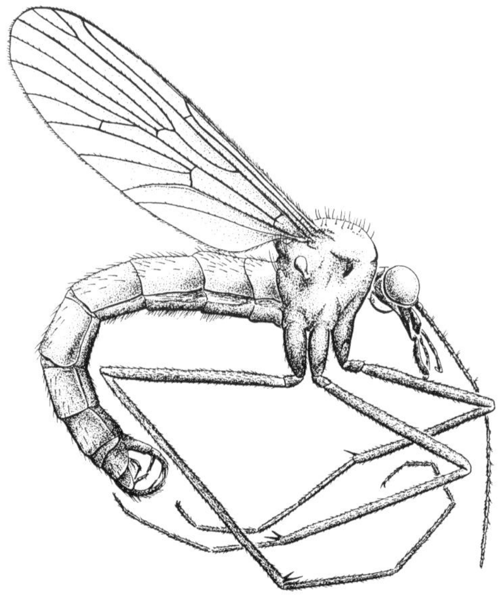

Fig. 89: *Macrochile baltica* Podenas, 1997 (Tanyderidae), holotype.

Plate 82: Flies (Insecta: Diptera) in Baltic amber III (Anisopodidae).

a Anisopodidae: Wood gnat, female with egg string, × 10.
b Anisopodidae: Wood gnat, wormlike larva, × 22.
c Anisopodidae: Wood gnat, pupa, × 17.
d Anisopodidae: Wood gnat, hatched imago with pupal exuvia, × 10.
e-h Anisopodidae: Wood gnats of the genus *Silvicola*, **e** × 10, **f** × 9, **g** × 9, **h** × 9.

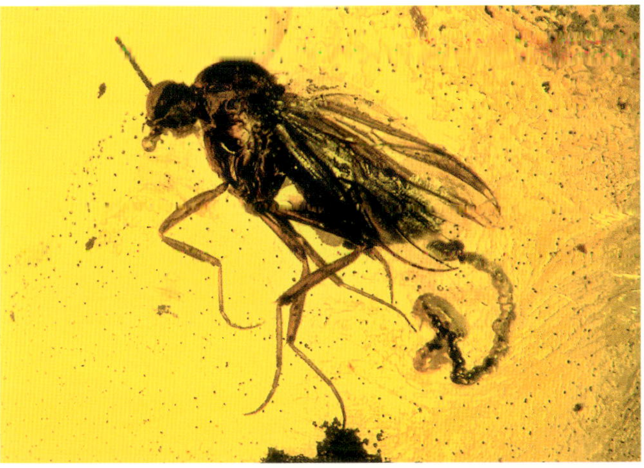
a

b

c

d

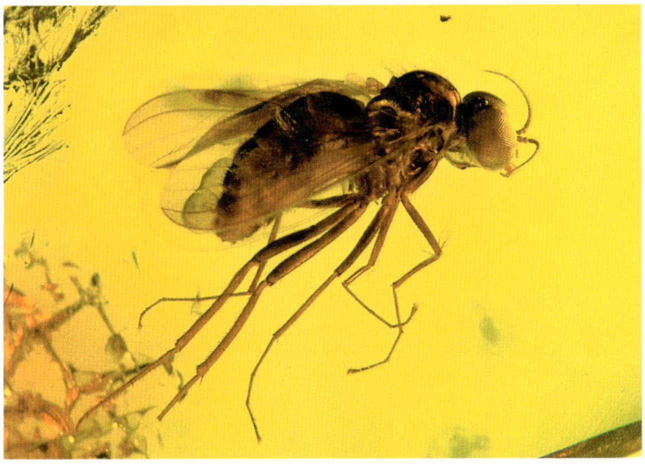
e

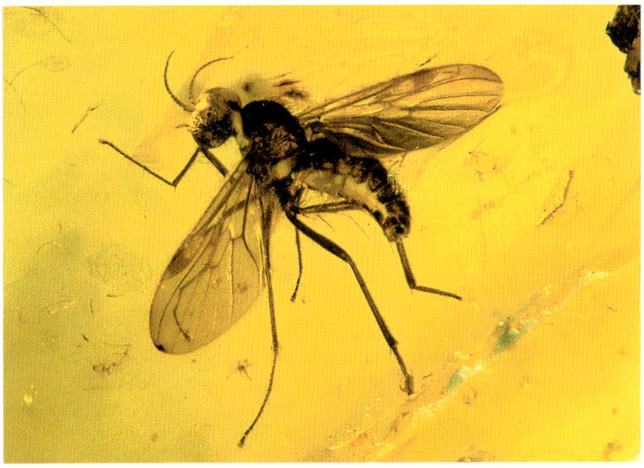
f

g

h

2.83 Culicomorpha in Baltic amber – Insecta: Diptera

The culicomorphous Nematocera are represented in Baltic amber by the two superfamilies Chironomoidea and Culicoidea. The Chironomoidea are represented by numerous Chironomidae (true midges) and Ceratopogonidae (biting midges), while the Simuliidae (blackflies) are rare and the Thaumaleidae do not occur at all (Chap. 2.81).

The publications by Szadziewski (1988, 1993) give a good overview of the numerous biting midges in Baltic amber, including its Bitterfeld deposits. Biting midges from various Cretaceous and Tertiary ambers have been examined, which allow conclusions to be drawn concerning the evolution of this group (e.g. the feeding habits of the blood-sucking females): Lebanese amber (Szadziewski 1996; Borkent 2000 a), Taimyr amber (Szadziewski 1996), French amber (Szadziewski & Schlüter 1992), Sachalin amber (Szadziewski 1990), New Jersey amber (Borkent 1995, 2000 b), Canadian amber (Borkent 1995) and Jordan amber (Szadziewski 2000).

The true midges (Chironomidae) are among the most common inclusions in Baltic amber and make up the most diverse and abundant family of Culicomorpha. The true midges of Baltic amber were first studied by Loew (1850, 1864) and Meunier (1904 b, 1916), and are currently under revision. Seredszus & Wichard (2002) demonstrated the chironomid subfamilies (Orthocladiinae – 90.2 %; Chironominae – 8.2 %; Podonominae – 0.9 %, Tanypodinae – 0.5 %, Diamesinae – <0.1 %, Buchonomyiinae – <0.1 %) and pointed out the special paleoecological importance of true midges.

The culicoid families Dixidae (dixid midges), Culicidae (mosquitoes), Chaoboridae (phantom midges) and Corethrellidae are noticeably rare in amber (Chap. 2.81). Hennig (1966) revised the Dixidae and defined four species. Five Culicidae, two Corethrellidae and one Chaoboridae species have been described to date:

Dixidae

Dixa minuta Meunier, 1906
Paradixa succinea (Meunier, 1906)
Paradixa filiforceps Hennig, 1966
Paradixa distans Hennig, 1966

Culicidae

Aedes damzeni Szadziewski, 1998
Aedes hoffeinsorum Szadziewski, 1998
Aedes perkunas Podenas, 1999
Aedes serafini Szadziewski, 1998
Culex erikae Szadziewski & Szadziewska, 1985

Corethrellidae

Corethrella prisca Borkent & Szadziewski, 1992
Corethrella miocenica Szadziewski et al., 1994

Chaoboridae

Mochlonyx sepultus Meunier, 1902

Most culicomorphous Nematocera are aquatic insects. Their larvae and pupae inhabit running or standing waters, or moist environments. An amber specimen examined by Hennig contains not only a dixid midge *(Paradixa succinea)*, but also the pupa of a blackfly, whose larvae and pupae live in running waters. The obvious and probable conclusion is that the larvae of *Paradixa succinea* also inhabited running waters.

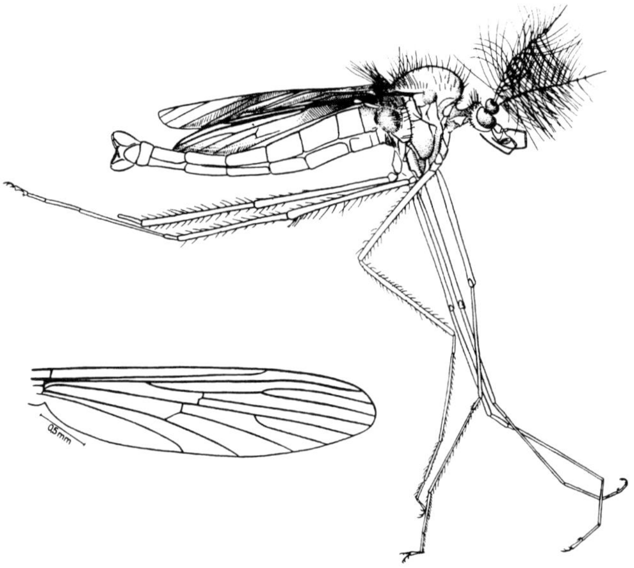

Fig. 90: *Mochlonyx sepultus* (Chaoboridae) (after Hennig 1966 d).
Fig. 91: Wing of the phantom midge.

Plate 83: Flies (Insecta: Diptera) in Baltic amber IV (Culicomorpha).

a Dixidae: Dixid moth, ×7.
b Chaoboridae: Phantom midge, ×12.
c Culicidae: Mosquito, ×10.
d Simuliidae: Blackfly, ×12.
e Ceratopogonidae: Biting midge, ×11.
f Ceratopogonidae: Biting midge, ×13.
g Chironomidae: True midge, ×11.
h Chironomidae: True midge, ×12.

a
b
c
d
e
f
g
h

2.84 Brachycera in Baltic amber – Insecta: Diptera

The suborder Brachycera is represented in Baltic amber by a wide range of families. Krzeminski & Evenhuis (2000) suspect that the Brachycera expanded in the Tertiary, when many of the families and subfamilies appear for the first time; however, the evolution of the group is likely more ancient since a diversity of lineages are well known from the Cretaceous (Grimaldi & Cumming 1999). A number of them occur in Eocene Baltic amber: Rachicerinae (Xylophagidae), Bombyliidae, Micropezidae, Pseudopomyzidae, Psilidae, Megamerinidae, Diopsidae, Conopidae, Lauxaniidae, Chamaemyiidae, Dryomyzidae, Sepsidae, Clusiidae, Acarthophthalmidae, Odiniidae, Anthomyzidae, Aulacigastridae, Asteiidae, Carnidae, Cryptochetidae, Heleomyzidae, Chyromyidae, Camillidae, Drosophilidae, Diastatidae. The successful adaptation and evolution of the Brachycera is attributed to the widespread availability of new, unoccupied ecological niches, which were primarily created by the spreading of angiosperms, but perhaps also by the development of mammals and birds in the Paleogene. The continuous production of leaves led to an increase in the amount of dead biomass and to the eutrophication of land and water, thus favoring all animals that fed directly or indirectly on decaying plant material.

The list of the Brachycera families in Baltic amber is based on the classification by Woodley (1989) and McAlpine (1989), taking into account the family lists from Spahr (1985):

Orthorrhapha
- **Xylophagidae** – Xylophagid flies
- **Xylomyidae**
- **Stratiomyidae**
- **Vermileonidae**
- **Rhagionidae** – Snipe flies
- **Tabanidae** – Horseflies, deerflies
- **Acroceridae** – Small-headed flies
- **Asilidae** – Robber flies
- **Therevidae** – Stiletto flies
- **Bombyliidae** – Bee flies
- **Dolichopodidae** – Long-legged flies
- **Empididae** – Dance flies

Cyclorrhapha (Aschiza)
- **Sciadoceridae**
- **Phoridae** – Scuttle flies
- **Syrphidae** – Syrphid flies
- **Pipunculidae** – Big-headed flies

Cyclorrhapha (Schizophora)

Nerioidea
- **Micropezidae** (+ **Calobatidae**) – Stilt-legged flies
- **Cypselosomatidae** (+ **Pseudopomyzidae**)

Diopsoidea
- **Psilidae** – Rust flies
- **Megamerinidae**
- **Diopsidae** – Stalk-eyed flies

Conopoidea
- **Conopidae** – Thick-headed flies

Tephritoidea
- **Lonchaeidae** – Lonchaeid flies
- **Pallopteridae**

Lauxanioidea
- **Lauxaniidae** – Lauxaniid flies
- **Chamaemyiidae** – Chamaemyiid flies

Sciomyzoidea
- **Dryomyzidae** – Dryomyzid flies
- **Sciomyzidae** – Marsh flies
- **Sepsidae** – Black scavenger flies

Opomyzoidea
- **Clusiidae**
- **Acartophthalmidae**
- **Odiniidae**
- **Anthomyzidae**
- **Aulacigastridae**
- **Asteiidae** – Asteiid flies

Carnoidea
- **Carnidae**
- **Milichiidae**
- **Cryptochetidae**
- **Chloropidae** – Frit and grass flies

Spaeroceroidea
- **Heleomyzidae** – Heleomyzid flies
- **Chyromyidae**

Ephydroidea
- **Camillidae**
- **Drosophilidae** – Small fruit flies
- **Diastatidae**

Plate 84: Flies (Insecta: Diptera) in Baltic amber V (Culicomorpha).

a Ceratopogonidae: Biting midge, head with antennae, ×55.

b Chironomidae: True midge, head with antennae, ×52.

a

b

2.85 Brachycera in Baltic amber – Insecta: Diptera

If the plants and animals in Baltic amber are compared to extant species, and the occurrence of their closest relatives examined, the results indicate that many of the present-day descendents are distributed in Southeast Asia. However, closer geographical analysis combined with sound phylogenetic knowledge of the taxa provides a more detailed picture of the global distribution. Southeast Asia is inhabited by numerous relict species. The diversity of these lineages and the complexity of their evolution were illustrated by HENNIG using well-studied amber Diptera as examples.

The xylophagid flies (Xylophagidae) in the subfamily Rachicerinae are represented by three taxa in Baltic amber (HENNIG 1967a):

Chrysothermis speciosa LOEW, 1850
Electra formosa LOEW, 1850
Lophyrophorus flabellatus MEUNIER, 1902

These can be compared to 34 extant Rachicerinae species, 18 of which are distributed in the Indo-Asian region and three in other regions: Spain (1 species), Neotropical region (10), Nearctic region (5).

The extant small-headed flies (Acroceridae), which occur not only in Southeast Asia, but also in other subtropical regions, have a similar geographical distribution. The genus *Eulonchiella* from Baltic amber belongs to the subfamily Philopotinae and is more closely related to extant Philopotinae genera in South Africa (HENNIG 1966f). The subfamily Acrocerinae was recorded from Baltic amber on the basis of the genus *Villalites*. According to detailed morphological and phylogenetic analysis, this genus is most closely related to the extant, Chilean genus *Villalus* (HENNIG 1966f).

In order to identify direct links to Eocene species, it must first be determined whether a monophyletic relationship can be assumed to exist between all of the geographically separated subgroups. If this is the case, the next step would be to resolve the phylogenetic relationships between the subgroups and the fossil forms.

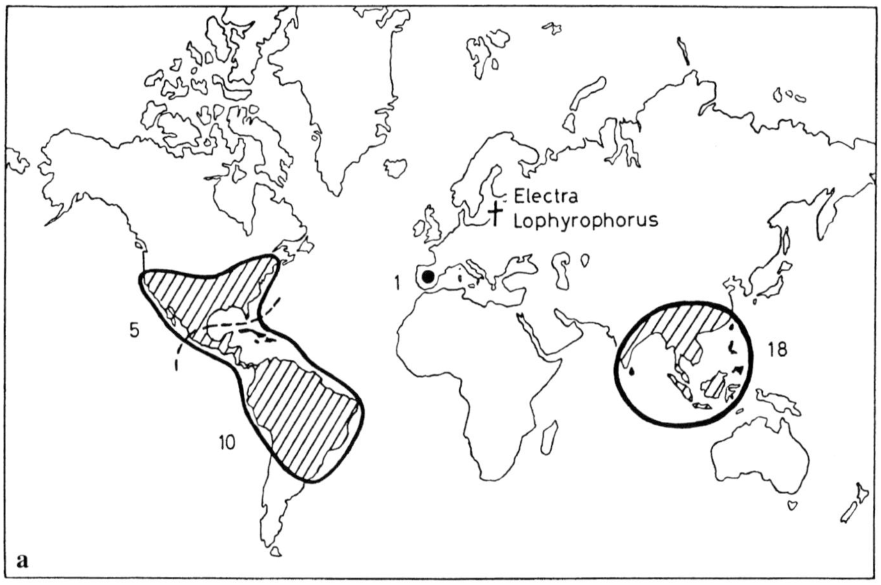

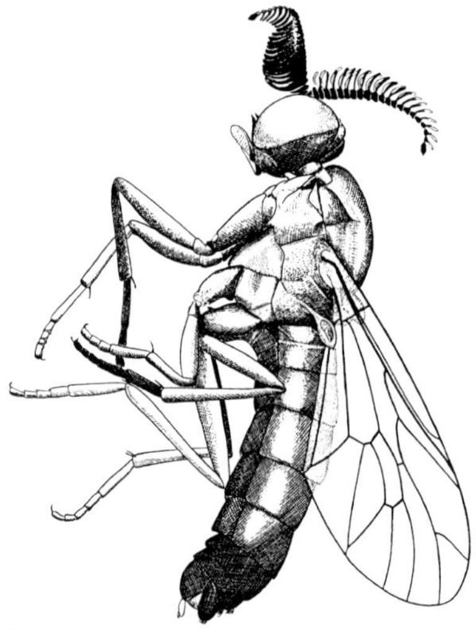

Fig. 92: **a** Geographical distribution of the Rachicerinae, **b** *Lophyrophorus flabellatus* (after HENNIG 1967 a).

Plate 85: Flies (Insecta: Diptera) in Baltic amber VI.

a Xylophagidae (Rachicerinae): Xylophagid fly, × 7.
b Xylophagidae (Rachycerinae): Head, × 15.
c Rhagionidae: Snipe fly, × 6.
d Rhagionidae: Head, mouthparts, × 22.
e Tabanidae: Horsefly, × 7.
f Tabanidae: Head, eyes, × 21.
g Acroceridae: Small-headed fly, × 9.
h Acroceridae: Small-headed fly, × 10.

a

b

c

d

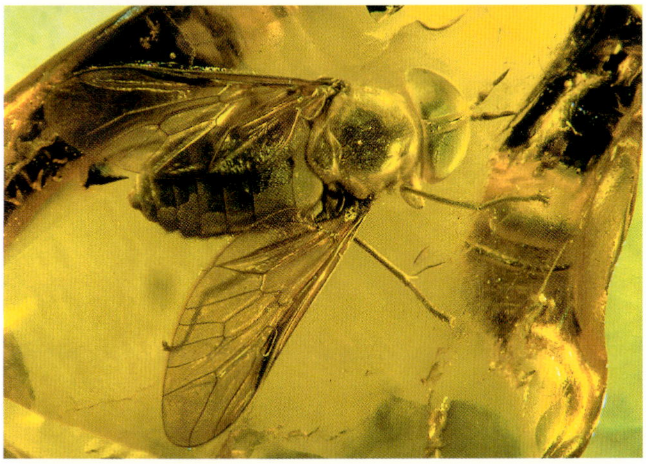
e

f

g

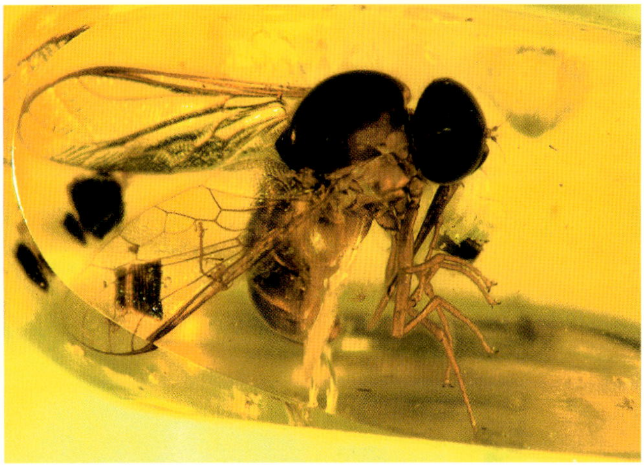
h

2.86 Brachycera in Baltic amber – Insecta: Diptera

The superfamily Empidoidea encompasses the families Empididae (dance flies) and Dolichopodidae (long-legged flies), both of which occur with great regularity in Baltic amber. The phylogenetic relationships and monophylety of the two taxa are still unclear. In particular, it cannot be ruled out that several subfamilies of the Empididae are more closely related to the Dolichopodidae than others. All subfamilies of the Empididae are represented in amber: Microphorinae, Atelestinae, Ocydromiinae, Hybotinae (Neozinae), Tachydromiinae (Corynetinae), Hemerodromiinae (+ Clinoceratinae, Ceratomerinae, Homalocerinae) and Empidinae (Hennig 1973). The Atelestinae have already been reported from Lower Cretaceous Lebanese amber on the basis of *Tricinites cretaceus* (Hennig 1970), *Atelestites senectus*, and *Phaetempis lebanensis* (Grimaldi & Cumming 1999), as well as from Canadian and New Jersey ambers (Grimaldi & Cumming 1999). Most Dolichopodidae families have also been confirmed for Baltic amber. The revision of the amber flies of both family groups will continue to be a primary objective of amber research, until more detailed knowledge of the Empididae and Dolichopodidae of Baltic amber helps unravel the phylogenetic relationships within the superfamily.

The Bombyliidae, or bee flies, are represented in Baltic amber by the Bombyliinae *(Paracorsomyza)*, Cyrtosiinae *(Proplatypygus, Proglabellula)* and Cylleniinae *(Palaeoamictus, Amictites, Glaesamictus)* (Hennig 1966b, 1967c, 1969). These groups are also the oldest fossil records of the Bombyliidae.

The Bombyliidae are cosmopolitan insects that occur all over the world, except in the cold regions of the southern and northern hemispheres. They are most diverse in the temperate zones, as is *Proglabellula* and its related genera from Baltic amber. In contrast, the amber genus *Paracorsomyza*, with the species *Paracorsomyza crassirostris* (Loew, 1850), is most closely related to the exclusively South African group *Corsomyza*, with its roughly 40 species (Fig. 93).

Due to the fact that the phylogenetic system of the Bombyliidae has long been incomplete, and the classification of the subfamilies of a preliminary nature (Mühlenberg 1971, Hennig 1973), it was previously difficult to assign fossil specimens to related, extant forms and to evaluate their associated biogeography. However, the recent cladistic studies by Yeates (1992, 1994) have opened the door for detailed studies of the fossil Bombyliidae.

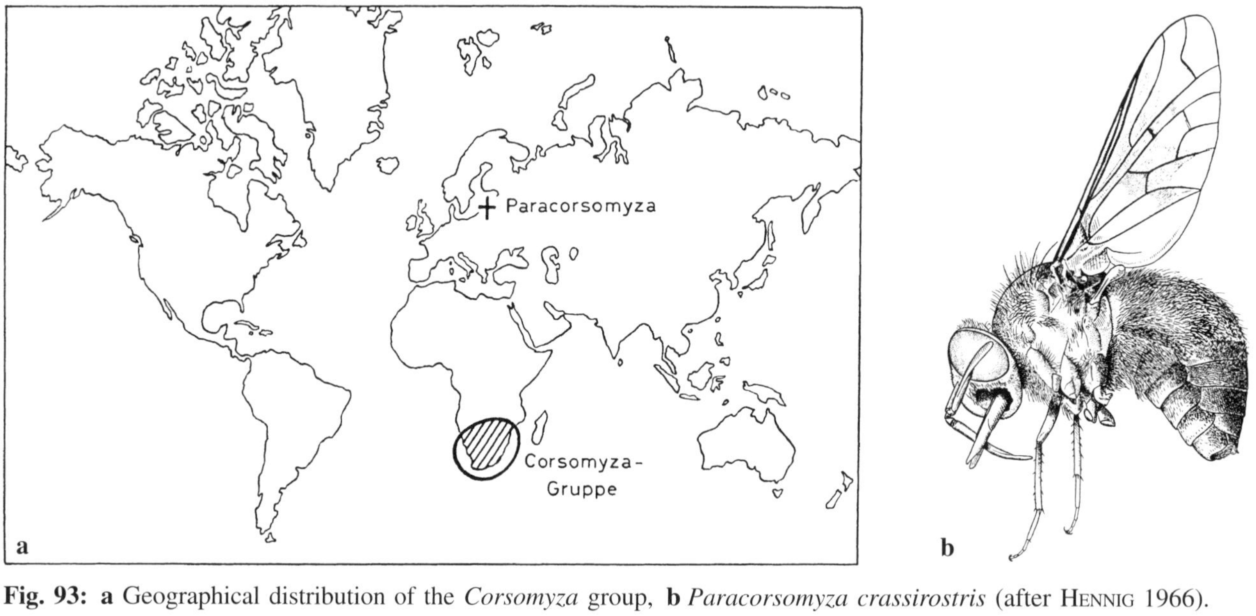

Fig. 93: **a** Geographical distribution of the *Corsomyza* group, **b** *Paracorsomyza crassirostris* (after Hennig 1966).

Plate 86: Flies (Insecta: Diptera) in Baltic amber VII.

- **a** Empididae: Dance fly, × 10.
- **b** Empididae: Head with eye and mouthparts, × 21.
- **c** Empididae: Dance fly, × 11.
- **d** Empididae: Head with eyes and antennae, × 23.
- **e** Dolichopodidae: Long-legged flies mating, × 10.
- **f** Empididae (Hybotinae): Dance fly, × 10.
- **g** Dolichopodidae: Long-legged fly, × 10.
- **h** Dolichopodidae: Head with green eyes, × 22.

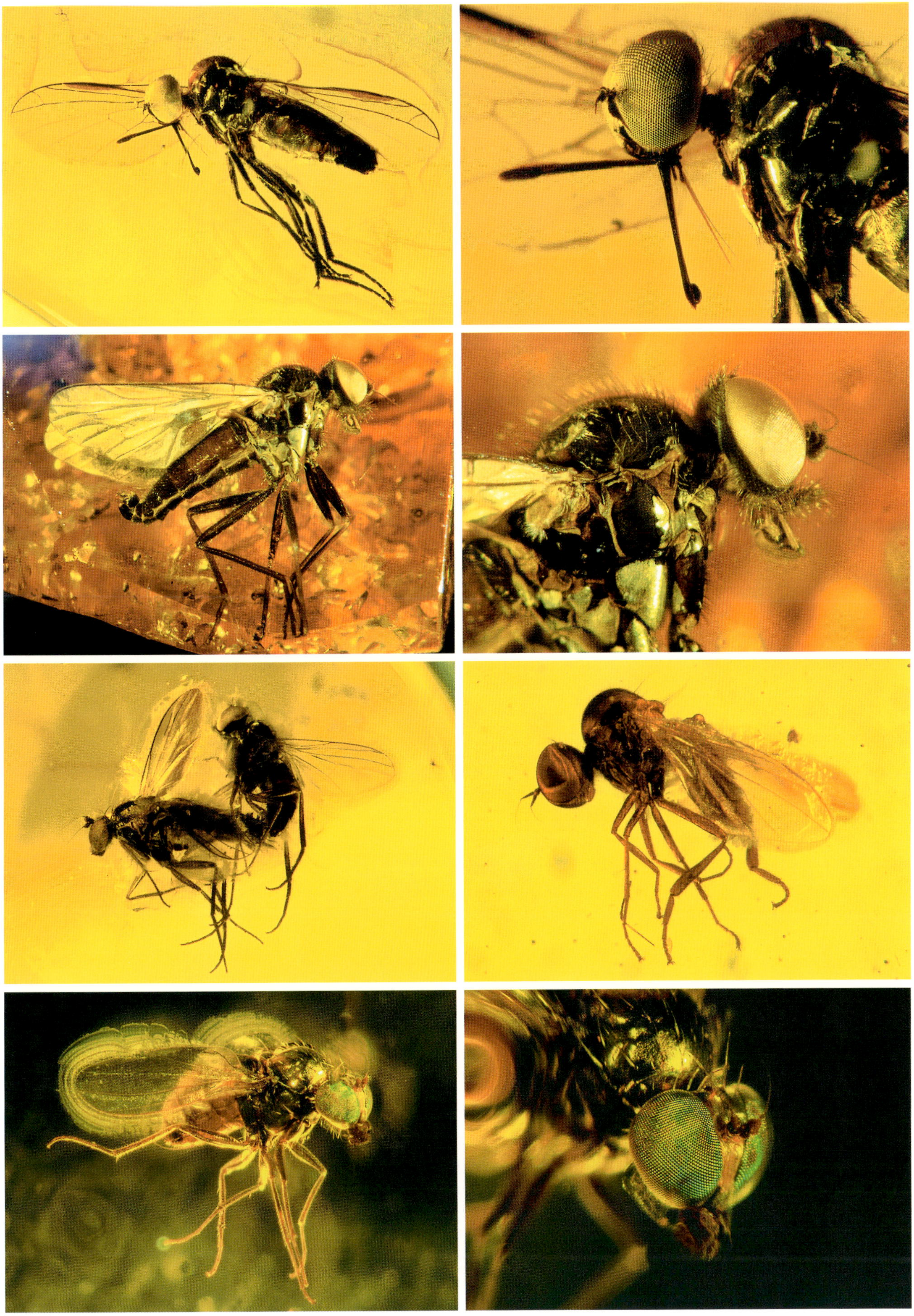

a b c d e f g h

2.87 Brachycera in Baltic amber – Insecta: Diptera

The Phoridae (Plate 87 d-h) are commonly referred to as "scuttle flies", due to the scurried manner in which they move. In some species, particularly those that are commensal in the nests of ants and termites, the wings are very short or completely reduced (as are the halteres). The family is represented in Baltic amber by some species (Brues 1923, 1939, Brown 1999).

According to the phylogenetic analysis by Hennig (1954, 1964), the record of the family Phoridae further indicates that both of its most closely related families, Sciadoceridae and Platypezidae (Clythiidae), also must have existed in the Eocene "amber forest". Evidence of the Platypezidae in amber is still considered to be uncertain, particularly since the questionable description of the species *Oppenheimiella baltica* Meunier, 1893 can no longer be verified due to the disappearance of the type material. Great significance was thus attributed to the report of a Sciadoceridae, which was described by Meunier and placed in the genus *Archiphora* by Hennig (1964): *A. robusta* (Meunier, 1905).

Only two extant species of Sciadoceridae are known: *Amphiphora patagonica* Schmitz, 1927 in southern Argentina (Patagonia) and *Sciadocera rufomaculata* White, 1917 in Tasmania and New Zealand. Little is known about the ecological requirements of the two extant species. Larvae of the Sciadoceridae may live in fungi, feeding on the fungi itself and on decaying plant material. It is conceivable that the larvae of the fossil species also led this kind of lifestyle, especially since the humid climate in the amber forest would have favored fungal vegetation.

"The evidence of a fossil species on the northern continents is what is so unexpected and significant about this newly discovered inclusion" (Hennig 1964). According to Hennig (1960), the trans-antarctic dispersal of the Sciadoceridae over the southern continents (South America, Australia/New Zealand) can virtually be ruled out. It is also very improbable that they came independently to South America and Tasmania/New Zealand from the northern continents. However, the two groups may merely be relicts of a global distribution at the time of the favorable, Eocene climate.

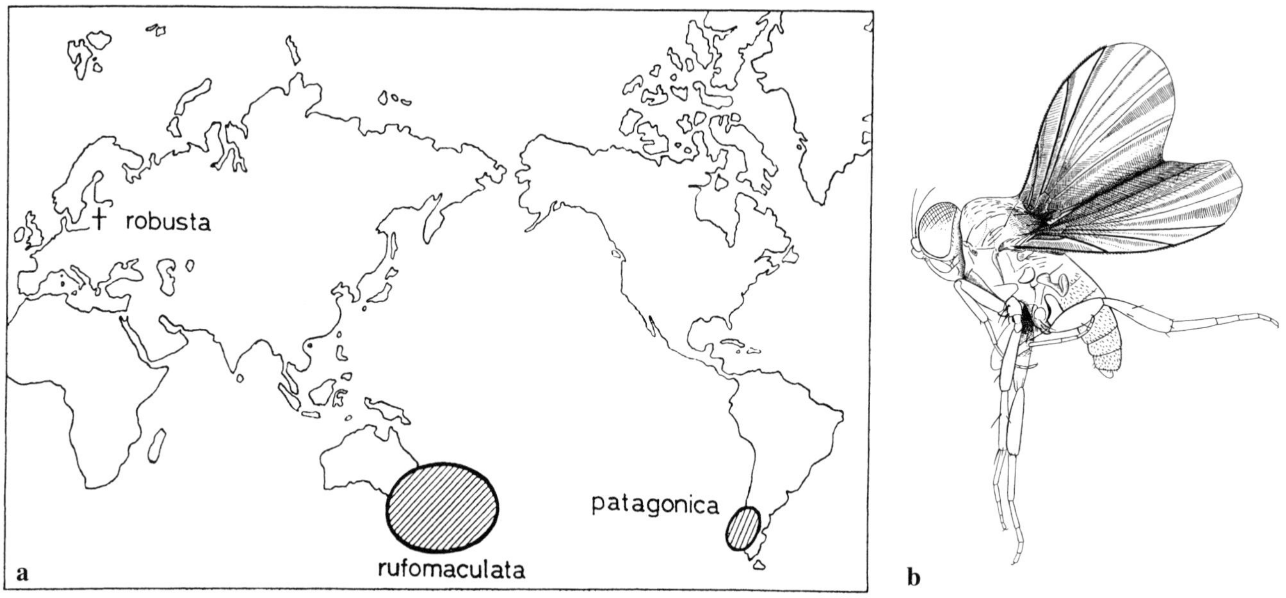

Fig. 94: a Geographical distribution of the Sciadoceridae, **b** *Archiphora robusta* (after Hennig 1964).

Plate 87: Flies (Insecta: Diptera) in Baltic amber VIII.

a Acalyptratae, ×6.
b Acalyptratae, ×8.
c Acalyptratae, ×9.
d Phoridae: Scuttle fly, ×17.
e Phoridae: Scuttle fly, ×14.
f Phoridae: Scuttle fly, ×15.
g Phoridae: Pair with wingless female, ×17.
h Phoridae: Wingless female, ×27.

a
b
c
d
e
f
g
h

2.88 Brachycera in Baltic amber – Insecta: Diptera

Syrphid flies (Syrphidae) are one of the hidden beauties of the insect world, attracting much less attention than beetles and butterflies due to their small size. As nectar and pollen-feeders, they are one of the primary pollinators. Their elegant flight patterns along flowery forest edges and glades make them easy to discover. Some syrphid flies mimic stinging bumble bees, bees and wasps as a form of protection, even against human beings. The small size of the syrphid flies becomes irrelevant in amber, overshadowed by the impressive colors and patterns that have been very well preserved in some fossil species. The fossil syrphid flies of Baltic amber were covered by MEUNIER and LOEW, and later on in great detail by HULL (1945, 1949, 1958).

Stalk-eyed flies (Diopsidae) have very wide-set eyes with characteristic eye stalks, the length of which varies among the Diopsinae. The primitive, African genus *Centrioncus* (Centrioncinae) does not have stalked eyes. The family Diopsidae is currently represented in Baltic amber by only one species. It is considered to be a representative of the stem group of the Diopsinae (HENNIG 1965):

Prosphyracephala succini (LOEW, 1873)

The distribution of the stalk-eyed flies extends from Africa to Madagascar, and from India to New Guinea. Based on its assumed ecological requirements and morphology, this Eocene amber species is most closely related to extant species restricted to areas of the Palearctic region that are considered to be refuges of past forest flora and fauna (Fig. 95 b).

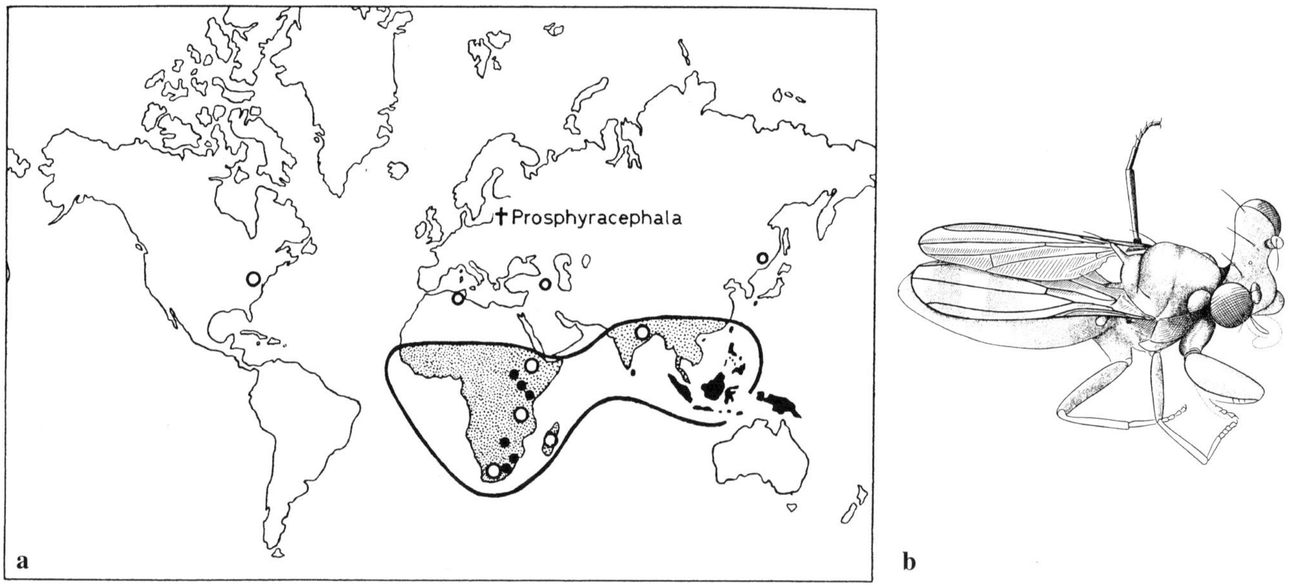

Fig. 95: a Geographical distribution of the Diopsidae, **b** *Prosphyracephala succini* (after HENNIG 1965).

Plate 88: Flies (Insecta: Diptera) in Baltic amber IX.

a-f Syrphidae: Syrphid flies, **a** × 10, **b** × 23, **c** × 9, **d** × 8, **e** × 10, **f** × 10.
g Diopsidae: Stalk-eyed fly, × 15.
h Pipunculidae: Big-headed fly, × 12.

a

b

c

d

e

f

g

h

2.89 Fleas in Baltic amber – Insecta: Siphonaptera

The Siphanoptera, or fleas, make up a separate order of the Endopterygota. Their phylogenetic position is not entirely certain. The Siphanoptera are considered as a sister group of either the Diptera or, more likely, the Mecoptera, and in fact the fleas may render the latter paraphyletic. Over 2,000 species are divided among sixteen families.

The adults are usually between two and three millimeters long, have a laterally flattened body and a characteristic keel-shaped head. Instead of compound eyes, fleas have atypical, lateral ocelli. Both males and females use the beak to suck blood from warm-blooded animals. Fleas have setae running in the same direction from the head to end of the abdomen, thus allowing them to maneuver unimpeded through the hair and plumage of their hosts. Typical spines (ctenidia) help retain the fleas on the host. The distribution of the spines on the head and body is a reliable characteristic for identification. Strong claws are used to cling to the host. The last two pairs of long legs are adapted for jumping. The biology of the fleas is extensively adapted to their hosts. Some leave the host soon after engorgement, while others are permanent inhabitants in the host's fur. Most fleas take blood from mammals, while a few are specialized on birds. Fleas are most diverse in the tropics and subtropics.

The discovery of a flea from Baltic amber in 1910 caused a minor sensation among the general public. Fleas were a much more common subject at that time and accordingly received mention in the daily press. Another flea was recorded from amber fifty years later and further specimens have even been found more recently, although these equally interesting discoveries have gone largely unnoticed.

Dampf (1911) and Peus (1968) described the two specimens and placed them in the same genus:

Palaeopsylla klebsiana Dampf, 1911
Palaeopsylla dissimilis Peus, 1968

The amber species belong to the family Hystrichopsyllidae, as does the extant species *Hystrichopsylla talpae*, which takes blood from moles. The fossil species in amber are also assumed to have infested the burrowing insectivores that appear to have been widely distributed in the Eocene, as indicated by specimens found in the Messel Pit and Geisel Valley.

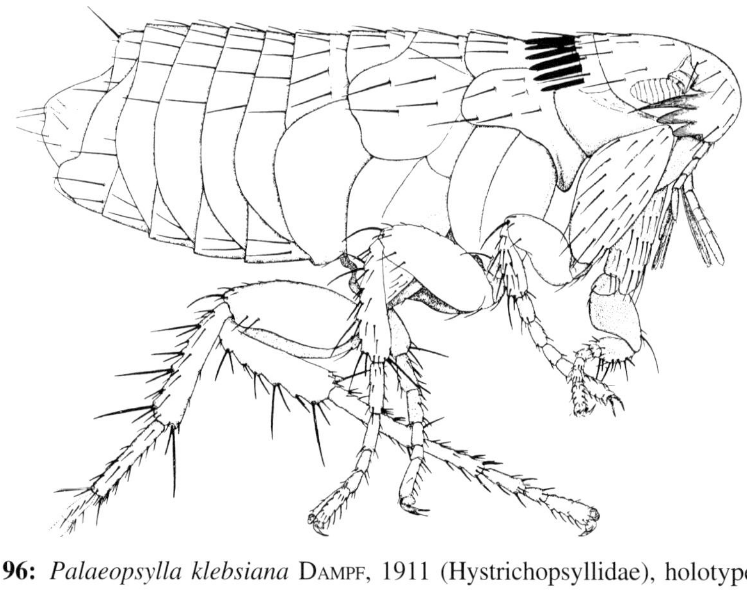

Fig. 96: *Palaeopsylla klebsiana* Dampf, 1911 (Hystrichopsyllidae), holotype.

Plate 89: Fleas (Insecta: Siphonaptera) in Baltic amber.

Hystrichopsyllidae: *Palaeopsylla dissimilis* Peus, 1968, holotype, ×92.

2.90 Lizards in Baltic amber – Vertebrata: Reptilia

The "amber lizard from Königsberg", which was described by KLEBS (1910) and given the name *Nucras succinea* by BOULENGER (1917), was long considered to be the only reptile found in Baltic amber. The fully intact lizard was the showpiece of the amber museum founded in Königsberg in 1889 by the STANTIEN & BECKER company. This collection was taken over ten years later by the Geology Institute of Königsberg University, where it remained until World War II.

The amber lizard was first mentioned by KLEBS (1889), who initially classified the specimen in the family Teiidae, in close relation to the extant genus *Knemidophorus*. Two years later, the unique amber fossil was presented to F. A. BOULENGER (British Museum of Natural History, London), who placed it in the lacertid genus *Nucras* of South Africa and pointed out its great similarity to the extant species *Nucras tessellata*. In a systematic study conducted later on, BOULENGER (1917) described the amber lizard as *Nucras succinea.*

LOVERIDGE (1942) included *Nucras succinea* in a list of South African copal lizards and expressed doubt about the authenticity of the inclusion, even though he had never examined it personally. LOVERIDGE (1957) mentioned in a footnote that the amber lizard *Nucras succinea* was embedded in Zanzibar copal. A number of authors (WERMUTH 1966, ESTES 1983, BÖHME 1984, SCHLEE 1990, POINAR 1992, SPAHR 1993 a) initially supported LOVERIDGE's authority and his assumption that the Königsberg lizard was actually a copal inclusion.

Other authors (KRZEMINSKA & KRZEMINSKI 1992, GRIMALDI 1996, KOSMOWSKA-CERANOWICZ et al.1997) reported that much of the Königsberg collection had been lost during World War II, including the amber lizard. However, RITZKOWSKI (1996) remarked that "a significant, albeit small portion" of the Königsberg collection survived World War II and is today deposited in the Institute of Geology and Paleontology of Göttingen University. As it turns out, the famous amber lizard is indeed among this salvaged material.

All doubt about the authenticity of the specimen was finally removed by a revision of *Nucras succinea* and the identification of co-occurring inclusions, i.e. in the same piece of amber (including stellate hairs). Together with other lizard remains from Baltic amber belonging to the same species, the Königsberg specimen was redescribed and placed in the newly established fossil genus *Succinilacerta* (BÖHME & WEITSCHAT 1998). A second, almost completely intact lizard has been found since then in Holocene coastal sediments near Danzig. Examination revealed (BORSUK-BIALYNICKA et al. 1999) that this juvenile of the Lacertidae is also representative of *Succinilacerta succinea* (BOULENGER, 1917) (Fig. 97).

Plate 90: Lizards (Vertebrata: Reptilia, Lacertidae) in Baltic amber.

a Lacertidae: *Succinilacerta succinea*, ventral view with hind legs and tail, ×4.
b Lacertidae: *Succinilacerta succinea*, exposed back with a view of the spinal column, ×10.
c Lacertidae: *Succinilacerta succinea*, toes of the left hind foot, ×12.

a

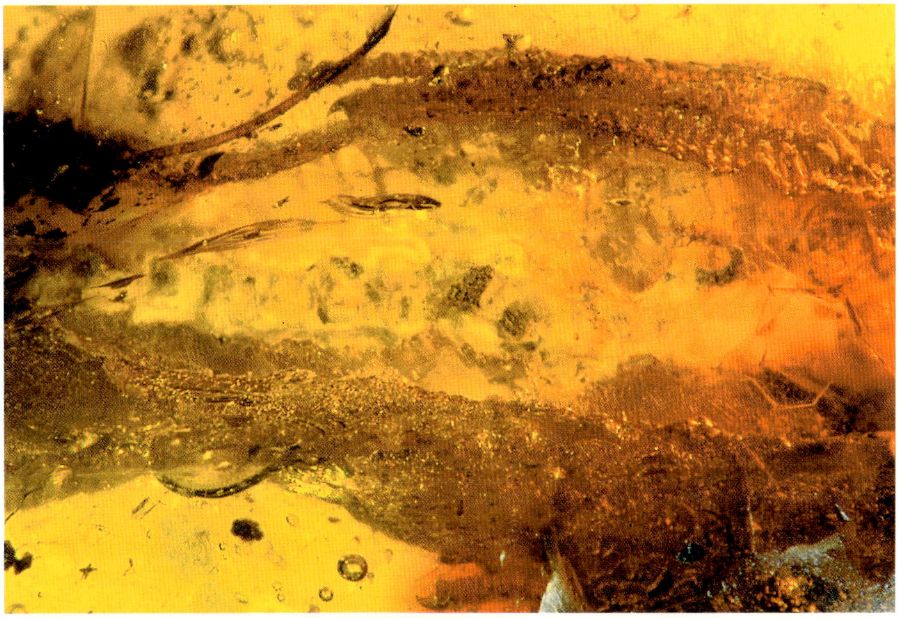
b

c

2.91 Bird feathers in Baltic amber – Vertebrata: Aves

The oldest bird feathers preserved in resin originate from Lower Cretaceous Lebanese amber (SCHLEE 1973). Together with other Mesozoic bird feathers, they are close in age to the oldest known feathers of *Archaeopteryx* from the Jurassic (Malm). Younger Baltic amber from the Tertiary contains well-preserved bird feathers. Most specimens are coverts that are similar in structure to present-day feathers. The size of these feathers can be used to estimate the size of the birds.

The microstructure of the contour feathers and particularly the down feathers is a characteristic that indicates phylogenetic relationships to extant families and genera. Comparative studies by BACHOFEN-ECHT (1936, 1944, 1949) suggest that the Tertiary amber forest was probably populated by passerines (*Sitta* – nuthatches, *Parus* – tits), woodpeckers and representatives of the genus *Momotus*, which today occur in Central and South America. These insectivorous birds nest in trees and fit in well with the picture of the Baltic amber forest.

Plate 91 shows the two types of bird feathers embedded in amber: the end segment of a down feather (Plate 91 a, b) and the base of a contour feather (Plate 91 c, d). The small vane from this inclusion consists of regularly arranged barbs connected to one another by overlapping, parallel barbules to form a continuous surface, although no hamuli are visible. The vane has down feathers on each side with barbules displaying successive nodes like a string of pearls.

The second amber specimen is a small down feather with long, curved, parallel barbs. They are densely covered on each side by short barbules, which become very soft towards the end, have no hamuli and therefore do not form a tight connection over their entire length. Down feathers develop before the contour feathers and are thus often found at the base of the contour feathers. They can also be located separately between the coverts as insulation.

Plate 91: Bird feathers (Vertebrata: Aves) in Baltic amber.

a Vane, × 35.

b Down of the counter feathers, × 30.

c End segment of a contour feather, × 10.

d Four barbs of a contour feather, × 40.

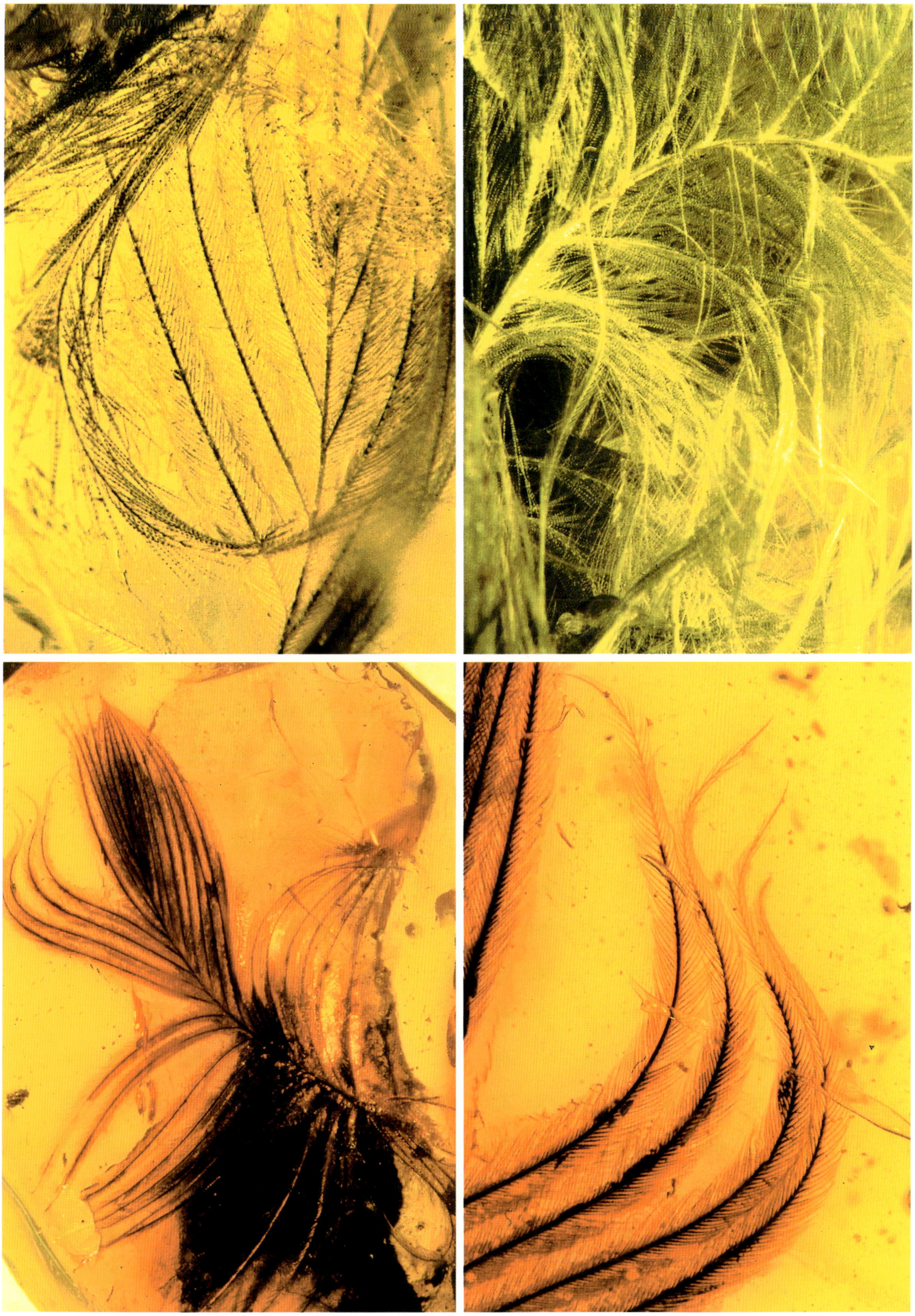

a b c d

2.92 Mammalian hair in Baltic amber – Vertebrata: Mammalia

Mammalian hair is much more common in Baltic amber than bird feathers. Most inclusions are tufts of hair consisting of two different types of hair: underhair, which is relatively thin (approx. 12 to 17 µm) and frizzy, and guard hair, which is longer, thicker (approx. 30 µm) and straighter. Eckstein (1890) and Lühe (1904) described hair inclusions and made an attempt to assign them to specific animals or animal groups. Eckstein believed he had identified a *Sciurus* species (squirrel) and a *Myoxus* species (dormouse). Voigt (1952 a, b) questioned the accuracy of this interpretation, pointing out that the genus *Sciurus* first appears in the Miocene. Lühe compared two hair inclusions with rodent hair (Myoxidae). Bachofen-Echt (1944, 1949) also showed an illustration of a tuft of hair from Baltic amber, suggesting it stemmed from a dormouse *(Glis)*.

Voigt (1952 a, b) subjected fossil mammalian hair to a detailed analysis and reported that the structure and definition of the cuticle are important criteria for distinguishing mammalian hair. In a comparison of cuticle samples from various extant mammalian groups, he concluded that most of the hair inclusions in Baltic amber stem from sciuromorph rodents. As the evolution of the rodents is placed in the later part of the Tertiary, Voigt felt that classification in extant genera (e.g. *Sciurus*) would be unjustifiable.

Voigt made a unique paleontological discovery in one of the hair inclusions shown here. He found the eggs of parasites attached to several hairs (Plate 92 a, b), which are unquestionably from animal lice (Phthiraptera). Voigt left the question open as to whether the eggs are from sucking lice (Anoplura) or biting lice (Mallophaga). In his opinion, however, the fact that sucking lice mainly parasitize rodents is a stronger indication that the eggs stem from this group. Apart from these eggs, no other fossil representatives of the suborders Phthiraptera and Mallophaga are known to exist.

In addition to the direct evidence of mammals in the Baltic "amber forest" provided by the records of hair, mammalian parasites point indirectly to the existence of their hosts. The most famous example in this context is the "amber flea" of the genus *Palaeopsylla* (Chap. 2.89), two species of which have been reported from Baltic amber. The Diptera include several blood-suckers (Culicidae, Tabanidae) that were also probably parasites on mammals.

Plate 92: Mammalian hair (Vertebrata: Mammalia) in Baltic amber.

a-b Hair with nits from animal lice (Phthiraptera), **a** ×20, **b** ×54.
c Tuft of hair, probably from a rodent, ×4.

a

b

c

3 References

ABDULLAH, M. (1965): New Anthicidae and Pyrochroidae (Coleoptera) from the Baltic amber (Oligocene). – The Entomologist 98: 38-42, London.

– (1965): *Palaeoasemum crowsoni* and *P. duffyi* a new genus and two new species of the Asemini (Coleoptera, Cerambycidae) from the Baltic amber. – Deutsche entomologische Zeitschrift, N. F. 14: 147-151, Berlin.

– (1969): *Dromius bakeri*, a new species of the Lebiini from the Klebs collection of the Baltic amber (Coleoptera, Carabidae). – Deutsche entomologische Zeitschrift, N. F. 16: 399-403, Berlin.

– (1975): The higher classification of the insect order Coleoptera including fossil records and a classified directory of the coleopterists and coleoptera collections of the world. – Zoologische Beiträge, N. F. 21: 363-461, Berlin.

ABDULLAH, M., ABDULLAH, A. (1967): *Crichtonia macleani* a new genus and species of the Hedobiini (Coleoptera: Anobiidae) from the Baltic amber. – Entomological News 78: 23-27.

– (1968): The discovery and probable phylogenetic significance of *Lathrobium (Palneolobrathium) whitei*, a new subgenus and species of Paederinae (Col., Staphylinidae) from Baltic amber. – The Entomologist's monthly Magazine 104: 1-3, Oxford.

ABEL, O. (1935): Vorzeitliche Lebensspuren. – Gustav Fischer Verlag Jena.

ACHTERBERG, C. VAN (1982): The fossil species of the subfamily Blacinae discribed by C. T. Brues (Hym.: Braconidae). – Entomologische Berichten 42: 91-96, Amsterdam.

AFZELIUS, J., DALLAI, R. (1994): Characteristics of the flagellar axoneme in Neoptera, Coleoptera and Strepsiptera. – J. Morphology 219: 15-20.

AGUIAR A. P., JANZEN, H.-W. (2000): An overview of fossil Stephanidae (Hymenoptera), with description of two new taxa from Baltic amber and key to species of *Electrostephanus* Brues. – Ent. Scand. **30**: 443-452.

ALEKSEEV, V. N. (1995): *Conostigmus antiquus* sp.n., a fossil ceraphronoid species from Baltic amber (Hymenoptera, Ceraphronoidea). – Amber & Fossils 1: 23-25.

ALEXANDER, C. P. (1927): Diptera. Fam. Tanyderidae. – Genera Insectorum (Hrsg.: P. WYTSMAN), 189: 1-13.

– (1931): Crane flies of the Baltic Amber (Diptera). – Bernstein-Forsch. 2: 1-135.

ANDER, K. (1942): Die Insektenfauna des Baltischen Bernsteins nebst damit verknüpften zoogeographischen Problemen. – Lunds Univ. Arsskrift 38: 3-82.

ANDERSEN, N. M. (1982): The semiaquatic Bugs (Hemiptera, Gerromorpha). Phylogeny, Adaptions, Biogeography and Classification. – Entomonograph 3: 1-455; Klampenborg, Dänemark.

– (1998): Water striders from the Paleogene of Denmark with a review of the fossil record and evolution of semiaquatic bugs (Hemiptera, Gerromorpha). – Biol. Skr. 50: 1-152.

– (2000): Fossil water striders in the Eocene Baltic amber (Hemiptera, Gerromorpha). – Insect Syst. Evol., 31: 257-284.

ANDRÉE, K. (1924): Ostpreußens Bernstein und seine Bedeutung, hauptsächlich für Wissenschaft, Kunstgewerbe und Industrie. – Ostdeutscher Naturwart 1924: 183-189, Breslau.

– (1927): Vom "Ostpreußischen Gold", dem Bernstein, im Allgemeinen und von der Klebs'schen Bernsteinsammlung und ihrer Bedeutung für Königsberg und die Bernsteinforschung im Besonderen. – Königsberger Universitätsbund. Jahresbericht 1926/27: 18-36.

– (1929): Bernsteinforschung einst und jetzt. Bernstein-Forsch. 1: 1-32, Berlin.

– (1937): Eine zweite Floh-Inkluse in baltischem Bernstein. – Bernstein-Forsch. 4: 131, Berlin.

– (1943): Miozäner Bernstein im Westbaltikum und an der Nordsee? Abalus, die Glaesarien oder Elekriden und der Eridanus der Alten. – PetermannsGeographische Mitteilungen, 5: 172-178.

– (1951): Der Bernstein. Das Bernsteinland und sein Leben. – Stuttgart (Kosmos).

ANDRÉE, K. & KEILBACH, R. (1936): Neues über Bernsteineinschlüsse. – Schrft. phys. ökon. Ges. Königsberg 69: 124-128.

ANSORGE, J. (1996): *Eotonisca tertiaria* MEUNIER, 1905 (Diptera, Psychodidae) aus dem Sächsischen Bernstein. – Studia dipterologica 3: 195-199.

– (1996): Insekten aus dem oberen Lias von Grimmen (Vorpommern, Norddeutschland). – Neue Paläontol. Abhandlungen 2: 1-132.

ARNOLD, V. (1998): Vergessene Einschlüsse – Blütenstaub in Baltischem Bernstein. – Mitt. Geol.-Paläont. Inst. Univ. Hamburg 81: 269-282.

ASPÖCK, H., ASPÖCK, U., HÖLZEL, H. (1980): Die Neuropteren Europas. – (2 Bd.) Goecke & Evers, Krefeld.

ASSMANN, A. (1870): Beiträge zur Insekten-Fauna der Vorwelt. – Z. Entomologie N.F. 1: 3-62, Breslau.

ATKINS, M. D. (1963): The Cupedidae of the world. – Canad. Entomologist 95: 140-162.

AX, P. (1984): Das Phylogenetische System. – G. Fischer, Stuttgart.

AZAR, D., NEL, A., SOLIGANC, M. (2000): A new Coniopterygidae from Lebanese amber. – Acta Geol. Hisp. 35: 31-36.

BACHOFEN-ECHT, A. (1928): Leben und Sterben im Bernsteinwald. – Palaeobiologica 1: 39- 50.

– (1934): Beobachtungen über im Bernstein vorkommende Spinnengewebe. – Biol. gen. 10: 179-184.

– (1935): Tod im Bernstein. – in: Abel, O., Vorzeitliche Lebenspuren: 601-619.

– (1936): Das Vorkommen von Federn im Bernstein. – Nova Acta Leopoldina 4: 341-347.

– (1942): Über die Myriapoden des Bernsteins. – Palaeobiologica 7: 391-403.
– (1944): Einschlüsse von Federn und Haaren im Bernstein. – Palaeobiologica 8: 113-119.
– (1949): Der Bernstein und seine Einschlüsse. – Wien (Springer), 204 S.

BAGNALL, R. S. (1914): Fossil Insect in Amber. On *Stenurothrips succineus* gen. et sp. nov., an interesting tertiary Thysanopteron. – Geological Magazine 6: 483-485, London.
– (1923, 1924, 1926): Fossil Thysanoptera I-IV. – Ent. Month. Mag. 59: 35-38, 60: 130-133, 251-252, 62: 16-17.
– (1929): Fossil Thysanoptera, V. – Tubulifera. – Ent. Month. Mag. 65:95-99.

BAKER, A. C. (1922): Two new aphids from Baltic amber. – J. Washington Acad. Sci. 12: 353-358.

BAKKENDORF, O. (1948): A comparison of a Mymarid from baltic amber with a recent species, *Petiolaria anomala* (Micro. Hym.). – Ent. Medd. 25: 213-218.

BALOGH, J. (1942): The Oribatid Genera of the World. – Budapest (Akadémiai Kiadó), 188 S.

BARFOD, J., JACOBS, F., RITZKOWSKI, S. (1989): Bernstein – Schätze in Niedersachsen. – Verlag Knorr & Hirth, 96. pp.

BARONI URBANI, C. (2000): Rediscovery of the Baltic amber ant genus *Prionomyrmex* (Hymenoptera, Formicidae) and its taxonomic consequences. – Eclogae geol. Helv. 93: 471-480.

BARONI URBANI, C., BOLTON, B., WARD, P. S. B. (1992): The internal phylogeny of ants (Hymenoptera: Formicidae). – Syst. Ent. 17: 301-329, Oxford.

BARTHEL, M., HETZER, H. (1982): Bernstein – Inklusen aus dem Miozän des Bitterfelder Raumes. – Z. angew. Geol. 28: 314-336; Berlin.

BECHLY, G. (1998): New fossil damselflies from Baltic amber, with description of a new species, a redescription of *Litheuphaea carpenteri* Fraser, and a discussion on the phylogeny of Epallagidae (Zygoptera: Caloptera). – Int. J. Odonatol. (Pantala), 1(1): 33-63.
– (in prep. 2001a): New dragonflies in amber (Insecta: Odonata: Anisoptera). – projected for Stuttg. Beitr. Naturk. B.
– (in prep. 2001b): New damselflies in amber (Insecta: Odonata: Zygoptera). – projected for Stuttg. Beitr. Naturk. B.

BECHLY, G., WICHARD, W. (in prep. 2001): New odonate larvae in Baltic amber (Insecta: Odonata). – projected for Stuttg. Beitr. Naturk. B.

BECHLY, G., WITTMANN, M. (2000): Two new tropical bugs (Insecta: Hemiptera: Thaumastocoridae – Xylastodorinae and Hypsipterygidae) from Baltic amber. – Stuttg. Beitr. Naturk. B 289: 1-11.

BECK, C. W. (1986): Spectroscopic investigations of amber. – Appl. Spectroscopy Rev. 22: 57-110.
– (1996): Zur Herkunftsbestimmung von Bernstein. – In: GANZELEWSKI, M., SLOTTA, R. (Hrsg): Tränen der Götter 59-61, Bochum.

BECK, C. W., WILBUR, E., MERET, S. (1964): Infrared spectra and origin of amber. – Nature 201: 256-257.

BECK, C. W., WILBUR, E., MERIT, S., KOSSOVE, D., KERMANI, K. (1965). The infrared spectra of amber and the identification of Baltic amber. – Archaeometry 9: 96-109.

BEIER, M. (1937): Pseudoscorpione aus dem baltischen Bernstein. – Festschrift für E. Strand 2: 302-316.
– (1947): Pseudoscorpione im baltischen Bernstein und die Untersuchung von Bernstein-Einschlüssen. – Mikroskopie (Wien) 1: 188-199.
– (1948): Phoresie und Phagophilie bei Pseudoscorpionen. – Österr. Zool. Z. 1: 441-497.
– (1955): Pseudoscorpione im baltischen Bernstein aus dem geologischen Staatsinstitut in Hamburg. – Mitt. geol. Staatsinstitut in Hamburg, 24: 48-54.
– (1963): Ordnung Pseudoscorpionidea. – Bestimmungsbücher zur Bodenfauna Europas 1: 1-313, Berlin.
– (1968): Mantodea (Fangheuschrecken). – In: Handbuch der Zoologie, 4 (2) 12: 1-47. Berlin.
– (1974): Blattariae (Schaben). – In: Handbuch der Zoologie, 4 (2) 13: 1-127; Berlin.

BENICK, L. (1943): Ein *Stenus* aus dem baltischen Bernstein: *St. (Parastenus) priscus* n. sp. (Coleoptera: Staphylinidae). – Arbeiten morphol. taxon. Entomologie 10: 101-104, Berlin.

BERENDT, G. C. (1830): Die Insekten im Bernstein. – Ein Beitrag zur Thiergeschichte der Vorwelt. – 38pp., Danzig.
– (1845-1856): Die im Bernstein befindlichen organischen Reste der Vorwelt. Vol. I-II, (Hrsg.), Berlin (Nicolai).

BERENDT, G. C. (1936): Memoire pour servir a l´histoire des blattes antediluviennes, traduit de l´allemand par M. HELLER. – Ann. Soc. Ent. France 5 : 539-546.

BERKELEY, M. J. (1948): On three species of Mould detected by Dr. Thomas in the amber of East Prussia. – Ann. Mag. Natural History, Ser. 2, 2: 380-383, London.

BERTRAND, H. P. I. (1972): Larves et nymphes des Coléoptères aquatiques du globe. – Abbeville (Paillart), 804 S.

BERVOETS, R. (1910): Diagnoses de quelques nouvelles espèces de Cicadines de l´ambre de la Baltique. – Ann. Mus. Nat. Hungar 8: 125-128.

BISCHOFF, H. (1927): Bernsteinhymenopteren. – Schr. phys.-ökon. Ges. Königsberg 56: 139-144.
– (1927): Biologie der Hymenopteren. – Biologische Studienbücher. 598 S.; Berlin (Springer).

BISMARCK, R. v. (1987): Bernstein – Gold des Nordens. – Wanderndes Museum der Christian Albrechts-Universität Kiel.

BOGNER, J. (1976): Die systematische Stellung von *Acoropsis* Conwentz, einer fossilen Aracee aus dem Bernstein. – Mitt. Bayr. Staatssamml. Paläontologie und Geologie, 16: 95-98, München.

BÖHME, W. (1984): Erstfund eines fossilen Kugelfingergeckos (Sauria: Gekkonidae: Sphaerodactylinae) aus Dominikanischem Bernstein (Oligozän von Hispaniola, Antillen). – Salamandra 20: 212-220, Bonn.

BÖHME, W., WEITSCHAT, W. (1998): Redescription of the Eocene lacertid lizard *Nucras succinea* BOULENGER, 1917 from Baltic amber and its allocation to *Succinilacerta* n.gen. – Mitt. Geol. Paläont. Inst. Univ. Hamburg. 81: 203-222.

Bolland, H. R., Magowski, W. L. (1990): *Neophyllobius succineus* n. sp. from Baltic amber (Acari: Raphignathoidea: Camerobiidae). – Entomologische Berichten 50: 17-21, Amsterdam.

Bollow, H. (1940): Die erste Helminide (Col. Dryop.) aus Bernstein. – Mitt. Münch. Ent. Ges. 30: 117-119.

Bölsche, W. (1927): Im Bernsteinwald. – Kosmos-Buchbeilage: 78 S.; Stuttgart (Franckh).

Borgmeier, T. (1968): A catalogue of the Phoridae of the world (Diptera, Phoridae). – Studia entomologica 11: 1-367, Petropolis, Rio de Janeiro.

Borkent, A. (1995): Biting Midges in the Crateceous Amber of North America (Diptera: Ceratopogonidae). – Backhuys Publ. Leiden.

– (2000a): Biting midges (Ceratopogonidae: Diptera) from Lower Cretaceous Lebanese amber with a discussion of the diversity and patterns found in other ambers. – In: Grimaldi, D. (Ed.): Studies on Fossils in Amber, with Particular References to the Cretaceous of New Jersey: 355-451.

– (2000b): Further biting midges (Diptera: Ceratopogonidae) from Upper Cretaceous New Jersey amber. – In: Grimaldi, D. (Ed.): Studies on Fossils in Amber, with Particular References to the Cretaceous of New Jersey: 453-472.

Borkent, A., Szadziewski, R. (1992): The first records of fossil Corethrellidae (Diptera). – Entomol. Scand. 22: 457-463.

Borsuk-Bialynicka, M., Lubka, M., Böhme, W. (1999): A lizard from Baltic amber (Eocene) and the ancestry of the crown group lacertids. – Acta Palaeontologica Polonica 44:349-382.

Botosaneanu L., Levanidova, I. M. (1987): The remarkable genus *Palaeagapetus* Ulmer, 1912 (Hydroptilidae). – Series Entomologica 39: 43-46, Lancaster (Junk).

Botosaneanu, L., Wichard, W. (1981): Trichoptera from the Baltic amber. [Summary]. – Series Entomologica 20: 31-32, The Hague (Junk).

Boulenger, G. A (1917): A revision of the Lizards of the genus *Nucras* Gray. – Ann. South African Museum 13: 195-216.

Brauns, A. (1964): Taschenbuch der Waldinsekten. – G. Fischer Verlag, Stuttgart.

Brechtel, F., Ehrmann, R., Detzel, P. (1996): Zum Vorkommen der Gottesanbeterin *Mantis religiosa* (Linne, 1758) in Deutschland. – Carolinea 54: 73-90.

Breyne, J. P. (1734): An account of a leaf of a plant impressed in a piece of amber. – Phil. Transact. 6: 232-233, London.

Brischke, D. (1886): Die Hymenopteren des Bernsteins. – Schr. naturf. Ges. Danzig 6: 278-279.

Brost, L., Reinicke, R. (1996): Bernstein in Südschweden – in: Ganzelewski, M., Slotta, R. (Hrsg.): "Bernstein – Tränen der Götter" 155-156.

Brothers D. J., Janzen J.-W. (1998): New generic synonymy in Scolebythidae, with redescription of both sexes of *Pristapenesia primaeva* Brues from Baltic amber (Hymenoptera: Chrysidoidea). – Proc. 1st Int. Palaeoent. Conf. Moscow 1998: 17-26.

Brown, W. L. (1954): Remaks on the internal phylogeny and subfamily classification of the family Formicidae. Insect Soc.1: 21-31.

Brown, B. V. (1999): Re-evaluation of the fossil Phoridae (Diptera). – J. Natural History 33: 1561-1573.

Brues, C. T. (1915): A synonymic catalogue of the Dipterous family Phoridae. – Bull. Wisconsin nat. Hist. Soc., 12 (1914): 85-152.

– (1923a): A fossil genus of Dinapsidae from Baltic Amber (Hymenoptera). – Psyche 30: 31-35.

– (1923b): Two new species of Phoridae from Baltic Amber. – Psyche 30: 59-62.

– (1923c): Some new fossil parasitic Hymenoptera from Baltic Amber. – Proc. Amer. Ac. Sci. Boston 58: 327-346.

– (1923d): Ancient insects, Fossils in amber and other deposits. – Scientific Monthly 17: 289-304; New York.

– (1926): A species of *Urocerus* from Baltic Amber. – Psyche 33: 168-170.

– (1933a): Progressive change in the insect population of forests since the Early Tertiary. – American Naturalist, 67: 385-406.

– (1933b): The parasitic Hymenoptera of the Baltic Amber. – Bernstein-Forsch. 3: 4-178.

– (1938): Ichneumonoidea, Serphoidea, and Chalcidoidea from Canadian amber. – Univ. Toronto Stud. Geol. 40: 27-44.

– (1939): Fossil Phoridae in Baltic Amber. – Bull. Mus. comp. Zool. Harvard 85: 411-436.

– (1939): New Oligocene Braconidae and Bethylidae from Baltic Amber (Hymenoptera). – Ann. ent. Soc. Amer. 32: 251-263.

– (1940a): Serphidae in Baltic Amber, with description of a new living genus. – Proc. Amer. Ac. Sci. Boston 73: 259-264.

– (1940b): Calliceratidae in Baltic Amber. – Proc. Amer. Ac. Sci. Boston 73: 265-269.

– (1940c): Fossil parasitic Hymenoptera of the family Scolionidae from Baltic Amber. – Proc. Amer. Ac. Sci. Boston 74: 69-70.

Buchardt, B. (1978): Oxygen isotop palaeotemperatures from the Tertiary period in the North Sea Area. – Nature 275: 121-123.

Buckton, G. B. (1883): Fossil Aphids. – In: Monograph of the British Aphides 4: 144-178, London.

Budrys, E. (1993): Digger wasps of the subfamily Pemphredoninae (Hymenoptera, Specidae) from the Baltic and the Taimyr amber. – Acta entomol. Lituanica 11: 34-56.

Burmeister, H. (1831-1932): Handbuch der Entomologie. Bd. 1+2, Berlin.

Burr, M. (1911): Dermaptera (earwigs) preserved in Amber from Prussia. – Trans. Linn. Soc. London 11: 145-150.

Buttel-Reepen, H. v. (1906): Apistica. – Mitt. zool. Mus. Berlin 3: 117-201.

Byers, G. W. (1979): *Hylobittacus*, a new genus of North American Bittacidae (Mecoptera). – J. Kansas Entomol. Soc. 52: 402-404.

CANO, R. J, POINAR, H. N, PIENIACEK, N. J, ACRA, A., POINAR, G. O. jr. (1993): Amplification and sequencing of DNA from a 120-135-million-year-old-weevil. – Nature 536-538,London.

CANO, R. J., POINAR, H. & POINAR jr. G.O. (1992): Isolation and partial characterization of DNA from *Problebeia dominicana* (Hymenoptera,: Apidae) in 25-40 million year old amber. – Medical Science Res., 20:249-251.

CARPENTER, F. M. (1927): Notes on a collection of amber ants. – Psyche 34: 30-32, Boston, Mass.

– (1931): The affinities of *Holcorpa maculosa* SCUDD. and other Tertiary Mecoptera. – J. N. York ent. Soc. 39: 409-410.

– (1935): Tertiary insects of the family Chrysopidae. – J. Paleont. 9: 259-271.

– (1954): The Baltic amber Mecoptera. – Psyche, 61: 31-40, Cambridge, Mass.

– (1955): An Eocene *Bittacus* (Mecoptera). – Psyche, 62: 39-41, Cambridge, Mass.

– (1956): The baltic amber snake-flies (Neuroptera). – Psyche 63: 77-81.

– (1976): Note on *Bittacus validus* in Baltic amber. – Psyche 82 (1975): 303.

CARPENTER, F. M., HULL, F. M. (1939): The fossil Pipunculidae. – Bernstein-Forsch. 4: 8-17.

CARVALHO, J. C. M. (1966): A new fossil genus and species of Miridae from the baltic amber (Hemiptera). – Revista Brasil. Biol. 26: 199-202.

CARVALHO, J. C. M., POPOV, V. (1984): A new genus and species of Mirid bug from the Baltic amber (Hemiptera, Miridae). – An. Acad. brasil. Cienc. 56: 203-205.

CARVER, M., GROSS, G. F., WOODWARD, T. E. (1991): Hemiptera – in: CSIRO (1991): The insects of Australia, Vol. 1+2: 429-509, Cornell Univ. Press, Ithaca, New York.

CASPARY, R. (1887): Einige neue Pflanzenreste aus dem samländischen Bernstein. – Schr. Physik.-Ökonom. Ges. Königsberg 27: 1-8.

CASPARY, R., KLEBS, R. (1907): Die Flora des Bernsteins und anderer fossiler Harze des ostpreußischen Tertiärs. – Abh. Königl. Preuß. Geol. Landesanst. 4: 1-182, Berlin.

CERNOVA, O. A. (1970): On the classification of fossil and recent Ephemeroptera. – Ent. Rev. 49: 71-81.

CHOPARD, L. (1936): Orthoptères fossiles et subfossiles de l'ambre et du copal. – Ann. Soc. ent. France. 105: 375-386.

– (1936): Un remarquable genre d'orthoptères de l'ambre de la Baltique. – Livre jubilaire de E. L. Bouvin 163-168, Paris.

CLOUDSLEY-THOMPSON, J. L. (1988): Evolution and Adaptation of Terrestrial Arthropods. – Springer, Berlin.

COCKERELL, T. D. A. (1909): Some European fossil bees. – The Entomologist 42: 313-317.

– (1910): Some insects in Baltic amber. – The Entomologist 43: 153-155.

– (1914): The fossil and recent Bombyliidae compared. – Bull. Am. Mus. nat. Hist. 33: 229-236; New York.

– (1917): Descriptions of fossil insects. – Proc. Biol. Soc Washington 30: 79-82.

– (1921): An Ortalid fly in British amber. – The Entomologist 54: 30-31.

COCKERELL, T. D. A., CLARK, E. (1918): A tipulid fly from Baltic amber. – Canad. Ent. 50: 115-116.

COLLESS, D. H., MCALPINE, D. K. (1991): Diptera – in: CSIRO (1991): The insects of Australia, Vol. 2, Cornell Univ. Press, Ithaca, New York.

COMMON, I. F. B. (1975): Evolution and classification of the Lepidoptera. – Ann. Rev. Entomology 20: 183-203.

CONWENTZ, H. (1886): Die Angiospermen des Bernsteins. – In: GÖPPERT, H. R., MENGE, A.: Die Flora des Bernsteins und ihre Beziehungen zur Flora der Tertiärformation und der Gegenwart, Bd. 2, 140 S., Danzig (Engelmann).

– (1886): Die Bernsteinfichte. – Ber. Dt. Botan. Ges. 4: 375-377, Berlin.

– (1890): Monographie der baltischen Bernsteinbäume. – Leipzig (Engelmann), 151 S.

CRAMPTON, G. C. (1926): The external anatomy of the primitive Tanyderid Dipteron *Macrochile spectrum* LOEW, preserved in Baltic amber. – Bull. Brooklyn ent. Soc. 21: 1-14.

CROWLY, T. & NORTH, G. (1991): Paleoclimatology. – Oxford University Press.

CROWSON, R. A. (1960): The phylogeny of Coleoptera. – Ann. Rev. Entomology 5: 111-134.

– (1965): Some thoughts concerning the insects of the Baltic amber. – Proc. 12th Int. Congr. Entomology London 1964: 133.

– (1973): On a new superfamily Artematopoidea of polyphagan beetles, with the definition of two new fossil genera from the Baltic amber. – J. nat. Hist.7: 225-238, London.

– (1975): The evolutionary history of Coleoptera, as documented by fossil and comparative evidence. – Congr. nazionale Ital. Entomol., 1974: 47-90, Firenze.

CSIRO (1991): The insects of Australia, Vol. 1, 2, Cornell Univ. Press, Ithaca, New York.

CULLINEY, T. W. (1983): Geologicai history and evolution of the honey bee. – Am. Bee Journal 123: 580-585, Hamilton, Ill.

CZECZOTT, H. (1961): Sklad i wiek flory bursztynow baltyckich. – Prace Muzeum Ziemi 4: 119-145, Warszawa.

DAHL, C. (1971): Trichocoridae (Dipt.) from the baltic amber. – Ent. Scand. 2: 29-40.

DAHL, C., ALEXANDER, C. P. (1976): A world catalogue of Trichoceridae Kertész, 1902 (Diptera). – Ent. Scand 7: 7-18.

DAMPF, A. (1911): *Palaeopsylla klebsani* n. sp., ein fossiler Floh aus dem baltischen Bernstein. – Schr. phys.-ökonom. Ges. Königsberg 51, 248-259.

DE SALLE, R., GATESY, J., WHEELER, W., GRIMALDI, D. A. (1992): DNA sequences from a fossil termite in oligomiocene dominican amber and their phylogenetic implications. – Science 257, 1933-1936.

DEMOULIN, G. (1954): Essai sur quelques Ephéméroptères fossiles adultes. – Vol. jubilaire Victor van Straelen, 1: 547-574; Bruxelles.

– (1955): Remarques critiques sur *Cronicus anomalus* (Pictet). Ephéméroptère de l´ambre oligocène de la Baltique. – Bull. Inst. R. Sci. Nat. Belg. 31 (4): 1-4.
– (1956): *Electrogenia dewalschei* n.gen. n.sp., Ephéméroptère fossile de l´ambre. – Bull. Ann. Soc. R. Entom. Belg 92: 95-100.
– (1958): Nouveau Schema de Classification des Archodonates et des Ephéméroptères. – Bull. Inst. R. Sci. Nat. Belg. 34 (27): 1-19.
– (1965): Contribution à la connaissance des Ephéméroptères de l´ambre oligocène de la Baltique. – Entom. Medd. 34: 143-153.
– (1968): Deuxième contribution à la connaissance des Ephéméroptères de l'ambre oligocène de la Baltique. – Bull. Inst. R. Sci. Nat. Belg., 46, 1-11 S.,
– (1968): Deuxième contribution à la connaissance des Ephéméroptères de l´ambre oligocène de la Baltique. – Dtsch. Ent. Z. 15: 233-276, Berlin.
– (1969): Remarques critiques sur la l'osition systematique des Baetiscidae et des Prosopistomatidae (Ephemeroptera). – Bull. Inst. R. Sci. Nat. Belg. 45, (27): 1-8.
– (1970): Contribution à la Connaissance des Ephéméroptères du Miocène. I. Siphurites explanatus Cockerell. – Bull. Inst. R. Sci. Nat. Belg. 46 (5): 1-4.
– (1970): Troisième contribution à la connaissance des Ephéméroptères de l´ambre oligocène de la Baltique. – Bull. Inst. R. Sci. Nat. Belg. 46 (2): 1-11.
– (1974): Quatrième Contribution à la Connaissance des Ephéméroptères de l'Ambre Oligocene de la Baltique. – Bull. Inst. R. Sci. Nat. Belg. 50 (Entomologie) 3: 1-4; Bruxelles.
– (1979): Les Avatars des Mesonetinae (Ephemeroptera). – Bull. Inst. R. Sci. Nat. Belg. 51 (Entomologie) 9: 1-6; Bruxelles.
Dessart, P. (1978): Revision de trois Conostigmus de l'ambre de la Baltique. (Hym. Ceraphronoidea Megaspilidae). – Bull. Ann. Soc. royal belge d'Entomologie 114: 50-58.
Dixon, H. N. (1922): Note on a moss in amber. – Journal of Botany, 60: 149-151; London.
Dlusskij, G. M. (1967): Ants of the genus Formica from the Baltic amber. – Paleontol. J. 2: 69-77.
Dobosz, R., Krzeminski, W, (2000): A new species of the Coniopterygidae (Neuroptera) from Baltic amber. – Polskie Pismo Entomol. 69: 219-224.
Dollfus, R. P. (1950): Liste des Némathelminthes connus à l'état fossile. – Compte rendu sommaire des Séances de la Soc. Geol. France, 5: 82-85; Paris.
Domke, W. (1952): Der erste sichere Fund eines Myxomyceten im baltischen Bernstein (*Stemonitis splendens* Rost. fa. succini fa. nov. foss.). – Mitt. Geol. Staatsinstitut in Hamburg, 21: 154-161.
Doutt, R. L. (1973): The fossil Mymaridae (Hymenoptera: Chalcidoidea). – The Pan-Pacific Entomologist 49: 221-228.
Dowens, J. A., Wirth, W. W. (1981): Ceratopogonidae. – In: McAlpine, J. F. et al. (Eds.): Manual of Nearctic Diptera 1: 393-421, Ottawa/Ontario.
Duisburg, H. von (1860): Urweltlicher Blütenstaub. – Neue Preußische Provinzial-Blätter, 3. Folge, 5: 294-298; Königsberg.
– (1862): Beitrag zur Bernsteinfauna. – Schr. Phys. Ökon. Ges. Königsberg 3: 31-36.
Dumpert, K. (1978): Das Sozialleben der Ameisen. – Parey Verlag, Berlin, Hamburg.
Dunger, W. (1974): Tiere im Boden – Die Neue Brehm Bücherei 327, Wittenberg.
Dybas, H. S. (1961): A new fossil feather- wing beetle from baltic amber (Coleoptera, Ptiliidae). – Fieldiana, (Zool.) Chicago 44: 1-9.

Eckstein, K. (1890): Tierische Haareinschlüsse im baltischen Bernstein. – Schrift. Naturf. Ges. Danzig, N.F. 7: 90-93.
Edwards, F. W. (1921): A note on the Dipterous subfamily Ditomyinae, with descriptions of new recent and fossil forms. – Ann. Mag. Nat. Hist., Ser. 9, 7: 431-437; London.
– (1924): Notes on the types of Diptera Nematocera (Mycetophilidae and Tipulidae) described by Mr. E. Brunetti. – Records of Indian Museum, 26: 291-307, Calcutta.
– (1925): British Fungus – Gnats (Diptera, Mycetophilidae). With a revised generic classification of the family. – Trans. Ent. Soc. London 1924: 505-670; London.
– (1928): Diptera. Fam. Protorhyphidae, Anisopodidae, Pachyneuridae, Trichoceridae. – In: Genera Insectorum (Hrsg.: P. Wytsman), 190: 1-41.
– (1932): Diptera. Fam. Culicidae. – In: Genera Insectorum (Hrsg.: P. Wytsman), 194: 1-258.
– (1940): Redefinitions and synonymy of some genera of amber fungus-gnats. (Diptera, Mycetophilidae). – Proc Roy entomol Soc London, Ser. B, 9: 120-126.
Ehrensberg, C. G. (1848): Fossil Infusoria in amber. – Ann. Mag. Nat. Hist., 2: 1-397; London.
Ehrmann, R. (1999): Gottesanbeterinnen in Kopal und Bernstein (Insecta: Mantodea). – Arthropoda 7: 2-8. Wernigerode.
Eisenbeis, G., Wichard, W. (1987): Atlas on the Biology of Soil Arthropods. – Springer Verlag, Berlin, Heidelberg, New York.
Enderlein, G. (1900): *Epipsocus ciliatus* Hagen, eine Psocide des Bernsteins und die recente peruanische *Epipsocus nepos* nov. sp. – Berliner ent. Zeitschrift, 45: 108-112; Berlin.
– (1905): Zwei neue beschuppte Copeognathen aus dem Bernstein. – Zool. Anz., 29: 576-580; Leipzig.
– (1906): The Scale winged Copeognatha (Monograph of the Amphientomidae, Lepidopsocidae and Lepidillidae in Relation to their Morphology and Taxonomy). – Spolia Zeylanica 4: 39-122; Colombo.
– (1907): Außereuropäische Copeognathen aus dem Stettiner Museum. – Zool. Jb. Syst. 24: 81-90.
– (1911a): Die fossilen Copeognathen und ihre Phylogenie. – Palaeontogr. 58: 279-360.
– (1911b): Die phyletischen Beziehungen der Lycoriiden (Sciariden) zu den Fungivoriden (Mycetophiliden) und Itonididen (Cecidomyiiden) und ihre systematische Gliederung. – Archiv Naturgesch., 77. Suppl. 3: 116-201; Berlin.

– (1921): Ein fossiler Simuliiden-Riese. – Zool. Anz. 53: 74-75; Leipzig.
– (1930): Die Klassifikation der Coniopterygiden auf Grund der rezenten und fossilen Gattungen. – Arch. klassif. phylogen. Ent. I. 2: 98-114.
ENGEL, M. S. (1995): A New Fossil Snake-Fly Species From Baltic Amber (Raphidioptera: Inocelliidae). – Psyche 102: 187-193.
– (1999): The first fossil of a pleasing lacewing (Neuroptera: Dilaridae). – Proc. Entomol. Soc. Washington 101: 822-826.
– (2001a): A monograph of the Baltic amber bees and evolution of the Apoidea (Hymenoptera). – Bull. Amer. Mus. Nat. Hist. 259: 1-192.
– (2001b): Monophyly and extensive extinction of advanced eusocial bees: Insights from an unexpected Eocene diversity. – Proc. Natl. Acad. Sci., U.S.A. 98: 1661-1664.
– (2001 in press): A new dustywing in Turonian amber from New Jersey, with a reassessment of *Glaesoconis* in Neocomian amber from Lebanon (Neuroptera: Coniopterygidae). – J. Kansas Entomol. Soc.
ERICHSON, U. (1996): Ribnitz-Damgarten – Standort der Bernsteinverarbeitung in Mecklenburg-Vorpommern und des Bernsteinmuseums – in: GANZELEWSKI, M., SLOTTA, R. (Hrsg.): "Bernstein – Tränen der Götter": 147-155.
ERMISCH, K. (1941): Mordelliden und Scraptiiden aus baltischem Bernstein. – Ent. Bl. 37: 177-185.
– (1943): Eine neue Mordellide und Scraptiide aus baltischem Bernstein. – Arb. morphol. taxon. Ent. 10: 64-68.
ERWIN, L. (1971): Fossil tachyine beetles from Mexican and baltic amber. With notes on a new synonymy of an extant group (Col. Carabidae). – Ent. Scand. 2: 233-236.
ESCHERICH, K. (1917): Die Ameise. – 2. Aufl., Vieweg, Braunschweig.
ESTES, R. (1983): Sauria terrestria, Amphisbaenia. – Handb. Paläoherpet. 10A, 1-249. G. Fischer Verlag, Stuttgart.
EVANS, G. O. (1992): Principles of Acarology. – CAB International, Cambridge.
EVANS, H. E. (1963): A new family of wasps. – Psyche 70: 7-16.
EVENHUIS, N. L. (1994): Catalogue of the fossil flies of the world (Insecta: Diptera). – Backhuys Publisher, Leiden.
EZHOVA, E. E., KOSTYASHOVA, Z. V. (1997): Specime of fossil oribatids (Acarina, Oribatida) from Baltic amber in the museums of Kaliningrad (Russia) – Metalla (Sonderheft) 66: 45-50.

FERRIS, G. F. (1941): Contributions to the knowledge of the Coccoidea (Homoptera) IX. A forgotten genus of the family Margarodidae from Baltic amber. – Microentomology, 6, Contr. 24: 6-10; Stanford/California.
FOELIX, R. F (1979): Biologie der Spinnen. – Thieme, Stuttgart.
FRAHM, J.-P. (1994): Die Identität von *Muscites hauchecornei* CASPARY & KLEBS (Musci) aus Baltischem Bernstein. – Nova Hedwigia 58: 239-243.
– (1996a): Laubmoose aus Baltischem Bernstein. – Palaeontographica, B, 241: 127-135.
– (1996b): Mosses newly recorded from Saxonian amber. – Nova Hedwigia 63: 525-527.
– (1999): Die Laubmoosflora des Baltischen und Bitterfelder Bernsteins. – Mitt. Geol.-Paläont. Inst. Univ. Hamburg 83: 219-238.
– (2000a): New and Interesting Records of mosses from Baltic and Saxonian Amber. – Lindbergia 25: 33-39.
– (2000b): Neue Laubmoosfunde aus Baltischem Bernstein. – Cryptogamie 21: 121-132.
– (2001a): Neue Laubmossfunde aus Sächsischem und Baltischem Bernstein. – Nova Hedwigia 72: 171-281.
– (2001b): *Hypnodontopsis confertus* comb. nov. from Baltic amber. – Tropical Bryology 20: 79-82.
FRAKES, L. A. (1979): Climates through Geologic time. – Elsevier, 310 S., Amsterdam.
FRANZ, H. (1976): Scydmaeniden (Coleoptera) aus dem baltischen Bernstein. – Verh. zool.-bot. Ges. Wien 115: 80-85.
– (1983): Fossile Scydmaenidae in Baltischem und Dominikanischem Bernstein (Coleoptera: Scydmaenidae). – Neue Ent. Nachr. 7: 25-29.
FUHRMANN, R., BORSDORF, R. (1986): Die Bernsteinarten des Untermiozäns von Sachsen. – Z. angew. Geol. 32: 309-316.

GANZELEWSKI, M. (1996): Aufbereitung und Verarbeitung von Bernstein im Samland bis 1945 – in: GANZELEWSKI, M., SLOTTA, R. (Hrsg.): Bernstein – Tränen der Götter: 215-236.
– (1996): Aussehen und Eigenschaften von "Bernstein" – in: GANZELEWSKI, M., SLOTTA, R. (Hrsg.): Bernstein – Tränen der Götter: 19-26.
– (1996): Entstehung und Lagerstätte des Baltischen Bernsteins – in: GANZELEWSKI, M., SLOTTA, R. (Hrsg.): Bernstein – Tränen der Götter: 11-18.
GANZELEWSKI, M., SLOTTA, R. (Hrsg.) (1996): Bernstein – Tränen der Götter – Deutsches Bergbau-Museum, Bochum 1996.
GARTY, J., GIELE, CH., KRUMBEIN, W. E. (1982): On the occurrence of Pyrite in a lichenlike inclusion in Eocene amber (Baltic). – Palaeogeogr., Palaeoclimat., Palaeoecology, 39: 139-147; Amsterdam.
GEBICKI, C., SZWEDO, J. (2000): *Kulickamia jantaris* gen. et sp.n. from Baltic amber (Hemiptera: Fulgoroidea: Cixiidae). – Polskie Pismo Entomol. 69: 167-173.
GERMAR, E. F. (1813): Insekten in Bernstein eingeschlossen, beschreiben aus dem academischen Mineralien-Cabinet zu Halle. – Mag. Entomologie (Hrsg.: GERMAR, E. F.) 1: 11-18.
GERMAR, E. F., BERENDT, G. C. (1856): Die im Bernstein befindlichen Hemipteren und Orthopteren der Vorwelt. – in: BERENDT, G. C. (1845-1856): Die im Bernstein befindlichen organischen Reste der Vorwelt. 2. Bd. I. Abth., Berlin.
GHIURCA, V (1996): Vergleich zwischen den fossilen Harzen Rumäniens und der Ostsee – in: GANZELEWSKI, M., SLOTTA, R. (Hrsg.): Bernstein – Tränen der Götter: 363-368.
GIEBEL, C. G. (1846): Paläozoologie. Entwurf einer systematischen Darstellung der Fauna der Vorwelt. – 359 S.; Merseburg (Garcke).

– (1856): Fauna der Vorwelt mit steter Berücksichtigung der lebenden Thiere monographisch dargestellt. 2. Gliederthiere 1. Abt. Insecten und Spinnen. – Leipzig 411 S.
– (1863): Wirbelthier und Insektenreste im Bernstein. – Zeitschrift ges. Naturwiss., 20: 311-321; Berlin, Halle.
GOECKE, H. (1943): Monographie der Schilfkäfer. II. Die fossilen Funde und ihre Bestimmung. – Nova Acta Leopoldina, N. F. 12: 339-380; Halle.
GOEPPERT, H. R. (1836): Fossile Pflanzenreste des Eisensandes von Aachen. – N. Acta Acad. C.L.C. Nat. Cur. 19.
– (1850): Monographie der fossilen Coniferen. – Leiden.
– (1852): Über die Braunkohlenflora des nordöstlichen Deutschland. – Z. Dt. Geol. Ges. 4: 484-496.
– (1853): Über die Bernsteinflora. – Monatsber. Königl. Preuß. Akad. Wiss. Berlin 1853: 450-477.
– (1883): Von den Bernstein-Coniferen, insbesondere auch in ihren beziehungen zu den Coniferen der gegenwart. – in: (Hrsg.) GOEPPERT, H. R., MENGE, A. (1883-86): Die Flora des Bernsteins und ihre Beziehungen zur Flora der Tertiärformation und der Gegenwart. – Bd. 1 (1883): VIII, 63 S.
GOEPPERT, H. R., BERENDT, G. C. (1845): Der Bernstein und die in ihm befindlichen Pflanzenreste der Vorwelt. – in: BERENDT, G. C. (1845-1956): Die im Bernstein befindlichen organischen Reste der Vorwelt. 1. Bd, I. Abth., Berlin.
GOEPPERT, H. R., MENGE, A. (1883-86): Die Flora des Bernsteins und ihre Beziehungen zur Flora der Tertiärformation und der Gegenwart. – Bd. 1 (1883): VIII , 63 S., Bd. 2 (1886): IX, 140 S.; Danzig (Engelmann).
GOLUB, V. B. (2001): *Archepopovia yurii* n. gen. n. sp. a new remarkable lace bug from Baltic amber, with some notes on phylogeny and classification of Tingidae. – Mitt. Geol.-Paläont. Inst. Univ. Hamburg, 85: 262-276.
GOLUB, V. B., POPOV, Y. A. (1998): Cantacaderid lace bugs from the Baltic Amber (Heteroptera: Tingidae, Cantacaderinae). – Mitt. Geol.-Paläolont. Inst. Univ. Hamburg, 81: 223-250.
GÖSSWALD, K. (1985): Organisation und Leben der Ameisen. – Wiss. Verlagsges., Stuttgart.
– (1989-1990): Die Waldameise. – 2 Vol. Aula-Verlag, Wiesbaden.
GOULET, H., HUBER, J. T. (1993): Hymenoptera of the World. Identification Guide to the Families. – Agriculture Canada Publ., Ottawa.
GRATSHEV, V. G., ZHERIKHIN, V. V. (1995): A new anthribid genus from the Baltic amber (Insecta: Coleoptera, Anthribidae). – Mitt. Geol.-Paläolont. Inst. Univ. Hamburg, 78: 149-157.
GRIMALDI, D. (2000): A diverse fauna of Neuropterodea in amber from the Cretaceous of New Jersey – In: GRIMALDI, D. (ed.): Studies on Fossils in Amber, with Particular Reference to the Cretaceous of New Jersey: 259-303; Backhuys Publishing House; Leiden.
GRIMALDI, D. A. (1989): The genus *Metopina* (Diptera: Phoridae) from cretaceous and tertiary ambers. – J. New York Entomol. Soc. 97: 65-72.
– (1991): Mycetobiine Woodgnats (Diptera: Anisopodidae) from the Oligo-Miocene Amber of the Dominican Republic, and Old Word Affinities. – Novitates 3014: 1-24, New York.
– (1996): Amber – Window to the Past. – Am. Museum of Natural History, New York.
GRIMALDI, D. A., BONWICH, B., DELANNOY, M., DOBERSTEIN, S. (1994): Electron microscopic studies of mummified tissues in amber fossils. – Novitates 3097: 1-31, New York.
GRIMALDI, D., CUMMING, J. (1999): Brachyceran Diptera in Cretaceous ambers and Mesozoic diversification of the Eremoneura. – Bull. Amer. Mus. Nat. Hist. 239: 1-124.
GROLLE, R. (1980a): Lebermoose im Bernstein. – 1. Feddes Repertorium, 91: 183-190; Berlin.
– (1980b): Lebermoose im Bernstein. – 2. Feddes Repertorium, 91: 401-407; Berlin.
– (1981a): *Nipponolejeunea* fossil in Europa. – Journal. Hattori Botanical Laboratory, 50: 143-157; Nichinan/Japan.
– (1981b): Was ist *Lejeunea schumannii* CASPARY aus dem baltischen Bernstein? – Occasional Papers Farlow Herbarium Cryptogamic Botany, 16: 101-110; Cambridge/Mass.
– (1982): *Trocholejeunea* fossil in Europa. – Journal. Hattori Botanical Laboratory, 51: 175-193, Nichinan/Japan.
– (1984): Die Lebermoosgattung *Cheilolejeunea* fossil in Mitteleuropa. – Feddes Repertorium, 95: 229-236; Berlin.
– (1985): Fossil *Spruceanthus* in Europe and two other Hepatics in Baltic amber. – Prace Muzeum Ziemi, 37: 79-85; Warszawa.
– (1985a): Monograph of Frullania in Baltic amber. – Prace Muzeum Ziemi, 37: 87-100; Warszawa.
– (1985b): Zwei weitere Lebermoose in Bernstein aus Bitterfeld (DDR). – Feddes Repertorium 96: 41-46; Berlin.
– (1988a): Bryophyte fossils in amber. – The Bryological Times 47: 4-5; Utrecht.
– (1988b): Die Lebermoosgattung *Notoscyphus* fossil in Europa. – Feddes Repertorium 99: 561-564; Berlin.
– (1989): Weitere Lebermoosfunde in Bitterfelder Bernstein, darunter *Radula oblongifolia* CASPARY mit Sporophyt. – Feddes Repertorium 100: 131-136.

HAGEDORN, M. (1906): Borkenkäfer des baltischen Bernsteins. – Schr. Phys.-ökon. Ges. Königsberg 47: 115-121.
– (1907): Fossile Borkenkäfer. – Dtsch. ent. Z. 1907: 259-261; Berlin.
HAGEN, H. (1848): Die fossilen Libellen Europas. – Stettin. ent. Z. 9: 6-13.
– (1854): Über die Neuropteren der Bernsteinfauna. – Verh. zool-bot. Vereins Wien, 4: 221-232.
– (1856): siehe PICTET-BARABAN, F. J., HAGEN, H. (1856).
– (1858): Monographie der Termiten. – Linnaea Entomologica 12: 1-342, Berlin.
HANDLIRSCH, A. (1906-1908): Die fossilen Insekten und die Phylogenie der rezenten Formen. – 1430 S., Leipzig.
– (1925): Palaeontologie. – In: SCHRÖDER, C. (Hrsg.): Handbuch der Entomologie, 3: 117-306.
HANDSCHIN, E. (1926a): Die Collembolen des baltischen Bernsteins. – Zool. Anz. 65: 179-182.

– (1926b): Revision der Collembolen des baltischen Bernsteins. – Ent. Mitt. 15: 161-185, 211-223, 330-342.
– (1926): Über Bernsteincollembolen. Ein Beitrag zur ökologischen Tiergeographie. – Rev. Suisse Zool. 33: 375-378.
HARDT, H. (1955): Der Bernstein. (2. Aufl.). – Die Neue Brehm-Büchcrei 128, Wittenberg.
HAUPT, H. (1944): Zur Kenntnis der Dryinidae III. (Hymenoptera-Sphecoidea). Über eine Bernstein-Dryinide. – Stettiner ent. Z. 105: 90-94.
– (1950): Die Käfer (Coleoptera) aus der eozänen Braunkohle des Geiseltales. – Geologica 6: 1-168.
HEER, O. (1849): Die Insektenfauna des Tertiärgebilde von Oenningen und von Radoboj in Croatien. Zweite Abtl.: Heuschrecken, Florfliegen, Aderflügler, Schmetterlinge und Fliegen. – N. Denkschr. Allg. Schweiz. Ges. ges. Naturwiss. 11 (1850): 264 pp + 17 plates, Zürich.
– (1865): Die Urwelt der Schweiz. – Zürich.
– (1869): Miocene baltische Flora. – Beitr. Naturk. Preußens 2: 104 S.; Königsberg.
HEIE, O. E. (1967): Studies on fossil aphids (Homoptera: Aphidoidea). – Spolis Zool. Musei Hauniensis 26: 1-273, Kopenhagen.
– (1968): An aphid identified with *Aphis transparens* GERMAR et BERENDT, 1856 and some other Baltic amber aphids in German collections. – Stuttg. Beitr. Naturk. 184: 1-7.
– (1969): The Baltic amber Aphidoidea of the Geologisches Staatsinstitut of Hamburg. – Mitt. Geol. Pal. Inst. Univ. Hamburg 38: 143-151.
– (1970): Notes on six little know tertiary aphids (Hem. Aphidoidea). – Ent. Scand. 1: 109-119.
– (1971): The rediscovered types of the fossil aphids described by GERMAR and BERENDT in 1856. – Dtsch. ent. Z., N. 18: 251-264.
– (1972): Some new fossil aphids from Baltic amber in the Copenhagen collection (Insecta, Homoptera, Aphididae). – Steenstrupia 2: 247-262.
– (1976): Taxonomy and phylogeny of the fossil family Elektraphididae STEFFAN, 1968 (Homoptera: Aphidoidea). – Ent. Scand. 7: 53-58.
– (1981): Morphology and Phylogeny of some Mesozoic Aphids (Insecta, Hemiptera). – Entmologica scand. Suppl. 15: 401-415.
– (1985): Fossil aphids. – In: Evolution and biosystematics of aphids. Proc. Int. Aphidological Symp. 1981: 101-134; Wroclaw, Warszawa.
HEISS, E. (1997): Erstnachweis einer Aneurinae aus dem Baltischen Bernstein: *Aneurus ancestralis* n. sp. (Heteroptera, Aradidae). – Carolinea, 55: 111-113.
– (1998): Revision der Familie Aradidae des Baltischen Bernsteins. I. Bisher beschriebene Taxa der Gattung *Aradus* und zwei neue Arten (Insecta, Heteroptera). – Mitt. Geol. Paläont. Inst. Univ. Hamburg. 81: 252-268.
– (2000): Revision der Familie Aradidae des Baltischen Bernsteins II. Drei neue Calisius – Arten (Insecta, Heteroptera). – Carolinea, 58:195-201.
HELM, O. (1877): Notizen über die chemische und physikalische Beschaffenheit des Bernsteins. – Archiv Pharmazie 211: 229-246.
– (1884): Mittheilungen über Bernstein. IX. Über die Holzreste im Bernstein und unter Bernstein. – Schr. Naturforsch. Ges. Danzig, 6: 127-132.
– (1885): Über die Herkunft des in alten Königs-Gräbern von Mykenae gefundenen Bernsteins und über den Bernsteinsäuregehalt verschiedener fossiler Harze. – Schrft. Naturforsch.Ges.Danzig, N.F.6: 234-239.
– (1886): Über die Insekten des Bernsteins. – Schr. Naturforsch. Ges. Danzig, 6: 267-277.
– (1896a): Beiträge zur Kenntnis der Insecten des Bernsteins. – Schr. naturforsch. Ges. Danzig, 9: 220-231.
– (1896b): Thierische Einschlüsse im Succinit. – Schr. Naturforsch. Ges. Danzig, 9: 88-89; Danzig.
– (1899): Bemerkenswerthe Käfereinschlüsse in Succinit. – Schr. Naturforsch. Ges. Danzig, 10: 37-38.
HENNIG, W. (1938): Die Gattung *Rhachicerus* WALKER und ihre Verwandten im Baltischen Bernstein (Dipt.). – Zool. Anz. 123: 33-41.
– (1939): Über einen Floh aus der Bernsteinsammlung des Herrn Scheele. – Arbeiten morphol. taxonom. Entomologie 6: 330-332, Berlin.
– (1941): Beiträge zur Kenntnis des Kopulationsapparates und der Systematik der Acalyptraten. – Arbeiten morph. taxon. Entomologie 8: 54-65; Berlin.
– (1950): Grundzüge einer Theorie der phylogenetischen Systematik. – Deutscher Zentralverlag, Berlin.
– (1954): Flügeläder und System der Dipteren. – Beitr. Ent. 4: 245-388.
– (1960): Die Dipterenfauna von Neuseeland als systematisches und tiergeographisches Problem. – Beitr. Ent. 10: 221-329.
– (1964): Die Dipteren – Familie Sciadoceridae im Baltischen Bernstein. – Stuttgart. Beitr. Naturk. 127: 1-10.
– (1965): Die Acalyptratae des Baltischen Bernsteins und ihre Bedeutung für die Erforschung der phylogenetischen Entwicklung dieser Dipteren – Gruppe. – Stuttg. Beitr. Naturk. 145: 1-215.
– (1966a): Phylogenetic Systematics. – Univ. Illinois Press.
– (1966b): Bombyliidae im Kopal und im Baltischen Bernstein (Diptera: Brachycera). – Stuttg. Beitr. Naturk. 166: 1-18.
– (1966c): Conopidae im Baltischen Bernstein (Diptera: Cyclorrhapha). – Stuttg. Beitr. Naturk. 154: 1-24.
– (1966d): Dixidae aus dem Baltischen Bernstein, mit Bemerkungen über einige andere fossile Arten aus der Gruppe Culicoidea. (Diptera: Nematocera). – Stuttg. Beitr. Naturk. 153: 1-16.
– (1966e): Einige Bemerkungen über die Typen der von GIEBEL 1862 angeblich aus dem Bernstein beschriebenen Insektenarten. – Stuttg. Beitr. Naturk. 162: 1-7.
– (1966f): Spinnenparasiten der Familie Acroceridae im baltischen Bernstein. – Stuttg. Beitr. Naturk. 165: 1-21.
– (1967a): Die sogenannten "niederen Brachycera" im baltischen Bernstein (Diptera Fam. Xylophagidae, Xylomyidae, Rhagionidae, Tabanidae). – Stuttg. Beitr. Naturk. 174: 1-51.
– (1967b): Neue Acalyptratae aus dem baltischen Bernstein (Diptera: Cyclorrhapha). – Stuttg. Beitr. Naturk. 175: 1-27.

– (1967c): Therevidae aus dem baltischen Bernstein mit einigen Bemerkungen über Asilidae und Bombyliidae (Diptera: Brachycera). – Stuttg. Beitr. Naturk. 176: 1-14.
– (1968): Ein weiterer Vertreter der Familie Acroceridae im baltischen Bernstein (Diptera: Brachycera). – Stuttg. Beitr. Naturk. 185: 1-6.
– (1969a): Die Stammesgeschichte der Insekten. – Frankfurt. a. M. (Kramer), 436 S.
– (1969b): Neue Übersicht über die aus dem baltischen Bernstein bekannten Acalyptratae (Diptera: Cyclorrhapha). – Stuttg. Beitr. Naturk. 209: 1-42.
– (1970): Insektenfossilien aus der unteren Kreide – II. Empididae (Diptera, Brachycera). – Stuttg. Beitr. Naturk. 214: 1-12.
– (1971): Die Familien Pseudomyzidae und Milichiidae im baltischen Bernstein (Diptera: Cyclorrhapha). – Stuttg. Beitr. Naturk. 233: 1-16.
– (1972): Insektenfossilien aus der unteren Kreide – IV. Psychodidae (Phlebotominae), mit einer kritischen Übersicht über das phylogenetische System der Familie und die bisher beschriebenen Fossilien (Diptera). – Stuttg. Beitr. Naturk. 241: 1-69.
– (1973): Diptera (Zweiflügler). – Handbuch der Zoologie, 4 (2) 31: 1-337; Berlin.
HENWOOD, A. A. (1992a): Exceptional preservation of dipteran flight muscle and the taphonomy of insects in amber. – Palaios: 203-212.
– (1992b): Soft-part preservation of beetles in Tertiary amber from the Dominican Republic. – Palaeontology, 35: 901-912.
HERCZEK, A. (1991a): *Amberofulvius dentatus*, a new genus and species of the subfamily Cylapinae (Heteroptera, Miridae). – Ann. Naturhist. Mus. Wien 92: 79-84.
– (1991b): *Mixocapsus eocenicus* gen.n., sp.n. (Miridae, Heteroptera) from Baltic amber. – Ann. Naturhist. Mus. Wien 92: 85-88.
– (2000): First Phylinae (Heteroptera: Miridae) from Baltic amber. – Polskie Pismo Entomol. 69: 143-153.
HERCZEK, A., GORCZYCA, J. (1991): A representative of the genus *Deraeocoris* in Baltic amber (Heteroptera, Miridae). – Ann. Naturhist. Mus. Wien A92: 89-92.
HERCZEK, A., POPOV, Y. A. (1992): A remarkable psallopinous bug from Baltic Amber (Insecta: Heteroptera, Miridae). – Mitt. Geol.-Paläolont. Inst. Univ. Hamburg, 73: 235-239.
– (1997a): New peculiar representatives of the Isometopinae from the Baltic amber (Heteroptera: Miridae). – Mitt. Geol.-Paläolont. Inst. Univ. Hamburg, 80: 189-195.
– (1997b): On the mirid genera *Archeofulvius* CARVALHO and *Balticofulvius* n.gen. from the Baltic amber (Heteroptere: Miridae, Cylapinae).- Mitt. Geol.-Paläolont. Inst. Univ. Hamburg, 80: 179-187.
– (1999): On plant bugs from the Baltic amber (Heteroptera: Miridae, Cylapinae) based on the old type species. – Mitt. Geol.-Paläont. Inst. Univ. Hamburg 83: 249-262.
– (2000): *Ambocylapus kulickae* gen. et. sp.n., a new plant bug from Baltic amber (Heteroptera: Miridae: Cylapinae). – Polskie Pismo Entomol. 69: 155-160.
HEVERS, J., LISKE, E. (1991): Lauernde Gefahr – Das Leben der Gottesanbeterinnen. – Staatl. Naturhist. Museum Braunschweig, 68 S.
HIEKE, F., PIETREZENIUK, E. (1984): Die Bernstein-Käfer des Museums für Naturkunde, Berlin (Insecta, Coleoptera). – Mitt. Zool. Mus., Berlin 60: 297-326.
HINTON, H. E. (1966): The spiracular gill of the fly *Eutanyderus* (Tanyderidae). – Aust. J. Zool. 14: 365-369.
HOFFEINS, Ch., HOFFEINS, H. W., WAGNER, R. (1997): Beschreibung einer Art der Gattung *Heleodromia* (Diptera, Empididae, Hemerodromiinae) aus dem Bitterfelder Bernstein. – Studia dipterologica 4: 441-446.
HOLL, F. (1829): Handbuch der Petrefactenkunde. – Hilscher, Dresden.
HÖLLDOBLER, B., WILSON, E. O. (1990): The Ants. – Springer Heidelberg, Berlin, New York.
– (1995): Ameisen – Die Entdeckung einer faszinierenden Welt. – Birkhäuser Verlag Basel.
HÖRNSCHEMEYER, T. (1998): New species of Electribius Crowson 1973 (Coleoptera: Artematopodidae) from Baltic amber. – Paläontologische Zeitschrift 72: 299-306.
HOPE, F. W. (1834): Observations on succinatic insects. – Trans. Ent. Soc. London 1: 133-147.
HORN, W. (1906): Über das Vorkommen von *Tetracha carolina* L. im preußischen Bernstein und die Phylogenie der *Cicindela*-Arten. – Dtsch ent. Z., 1906: 329-336; Berlin.
– (1907): Brullés "Odontochila aus dem baltischen Bernstein" und die Phylogenie der Cicindeliden (Col.). – Dtsch. ent. Z., 1907: 461-466; Berlin.
HUBBARD, M. D. (1987): Ephemeroptera. Fossilium Catalogus. I: Animalia 129: 1-99.
– (1990): Mayflies of the World – A Catalog of the Family and Genus Group Taxa (Insecta: Ephemeroptera). – Flora & Fauna Handbook 8: 1-119, Gainesville.
HULL, F. M. (1945): A revisional study of the fossil Syrphidae. – Bull. Mus. comp. Zool. Harvard 95: 215-355.
– (1949): The morphology and inter-relationship of the genera of Syrphid flies, recent and fossil. – Trans. Zool. Soc. London, 26 (1947-1950), 4: 257-408; London.
– (1958): Tertiary flies from Colorado and the Baltic amber. – Psyche 64: 37-45.
HUMMEL, D. (1958): Kunststoff-, Lack- und Gummianalyse. Chemische und infrarotspektroskopische Methoden, II. – München.
HUNGER, R. (1977): The magic of amber. – 131 S.; London (NAG Press).
HUSTACHE, A. (1942): Un Curculionide de l'ambre de la Baltique. – Bull. Mens. Soc. Linné. Lyon 11: 108-109.

ILLIES, J. (1965): Phylogeny and Zoogeography of the Plecoptera. – Ann. Rev. Entomology, 10: 117-140; Palo Alto.
– (1967): Die Gattung *Megaleuctra* (Plecopt., Ins.). Beitrag zur konsequent-phylogenetischen Behandlung eines incertae-sedis-Problems. – Z. Morph. Ökol. Tiere 60: 124-134, Berlin.
– (1968): Ephemeroptera (Eintagsfliegen). – In: Handbuch der Zoologie, 4 (2) 2: 1-63; Berlin.

JABLOKOV-CHNZORJAN, S. M. (1960): Novye zestkokrylye iz baltijskogo jantarja. – Paleonto-logiceskij Zurnal Akademii Nauk SSSR, 1960: 90-101; Moskva.
– (1961): Novye zestkokrylye semejstva Elateridae iz baltijskogo jantarja. – Paleontologiceskij Zurnal Akademii Nauk SSSR, 1961: 84-97; Moskva.
– (1961): Predstaviteli semejstva Helodidae (Coleoptera) iz baltijskogo jantarja. – Paleontologiceskij Zurnal Akademii Nauk SSSR, 1961: 108-116; Moskva.
– (1962): Predstaviteli Sternoxia (Coleoptera) iz baltijskogo jantarja. Semejstva Throscidae BACH, 1849 (Trixagidae sensu Crowson). – Paleontologiceskij Zurnal Akademii Nauk SSSR, 1962: 81-89, Moskva.
JACOBI, A. (1938): Eine neue Bernsteinzikade (Homopt.). – Sber. Ges. naturf. Fr. Berlin 15: 188-189.
JACOBS, W., RENNER, M. (1988): Biologie und Ökologie der Insekten. – G. Fischer, Stuttgart.
JACOT-GUILLARMOD, C. F. (1970): Catalogue of the Thysanoptera of the World – Ann. Cape prov. Mus.
JOHANSON, K. A., WICHARD, W. (1997): Caddis flies of Baltic Amber. 4. New descriptions of *Palaeohelicopsyche* (Trichoptera, Helicopsychidae). – Mitt. Münch. Ent. Ges. 87: 101-108.
JORDAN, K. H. C. (1944): Eine neue Miride in Bernstein, *Electrocoris fuscus* n. gen. et n. sp. (Hemiptera Heteroptera: Miridae). – Arbeiten morph. Taxon. Entomologie, 11: 132-134, Berlin.
– (1944): *Oligocoris bidentata* n. gen. et n. sp., eine Miride aus dem ostpreußischen Bernstein (Hemiptera – Heteroptera). – Arbeiten morph. taxon. Entomologie, 11: 8-10, Berlin.
– (1952): *Nabis succini* n. sp. eine Nabide aus dem Bernstein Ostpreußens (Hemiptera Heteroptera). – Beitr. Ent. 2: 455-457.
– (1953): Eine weitere fossile Notonectide von Rott im Siebengebirge. – Zool. Anz. 150: 245-249. Leipzig.
– (1972): Heteroptera (Wanzen). – Handbuch der Zoologie, 4 (2) 20: 1-113; Berlin.
JUST, J. (1974): On *Palaeogammarus* ZADDACH, 1864, with a description of a new species from western Baltic amber. – Steenstrupia 3, 93-99, Copenhagen.

KAISER, P. (1890): Die fossilen Laubhölzer. I. Nachweise und Belege. – Wissenschaftl. Beilage Jb. Realprogymnasiums zu Schönebeck: 1-46.
KALTENBACH, A. (1968): Embiodea (Spinnfüsser). – In: Handbuch der Zoologie, 4 (2) 28: 1-29; Berlin.
KAPLAN, A. A., GRIGIALIS, A. A., STRELNIKOVA, N. I., GLIKMAN, L. S. (1977): Stratigraphy and correlation. – Sovietskaja Geologija, 4: 30-43, Moskau.
KARSCH, F. (1884): Neue Milben im Bernstein. – Berliner Ent. Z. 28: 175-176.
KATHIRITHAMBY, J. (1989): Review of the order Strepsiptera. – System. Entomol.14: 41-92.
KATINAS, V. (1971): Amber and amber-bearing deposits of the southern Baltic area. – Trans. Lithuanian scient. research geol. Survey Inst. 20: 1-156.
– (1988): Fossil resin studies by chemical and physical methods. – VI. Meeeting Amber and Amber-bearing Sediments (MS).
KEFFERSTEIN, C. (1834): Die Naturgeschichte des Erdkörpers in ihren ersten Grundzügen dargestellt, (2. Teil). – Fleischer, Leipzig.
KEILBACH, R. (1939): Neue Funde des Strepsipterons *Mengea tertiaria* MENGE im baltischen Bernstein. – Bernstein-Forsch. 4: 1-7.
– (1982): Bibliographie und Liste der Arten tierischer Einschlüsse in fossilen Harzen sowie ihrer Aufbewahrungsorte. – Dtsch. Ent. Z. 29: 129-286, 301-491, Berlin.
KELNER-PILLAULT, S. (1969): Les Abeilles fossiles. – Mem. Soc. Entomol. Italiana 48: 519-534; Genova.
– (1970): L'ambre balte et sa faune entomologique avec déscription de deux Apoides nouveaux. – Ann. Soc. ent. France, 6: 3-24; Paris.
– (1970): Une Mélipone (s.l.) de l'ambre Balte (Hym. Apidae). – Ann. Soc. ent. France, 6: 437-441.
– (1974): Etat d'evolution des Apides de l'ambre Balte. – Ann. Soc. ent. France, 10: 623-634.
KINZELBACH, R. (1971): Morphologische Befunde an Fächerflüglern und ihre phylogenetische Bedeutung (Insecta: Strepsiptera). – Zoologica, 41, 119: 1-256; Stuttgart.
– (1971): Strepsiptera (Fächerflügler). – In: Handbuch der Zoologie, 4 (2) 24: 1-73; Berlin.
– (1983): Fächerflügler aus dem dominikanischen Bernstein (Insecta: Strepsiptera: Myrmecolacidae). – Verhandl. Naturwiss. Verein Hamburg, 26: 29-36; Hamburg.
– (1990): The systematic position of Strepsiptera (Insecta). – Am. Entomol. 35: 292-303.
KINZELBACH, R., LUTZ, H. (1985): A stylopid larva from the eocene – a spotlight on the phylogeny of the stylopids (Strepsiptera). – Ann. Entomol. Soc. America. 78: 600-602.
KINZELBACH, R., POHL, H. (1994): The fossil Strepsiptera (Insecta: Strepsiptera). – Ann. Entomol. Soc. America 87: 59-70.
KIRCHHEIMER, F. (1937): Beiträge zur Kenntnis der Flora des baltischen Bernsteins I. – Beihefte zum Botanischen Centralblatt, Abt. B, 57: 441-484; Kassel.
KLAUSNITZER, B. (1975): Probleme der Abgrenzung von Unterordnungen bei den Coleoptera. – Ent. Abh. Staatl. Mus. Tierk. Dresden 40: 269-275.
– (1976): Neue Arten der Gattung *Helodes* LATREILLE aus Bernstein (Coleoptera, Helodidae). – Reichenbachia 16, 53-61, Dresden.
– (1982): Wunderwelt der Käfer. – Herder, Freiburg.
– (1996): Käfer im und am Wasser. – Die Neue Brehm-Bücherei 567 (2. Aufl.), Magdeburg.
KLEBS, R. (1886a): Beitrag zur Kenntnis fossiler Conchylien Ostpreußens. – Malakozoologische Blätter, 8: 149-160; Kassel.
– (1886b): Gastropoden im Bernstein. – Jb. Königl.-preuß. Geol. Landesanstalt, 1885: 366-394; Berlin.
– (1889): Über die Fauna des Bernsteins. – Tageblatt Vers. Dtsch. Naturf. Ärzte, 62: 268-271.
– (1890): The fauna of amber. – Ann. Mag. Nat Hist., incl. Zool., Bot., Geol. 6: 486-491; London.

– (1910): Über Bernsteineinschlüsse im allgemeinen und die Coleopteren meiner Bernsteinsammlung. – Schr. phys. ökon. Ges. Königsberg 51: 217-242.

KLEINE, R. (1940): Eine Lycide aus dem baltischen Bernstein. – Ent. Blätter 36: 179-180.

KLEINOW, W. (1966): Untersuchungen zum Flügelmechanismus der Dermapteren. – Z. Morph. Ökol. Tiere 56: 363-416.

KLIMASZEWSKI, S. M. (1993): New species of Eocene psyllids representing the tribe Paleopsylloidini BECKER-MIGDISOVA (Homoptera, Aphalaridae). – Acta Biol. Silesiana. 22: 9-18.

– (1997): New psyllids from the Baltic amber (Insecta: Homoptera, Aphalaridae). – Mitt. Geol.-Paläolont. Inst. Univ. Hamburg, 80: 157-171.

KLYUGE, N. Y. (1986): A reccnt May fly species (Ephemeroptera, Heptageniidae) in Baltic amber. – Paleontological Journal, 20: 106-107; Washington.

KOCH, C. L., BERENDT, G. C. (1854): Die im Bernstein befindlichen Crustaceen, Myriapoden, Arachniden und Apteren der Vorwelt. – in: BERENDT, G. C. (1845-1856): Die im Bernstein befindlichen organischen Reste der Vorwelt. 1. Bd, 2. Abth., Berlin.

KOENEN, A. v. (1865): Die Fauna der unteroligozänen Tertiärschichten von Helmstedt bei Braunschweig. – Z. Dtsch. Geol. Ges., 17:459-534.

KOHRING, R. (1998): REM-Untersuchungen an harzkonservierten Arthropoden. – Entomol. Gener. 23: 95-106.

KOHRING, R., WEITSCHAT, W. (1994): Weichteil-Erhaltung an Bernstein-Insekten. – Abstract Jahrestagung Paläont. Ges., 64, Budapest.

KOLBE, N. J. (1883): Neue Beiträge zur Kenntnis der Psociden der Bernsteinfauna. – Stettin. ent. Z. 44: 186-191.

KOLIBAC, J., GERSTMEIER, R. (1997): Description of *Eurymetopum wachteli* sp.n. from the Baltic amber (Coleoptera, Cleridae, Hydnocerinae). – Mitt. Münch. Ent. Ges. 87: 97-100.

KOLLER, J., BAUMER, B., BAUMER, U. (1997): Die Untersuchung von Bernstein, Bernsteinölen und Bernsteinlacken – Metalla (Sonderheft) 66: 85-102.

KÖNIGSMANN, E. (1978): Das phylogenetische System der Hymenoptera. Teil 3: "Terebrantes" (Unterordnung Apocrita). – Dtsch ent. Z. 25: 1-55; Berlin.

– (1978): Das phylogenetische System der Hymenoptera. Teil 4: Aculeata (Unterordnung Apocrita). – Dtsch ent. Z. 25: 365-435; Berlin.

KONOW, F. W. (1897): Ueber fossile Blatt- und Halmwespen. – Ent. Nachr. Berlin 23: 36-38.

KORNILOWITCH, N. (1903): Hat sich die Struktur quergestreifter Muskeln bei Insekten, die wir im Bernstein antreffen, erhalten? – S. Naturf. Ges. Dorpat 13: 198-206.

KORSCHEFSKY, R. (1939): Abbildungen und Bemerkungen zu vier Schaufuß'schen Coleopteren aus dem deutschen Bernstein. – Arbeiten Morph. Taxon. Entomologie, 6: 11-12; Berlin.

KOSMOWSKA-CERANOWICZ, B. (1984): Plant Inclusions in Amber. – In: Amber in the Nature (Guide and Catalogue of the Exhibition): 24-26; Warsaw.

– (1987): Mineralogical-petrographic characteristics of the Eocene amber-bearing sediments in the area of Chlapowo and the Paleogene sediments of northern Poland. – Bull. Inst. Geol. 356: 29-50, Warszawa.

– (1991/92): Spuren des Bernsteins. – Katalog, Museum der Erde, Polnische Akademie der Wissenschaften, 102 pp.

– (1996): Bernstein – die Lagerstätte und ihre Entstehung – in: GANZELEWSKI, M., SLOTTA, R. (Hrsg.): Bernstein – Tränen der Götter: 161-168.

– (1996): Die tertiären und quartiären Bernsteinvorkommen in Polen – in: GANZELEWSKI, M., SLOTTA, R. (Hrsg.): Bernstein – Tränen der Götter: 299-310.

KOSMOWSKA-CERANOWICZ, B., KOHLMAN-ADAMSKA, A., GRABOWSKA, I. (1997): Erste Ergebnisse zur Lithologie und Palynologie der bernsteinführenden Sedimente im Tagebau Primorskoje – Metalla (Sonderheft) 66: 5-17.

KOSMOWSKA-CERANOWICZ, B., KRUMBIEGEL, G. (1989): Geologie und Geschichte des Bitterfelder Bernsteins und anderer fossiler Harze. – Hall. Jb. Geowiss. 14: 1-25.

KOSMOWSKA-CERANOWICZ, B., KULICKA, R. (1995): Amber molars. – Amber & Fossils 1: 38-41.

KOSMOWSKA-CERANOWICZ, B., KULICKA, R., GIERLOWSKA, G. (1997): Nowe znalezisko jaszczurki w bursztynie baltyckim. – Przegl. Geol. 45: 1028-1030, Warszawa.

KOSMOWSKA-CERANOWICZ, B., MIERZEJEWSKI, P. (1978): L'Ambiente naturele nel quale si formava l´ambra. – Ambra oro del Nord: 34-43, Venezia.

KOTEJA, J. (1984): The Baltic amber Matsucoccidae (Homoptera, Coccinea). – Annales zoologici 37: 437-496; Warszawa.

– (1985): Czerwce (Homoptera, Coccinea) bursztynu baltyckiego. – Wiadomosci entomologiczne, 6: 195-205; Warszawa.

– (1986): *Matsucoccus saxonicus* sp. n. from Saxonian amber. (Homoptera, Coccinea). – Dtsch ent. Z. 33: 55-63; Berlin.

– (1986): Stellata Hairs – index fossils of ambers. – Inclusion 5: 4-8; Krakow.

– (1986): Matsucoccidae (Homoptera, Coccinea), living fossils. – Boll. Lab. Ent. Filippo Silvestri, 43 (1986), Suppl.: 41-44; Portici.

– (1987a): Current state of coccid paleontology. – Boll. Lab. Ent. Filippo Silvestri, 43 (1986), Suppl. 41: 29-34; Portici.

– (1987b): *Palaeonewsteadia huaniae* gen. et sp.n. (Homoptera, Coccinea, Ortheziidae) from Baltic amber. – Polskie Pismo entomologiczne, 57: 235-240; Warszawa.

– (1987c): *Protorthezia aurea* gen. et sp.n. (Homoptera, Coccinea, Ortheziidae) from Baltic amber. – Polskie Pismo entomologiczne, 57: 241-249; Warszawa.

– (1988a): Eriococcid crawlers (Homoptera, Coccinea) from Baltic amber. – Bull. Ent. Pol. 58: 503-524, Wroclaw.

– (1988b): *Sucinikermes kulickae* gen. et sp. n. (Homoptera, Coccinea) from Baltic amber. – Bull. Ent. Pol. 58: 525-535, Wroclaw.

– (1998c): Essays on coccids (Homoptera): Sudden death in amber? – Polskie Pismo Entomol. 67: 185-218.

– (2000): Advances in the study of fossil coccids (Hemiptera: Coccinea). – Polskie Pismo Entomol. 69: 187-218.

KOTEJA, J., ZAK-OGAZA, B. (1988a): *Arctorthezia antiqua* sp.n. (Homoptera, Coccinea) from Baltic amber. – Ann. Zool. 41: 321-328; Warszawa.

KOTEJA, J., ZAK-OGAZA, B. (1988b): *Newsteadia succini* sp.n. (Homoptera, Coccinea) from Baltic amber. – Ann. Zool. 41: 329-334; Warszawa.

KOVALEV, O. V. (1994): Paleontological History, Phylogeny, and Systematics of Brachycleistogastromorpha, Infraorder N., and Cynipomorpha Infraorder N. (Hymenoptera) with Descriptions of New Fossil and Recent Families, Subfamilies, and Genera. – Ent. Obozreniye 73: 385-426.

– (1995): New taxa of fossil cynipoids (Hymenoptera, Cynipoidea) from the Createaceous and Palaeogene. – Amber & Fossils 1: 9-16.

KOZLOV, M. V. (1987): New moth like Lepidoptera from the Baltic amber. – Paleont. J., 21, 4: 56-65; New York.

KRANTZ, G. W. (1978): A Manual of Acarology. – 2 Ed., Oregon State University Bookstores, Corvallis.

KRAUS, O. (1995): System der Insekten. – in: GEWECKE, M.: Physiologie der Insekten, 10: 429-436, G. Fischer, Stuttgart

KRÄUSEL, R. (1919): Die fossilen Koniferenhölzer (unter Ausschluß von *Araucanoxylon* KRAUS). – Palaeontogr. 62 (1916-1919): 185-275; Stuttgart.

– (1949): Die fossilen Koniferen-Hölzer (unter Ausschluß von *Araucarioxylon* KRAUS). II. – Palaeontogr., 89: 83-203; Stuttgart.

KRISHNA, K. (1961): A generic revision and phylogenetic study of the family Kalotermitidae (Isoptera). – Bull. Amer. Mus. Nat. Hist. 122: 303-408.

KRIVOLUCKIJ, D. A., DRUK, A. Y. (1986): Fossil oribatid mites. – Ann. Rev. Entomology 31: 533-545; Palo Alto.

KROMBEIN, K. V. (1986): Three cuckoo wasps from Siberian and Baltic amber (Hymenoptera: Chrysisidae: Amiseginae and Elampinae). – Proc. Ent. Soc. Washington 88: 740-747.

KRUCKOW, T. (1962): Eine echte Bernstein-Eidechse. – Der Aufschluß, 13: 267-270; Göttingen.

KRÜGER, L. (1913): Osmylidae. Beiträge zu einer Monographie der Ncuropteren Familie der Osmyliden. II-IV. – Stett. Ent. Z., 74: 3-123, 193-224, 225-294.

– (1922):Hemerobiidae. Beiträge zu einer Monographie der Neuropteren. – Stett. Ent. Z. 83: 138-172.

– (1923): Neuroptera succinica baltica. Die im baltischen Bernstein eingeschlossenen Neuropteren des Westpreussischen Provinzial – Museums in Danzig. – Stett. Ent. Z. 84: 68-92.

– (1923): Sisyridae. Beiträge zu einer Monographie der Neuropteren-Familie der Sisyridae. Stett. Ent. Z. 84: 25-66.

KRUMBIEGEL, G., KOSMOWSKA-CERANOWICZ, B. (1989): Geologie und Geschichte des Bitterfelder Bernsteins und anderer fossiler Harze. – Hallesches Jb. Geowiss. 14: 1-25.

KRUMBIEGEL, G., KRUMBIEGEL, B. (1994): Bernstein – fossile Harze aus aller Welt. – Fossilien, Sonderband 7, 110, S., Korb.

– (1996): Bernstein im weiteren Sinne – Die Akzessorischen Harze – in: GANZELEWSKI, M., SLOTTA, R. (Hrsg.): Bernstein – Tränen der Götter: 27-30.

– (1996): Bernsteinlagerstätten und -vorkommen in aller Welt – in: GANZELEWSKI, M., SLOTTA, R. (Hrsg.): Bernstein – Tränen der Götter: 31-46.

– (1996): Pflanzliche und tierische Organismen im Bernstein – Biologische Indikatoren der Erdgeschichte – in: GANZELEWSKI, M., SLOTTA, R. (Hrsg.): Bernstein – Tränen der Götter:47-58.

– (2001): Faszination Bernstein. – Goldschneck-Verlag, Korb.

KRZEMINSKI, W. (1985a): A representative of Trichoceridae (Diptera Nematocera) from Baltic amber (in the collection of the Museum of the Earth in Warsaw). – Prace Muzeum Ziemi 37:119-121.

– (1985b): Limoniidae (Diptera Nematocera) from Baltic amber (in the collection of the Museum of the Earth in Warsaw) Part. I Subfamily Limoniinae. – Prace Muzeum Ziemi 37: 113-117.

– (1998a): *Limonia* (s. str.) *skalskii* sp.n. from Baltic amber (Diptera: Limoniidae). – Polskie Pismo Entomol. 67: 299-301.

– (1998b): Hexatoma (Eriocera) serafini sp.n. from Baltic amber (Diptera: Limoniidae). – Polskie Pismo Entomol. 67: 309-310.

– (2000a): Fossil Tipulomorpha from Baltic amber (Upper Eocene): a new species of Dactylolabis (Idiolabis) (Diptera: Limoniidae). – Polskie Pismo Entomol. 69: 247-250.

– (2000b): A new species and other representatives of the genus *Dicranomyia* (Diptera: Limoniidae) in the collection of the Museum of Amber Inclusions, University of Gdansk. – Polskie Pismo Entomol. 69: 347-353.

KRZEMINSKI, W., EVENHUIS, (2000): Review of the Diptera palaeontological records. – in: PAPP, L., DARVAS, B. (eds.): Contributions to a Manual of Palaearctic Diptera. Vol. 1 General and Applied Dipterology: 535-564, Science Herald, Budapest.

KRZEMINSKA, E., KRZEMINSKI, W. (1992): Les fantomes de l'ambre. – Musée d'Histoire naturelle, 142 S., Neuchâtel.

KUCHARSKA, M , KWIATKOWSKI A. (1978): Research methods for the chemical composition of amber and problems concerning the origin of amber. – Prace Muzeum Ziemi 29: 149-156 (in Polish).

KUKALOVA-PECK, J., LAWRENCE, J. F. (1993): Evolution of the hind wing in Coleoptera. – Canad. Entomol. 125: 181-258.

KULICKA, R. (1977): *Mengea tertiaria* w bursztynie baltyckim (w zbiorach Muzeum Ziemi Pan). – Przeglad geol. 25: 32-33; Warszawa.

– (1979): *Mengea mengei* sp. n. from the Baltic amber. – Prace Muzeum Ziemi, 32: 109-112; Warszawa.

– (1984): Zbior inkluzji zwierzceych w bursztynie Muzeum Ziemi Pan. – Przeglad zool., 28: 387-389 (97-99); Wroclaw.

– (1985): Inkluzje zwierzece w bursztynie baltyekim w zbiorach Muzeum Ziemi Pan w Warszawie. – Wiadomosci Entomologiczne, 6: 179-186; Warszawa.

– (1990): The list of animal inclusions in Baltic amber from collection of the Museum of the Earth in Warsaw. – Prace Muzeum Ziemi 41: 144-146; Wroclaw.

Kulicka, R., Herczek, A. Popov, Y. A. (1997): Heteropteran in the Baltic amber. – Prace Muzeum Ziemi 44: 19-22.

Kulicka, R., Krzeminski, W., Szadziewski, R. (1985): Kolekcja muchówek (Diptera Nematocera) w bursztynie baltyckim ze zbiorów Muzeum Ziemi PAN x Warszawie. – Prace Muzeum Ziemi 37: 105-111.

Kulicka, R., Slipinski, A. (1996): A review of the Coleoptera inclusions in the Baltic amber. – Prace Muzeum Ziemi 44: 5-8.

Kulicka, R., Wegierek, P. (1996): Aphid species (Homoptera: Aphidinea) from the collection of the Baltic amber in the Museum of the Earth, Polish Academy of Sciences, Warsaw (Part three). – Prace Museum Ziemi, 44: 41-44.

Kullmann, E., Stern, H. (1975): Leben am seidenen Faden. – Bertelsmann, Gütersloh.

Kunow, G. (1872): Zwei Schnecken im Bernstein. – Schr. phys.-ökon. Ges. Königsberg 13: 150-154.

Kutscher, M., Koteja, J. (2000): Coccids and aphids (Hemiptera: Coccinea, Aphidinea), prey of ants (Hymenoptera: Formicidae): evidence from Bitterfeld amber. – Polskie Pismo Entomol. 69: 179-185.

Lakowitz, N. (1882): Die Bernsteinbäume. – Forstlich- naturwiss. Z., 1: 244-248; München.

Langenheim, J. H. (1969): Amber: A botanical inquiry. – Science 163:1157-1169.

Langenheim, R. L., Beck, C. W. (1965): Infrared spectra as a means of determinig botanical source of amber. – Science 149: 52-55.

Lanterno, E. , Besuchet, C. (1966): L'ambre et ses insectes. – Rev. Mens. Mus., Coll. Ville de Geneve, 7: 8-10.

Larsson, S. G. (1962): The Copenhagen collection of amberfossils. – Entomologiske Meddelelser, 31: 323-326, Kobenhavn.

– (1965): Reflections on the Baltic amber inclusions. – Entomologiske Meddelelser, 34: 135-142; Kobenhavn.

– (1978): Baltic Amber – a Palaeobiological Study. – Entomonograph, 1: 192 S.; Klampenborg/Denmark.

Lawrence, J. F., Britton, E. B. (1991): Coleoptera (Beetles). – in: Csiro: The insects of Australia, Vol. 2, Cornell Univ. Press, Ithaca, New York.

Lawrence, P. N. (1985): Ten species of Collembola from Baltic amber. – Prace Muzeum Ziemi 37: 101-104; Warszawa.

Legg, G., Jones, R. E. (1988): Pseudoscorpiones. – Brill/ Backhuys, Leiden/Köln.

Lengerken, H. v. (1913): Etwas über den Erhaltungszustand von Insekteninklusen im Bernstein. – Zool. Anz. 41: 284-286.

Liedtke, W. (1975): Käfer des baltischen Bernsteins. – Diplomarbeit Univ. Kiel: 68 S.

Lindquist, E. E. (1984): Current theories on the evolution of mjar groups of Acari and on their relationships with other groups of Arachnida, with consequent implications for their classification. – In: D. A. Griffiths, C. E. Bowman (eds), Acarology VI, John Wiley, Sons, New York.

Loew, H. (1845): Dipterologische Beiträge, Teil 1. – Schulprogramm Kgl. Friedrich-Wilhelm-Gymnasium, Posen, 1845: 1-52.

– (1850): Über den Bernstein und die Bernsteinfauna. – Progr. Realschule Meseritz, Berlin, 3-44.

– (1861): Über die Dipterenfauna des Bernsteins. – Vers. Dtsch. Naturf. 1860, Amtl. Ber. 35: 88-98 Königsberg.

– (1864): On the Diptera or twowinged insects of amberfauna. – Amer. J. Sci. Arts, 2. Ser. 37: 305-324.

– (1873): Über die Arten der Gattung *Sphyracephala* Say. – Z. ges. Naturwiss. 42: 101-105.

Lourenco, W. R., Weitschat, W. (1996): More than 120 years after its description, the enigmatic status of the baltic amber scorpion *'Tityus eogenus'*, Menge, 1869 can finally be clarified. – Mitt. Geol.-Paläont. Inst. Univ. Hamburg 79: 183-193.

– (2000): New fossil scorpions from the Baltic amber – implications for Cenozoic biodiversity. – Mitt. Geol.-Paläont. Inst. Univ. Hamburg 84: 247-260.

Loveridge, A. (1942): Scientific results of a fourth expedition to forested areas in East, Central Africa. – Bull. Mus. Comp. Zool. 91: 237-373, Cambridge/Mass.

– (1957): Check list of the reptiles and amphibians of East Africa (Uganda; Kenya; Tanganyika; Zanzibar). – Bull. Mus. Comp. Zool. 117: 153-362, Cambridge/Mass.

Lucks, R. (1927): *Palaeogammarus balticus,* nov. sp., ein neuer Gammaride aus dem Bernstein. – Schr. Nat. Ges. Danzig 8, 1-13.

Lühe, M. (1904): Säugetierhaare im Bernstein. – Schr. phys.-ökon. Ges. Königsberg 45, 62-63.

MacKay, M. R. (1969): Microlepidopterous larvae in Baltic amber. – Canad. Entomol. 101: 1173-1180.

MacLeod, E. G. (1970): The Neuroptera of the Baltic amber. I. Ascalaphidae, Nymphidae, and Psychopsidae. – Psyche 77: 147-180.

Mägdefrau, K. (1957): Flechten und Moose im baltischen Bernstein. – Ber.. Dtsch. Botan. Ges., 70: 433-435; Stuttgart.

Maidanovitsch, I.A. & Makarenko, D.E.(1988): Geologie und Genese bernststein-führender Sedimente des Ukrainischen Polesje. – Kiev, (in Russ. language).

Malicky, H. (1984): Ein Beitrag zur Autökologie und Bionomie der aquatischen Netzflüglergattung *Neurorthus* (Insecta, Neuroptera, Neurorthidae). – Arch. Hydrobiol. 101: 231-246.

Maneval, H. (1938): Trois Serphoides de l'ambre de la Baltique. – Rev. franc. Entom. 5: 107-116; Paris.

Manning, F. J. (1952): Recent and fossil honey bees: Some aspects of their cytology, phylogeny and evolution. – Proc. Linnean Society London 163: 3-8.

– (1960): A new fossil bee from Baltic amber. – Verh. XI. Intern. Kongr. Entomol. Wien 1: 306-308.

Manton, S. M. (1977): The Arthropoda. – Clarendon, Oxford.

Martens, J. (1978): Weberknechte, Opiliones. – Die Tierwelt Deutschlands 64: 1-464.

MATILE, L. (1979): Un nouveau genre de Keroplatidae de l'ambre oligocène de la Baltique. (Diptera: Mycetophiloidea). – Rev. franc. Entom, N.S. 1: 36-41; Paris.
– (1980): Identité du genre fossile de l'ambre balte Archaeumacrocera et du récent nord-americain Hesperodes (Dipt. Mycetophiloidea Keroplatidae). – L'Entomologiste, 36: 65-70; Paris.
– (1983): Systématique phylogénétique: application à l'étude des Diptères de la famille des Keroplatidae. – Bull. Soc. Ent. France, 88: 406-414; Paris.
MAYR, L. (1868): Die Ameisen des baltischen Bernsteins. – Beitr. Naturk. Preussens 1: 1-102. Königsberg.
MCALPINE, J. F. (1989): Phylogeny and Classification of the Muscomorpha. – in: MCALPINE, J. F. (Editor): Manual of Nearctic Diptera. Vol. 3, Hull, Quebec.
MCCAFFERTY, W. P. (1987): New fossil Mayfly in Amber and its Relationships among extant Ephemeridae (Ephemeroptera). – Ann. Ent. Soc. America, 80: 472-474.
MEINANDER, M. (1972): A revision of the family Coniopterygidae (Planipennia). – Acta zool. Fenn. 136: 1-357.
– (1975): Fossil Coniopterygidae (Neuroptera). – Notul. Ent. Helsingfors 55: 53-57.
– (1981): A review of the genus *Coniopteryx* (Neuroptera, Coniopterygidae). – Ann. Ent. Fennici, 47: 97-110; Helsinki.
– (1990): The Coniopterygidae (Neuroptera, Planipennia). A check-list of the species of the world, descriptions of new species and other new data. – Acta Zool. Fennica, 195: 1-95.
MENGE, A. (1855): Über die Scherenspinnen, Chernetidae. – Neueste Schr. naturforsch. Ges. Danzig, 5: 143.
– (1856): Lebenszeichen vorweltlicher im Bernstein eingeschlossener Tiere. – Progr. Petrischule Danzig, 1-32.
– (1858): Beitrag zur Bernsteinflora. – Neueste Schr. naturforsch. Ges. Danzig 6: 1-18.
– (1866): Über ein Rhipidopteron und einige Helminthen im Bernstein. – Schr. naturf. Ges. Danzig N. F. 1: 1-8.
– (1869): Über einen Scorpion und zwei Spinnen im Bernstein. – Schr. naturf. Ges. Danzig N. F. 2: 1-9.
– (1872): Über eine im Bernstein eingeschlossene *Mermis*. – Schr. naturf. Ges. Danzig. N. F. 3: 1-2.
METCALF, Z. P., WADE, V. (1966): A catalogue of the fossil Homoptera (Homoptera: Auchenorrhyncha). – Contr. Ent. Dep., North Carolina agricult. Exp. Station, Paper 2049, Raleigh.
MEUNIER, F. (1894): Note complementaire sur les Platypezidae de l'ambre. – Bull Soc. zool. France 22-24.
– (1894): Note sur les Mycetophilidae fossiles de l'ambre tertaire. – Wien. ent. Ztg. 62-64.
– (1895): Les Dolichopodidae de l'ambre tertiare. – Ann. Soc. scient. Bruxelles 19: 5-6.
– (1899): Etudes de quelques Dipteres de l'ambre tertiaire. – Bull. Soc. Ent. France 1899: 334-335, 358-359, 392-393.
– (1900): Über die Mycetophiliden (Sciophilinae) des Bernsteins. – Ill. Z. Ent. Germar 5: 68-70.
– (1901): Nouvelles recherches sur quelques Cecidomyidae et Mycetophilidae de l'ambre. – Ann. Soc. scient. Bruxelles 25: 183-203.
– (1901): Über. die Syrphiden des Bernsteins. – Allg. Z. Ent. 6: 70-72.
– (1902): Description de quelques diptères de l'ambre. – Ann. Soc. scient. Bruxelles. 26: 96-104.
– (1902): Études de quelques Diptères de l'ambre. – Ann. Sci. nat. Zool. 16: 395-405.
– (1902): Les Cecidomyidae de l'ambre de la Baltique. – Marcellia. Riv. int. de Cecidologia, Padova 100-105.
– (1902): Les Culicidae de l'ambre. – Rev. Sci. Bourbonnais 15: 199-200.
– (1903): Beitrag zur Fauna der Bibioniden, Simuliiden und Rhyphiden des Bernsteins. – Jb. geol. Landesanst. Berlin (1903) 24: 391-404.
– (1903): Beitrag zur Syrphidenfauna des Bernsteins. – Jb. geol. Landesanst. Berlin für 1903: 201-210.
– (1904a): Contribution à la faune des Helomyzinae de l'ambre de la Baltique. – Feuille jeunes Naturalistes 35: 21-27.
– (1904b): Monographie des Cecidmyidae, des Sciaridae, des Mycetophilidae et des Chironomidae de l'ambre. – Ann. Soc. scient. Bruxelles 28: 12-275.
– (1905): Contribution à la faune des acalyptères Agromyzinae de l'ambre. – Ann. Soc. scient. Bruxelles 29: 89-94.
– (1905): Sur deux Mymaridae de l'ambre de la Baltique. – Miscellanea Entomologica 13: 1-4, Narbonne.
– (1906): Les Dolichopodidae de l'ambre de la Baltique. – C.R. Ac. Sci. Paris 617-618.
– (1906): Monographie des Tipulidae et des Dixidae de l'ambre de la Baltique. – Ann. Sci. nat. Zool. 4: 349-401.
– (1908): Les Asilidae de l'ambre de la Baltique. – Bull. Soc. ent. France 2: 18-20.
– (1909): Monographie der Leptiden und Phoriden des Bernsteins. – Jb. geol. Landesanst. 30: 64-90.
– (1909): Sur quelques Diptères (Xylophagidae, Therevidae, Arthropidae, Stratiomyiidae, Tanypezinae et Ortalinae) de l'ambre de la Baltique. – Ann. Soc. scient. Bruxelles 32 (1908): 258-266.
– (1912): Coup d'oil rétrospectif sur les Diptères du succin de la Baltique. – Ann. Soc. scient. Bruxelles 36: 160-186.
– (1916): Sur quelques diptères (Bombylidae, Leptidae, Dolichopodidae, Conopidae et Chironomidae) de l'ambre de la Baltique. – Tijdschr. Ent. 59: 274-286.
– (1917): Beitrag zur Monographie der Mycetophiliden und Tipuliden des Bernsteins. – Jb. geol. Landesanst. Berlin 68 : 477-493.
– (1917): Über einige Mycetophiliden und Tipuliden des Bernsteins nebst Beschreibung der Gattung *Palaeotanypeza* (Tanypezinae) derselben Formation. – N. Jb. Mineral. 73-106.
– (1917): Un Pompilidae de l'ambre de la Baltique. – Tijdschrift voor Entomologie, 60: 181-184.
– (1922-23): Nouvelle contribution a la monographie des "Mycetophilidae" (Ceroplatinae, Mycetophilinae et Sciophilinae) de l'ambre de la Baltique. – Rev. scient. Bourbon. Centre de la France, 1922: 114-120, 1923: 14-34; Moulins.
MEY, W. (1985): Die Köcherfliegen des Sächsischen Bernsteins (I). (Trichoptera). – Dtsch. ent. Z. 32: 275-278, Berlin.

– (1986): Die Köcherfliegen des Sächsischen Bernsteins (II). (Trichoptera). – Dtsch. ent. Z. 33: 241-248, Berlin.
– (1988): Die Köcherfliegen des Sächsischen Bernsteins (III). (Trichoptera). – Dtsch. ent. Z. 35: 299-309, Berlin.
MEYER, A. B. (1887): Notiz über im Ostsee-Bernstein eingeschlossene Vogelfedern. – Schr. naturf. Ges. Danzig N. F. 6: 206-208.
MEYN, I. (1876): Der Bernstein der norddeutschen Ebene auf zweiter, dritter, vierter, fünfter und sechster Lagerstätte. – Z. Dtsch. Geol. Ges., 28:172-198.
MICHELSEN, V. (2000): Oldest authentic record of a fossil calyptrate fly (Diptera): a species of Anthomyiidae from early Coenozoic Baltic amber. – Studia dipterologica 7: 11-18.
MIERZEJEWSKI, P. (1976a): Scanning electron microscope studies on the fosslization of Baltic amber spiders (preliminary note). – Ann. Med. Sect. Polish Acad. Sci., 21: 81-82.
– (1976b): On application of scanning electron microscope to the study of organic inclusions from Baltic amber. – Ann. Soc. Geol. Pologne 46: 291-295.
– (1978): Electron microscopy study on the milky impurities covering arthropod inclusions in the Baltic amber. – Prace Muzeum Ziemi, 28, Prace geologiczne: 79-84; Warszawa.
MOENKE, H. (1962): Spectralanalyse von Mineralien und Gesteinen. – Leipzig.
MÜHLENBERG, M. (1971): Phylogenetisch-systematische Studien an Bombyliidae (Diptera). – Z. Morph. Tiere 70: 73-102, Berlin.

NAGEL, P. (1980): Aspects of the Evolution of Myrmecophilous Adaptions in Paussinae (Coleoptera, Carabidae). – Misc. Pap. Agric. Univ. Wageningen, 18: 15-34.
– (1987): Fossil Ant Nest Beetles (Coleoptera, Carabidae, Paussinae). – Ent. Arb. Mus. Frey, 35/36: 137-170.
NEL, A., JARZEMBROWSKI, E. (1997): New fossil Sisyridae and Nevrorthidae (Insecta: Neuroptera) from Eocene Baltic amber and Upper Miocene of France. – Eur. J. Entomol. 94: 287-294.
NEUBAUER, M. (1994): Die Bernsteinverbreitung in glazialen Ablagerungen insbesondere von Nordwestdeutschland. – Unveröfftl. Meldearbeit der Universität Bremen.
NEW, T. R. (1989): Planipennia (Lacewings). – In: Handbuch der Zoologie, 4 (2) 30: 1-132; Berlin.
NEW, T. R., TEISCHINGER, G. (1993): Megaloptera (Alderflies, Dobsonflies). – In: Handbuch der Zoologie, 4 (2) 33: 1-97; Berlin.
NIELSEN, E. S., COMMON, I. F. B. (1991): Lepidoptera. – in: CSIRO (1991): The insects of Australia, Vol. 2, Cornell Univ. Press, Ithaca, New York.
NOETLING, F. (1883): Über das Alter der samländischen Tertiärformation. – Z. dtsch. Geol. Ges., 35, 671-694, Berlin.
– (1888): Die Fauna des samländischen Tertiärs Teil 2 (Gastropoda, Pelycypoda, Bryozoa). – Abh. Geol. Spezialk. Preußen u. Thüring. Staaten, 6: 1-109.
OBENBERGER, J. (1957): Eine neue Buprestidenart aus dem baltischen Bernstein nebst Bemerkungen über einige fossile Buprestiden. (Coleoptera: Buprestidae). – Beitr. Entomologie, 7: 308-316; Berlin.
ODIN, G.S. & LUTTERBACHER, H. (1992). The Age of Paleogene Stage Boundaries. – N.Jb.Geol.Paläont.Abh., 186: 21-48.
OESLER, R. (1943): Über einige Copeognathengenera. – Stett. Ent. Z., 104: 1-14.
OHL, M. (1995): Eine neue Bethylidae der Gattung *Lytopsenella* KIEFER 1911 aus dem Baltischen Bernstein (Hymenoptera: Chrysidoidea). – Paläont. Z. 69: 409-416.
OLFERS, E. W. M. (1907): Die "Ur-Insecten". (Thysanura und Collembola im Bernstein). – Schr. phys.-ökon. Ges. Königsberg, 48: 1-40.
– (1911): Ein neuer Thysanure im Bernstein. – Berliner ent. Z. 56: 151-152.
OSTEN SACKEN, C. R. (1864): On the Diptera or two-winged insects of the Amber-fauna. – Am. J. Sci. 37: 305-324.
– (1881): A relic of the tertiary period in Europe, *Elephantomyia*, a genus of Tipulidae. – Mitt. Münch. ent. Vers. 5: 152-154.
OSWALD, J. D. (1998): Annotated catalogue of the Dilaridae (Insecta: Neuroptera) of the world. – Tijdschr. Entomol. 14:115-128.
OWEN, R. M, REA, D. K (1985): Sea floor hydrothermal activity links climate to tectonics: The Eocene carbon dioxide greenhouse. – Sciene, 227, 166-169.

PANTSCHENKO, V. I. & KVASNICA, V. N. (1982): Bernstein Wolhyniens. – Kiev, (in Russ. language).
PARSONS, C. T. (1939): A ptiliid beetle from Baltic amber in the Museum of comparative Zoology. – Psyche, 46: 62-64.
PETRUNKEVITCH, A. (1942): A study of amber spiders. – Trans. Conn. Acad. Arts. Sci., 34, 119-464, New Haven, Connecticut.
– (1950): Baltic amber spiders in the museum of comparative Zoology. – Bull. Mus. comp. Zool. Harvard 103: 259-337.
– (1958): Amber spiders in European collections. – Trans. Connec. Ac. Arts Sci. 41: 97-400.
– (1963): Chiapas Amber Spider. –Univ. Cal. Publ. Ent. 31: 1-40.
– (1971): Chiapas Amber Spider II. –Univ. Cal. Publ. Ent. 63: 1-44.
PEUS, F. (1968): Über die beiden Bernsteinflöhe (Insecta, Siphonaptera). – Paläont. Z. 42: 62-72.
PEYERIMHOFF, P. DE (1909): Le *Cupes* de l'ambre de la Baltique (Col.). – Bull. Soc. Ent. France, 1909: 57-60, Paris.
PFAU, H. K. (1975): Zwei neue Kleinlibellen (Odonata, Zygoptera – möglicherweise Platycnemididae) aus dem baltischen Bernstein. – Stuttg. Beitr. Naturk. A 270: 1-7, Stuttgart.
PFEIFFER, H. (1956): Zur Ökologie und Larvalsystematik der Weberknechte. – Mitt. Zool. Mus. Berlin 32: 59-104.
PICTET, F. J. (1845): Résultat de ses recherches sur les insects fossiles de l'ordre des Névroptères contenus dans l'ambre. – Act. soc. Helv. sc. nat. 30: 69-75.

– (1856): siehe PICTET-BARABAN, F. J., HAGEN, H. (1856).

PICTET-BARABAN, F. J., HAGEN, H. (1856): Die im Bernstein befindlichen Neuropteren der Vorwelt (mit Zusätzen von A. MENGE). – in: BERENDT, G. C. (1854-1856): Die im Bernstein befindlichen organischen Reste der Vorwelt. 2 Vols., Berlin (Nicolai), in: Vol. 2, II. Abhl.

PIELINSKA, A. (1990): The list of plant inclusions in Baltic amber from collection of the Museum of the Earth in Warsaw. – Prace Muzeum Ziemi, 41: 147-148; Wroclaw.

– (1997): Inclusions of wood in the amber collections of the Museum of the Earth in Warsaw – Metalla (Sonderheft) 66: 25-28.

PIROZYNSKI, K. A. (1976): Fossil Fungi. – Ann. Rev. Phytopathology, 14: 237-246.

PITON, L. E. (1938): *Succinotettix chopardi* PITON, Orthoptère (Tetricinae) inédit de l'ambre de la Baltique. – Bull. Soc. ent. France 43: 226-227.

– (1940): Un Longicorne nouveau de l'ambre de la Baltique. – Bull. Soc. Ent. France, 45: 63-64; Paris.

– (1940): Une Blatte nouvelle de l'ambre de la Baltique. – Bull. Mens Soc. Linné. Lyon, 9: 68-69.

PODENAS, S. (1997): New *Macrochile* LOEW, 1850 (Diptera, Tanyderidae) from the Baltic amber. – Mitt. Geol.-Paläolont. Inst. Univ. Hamburg, 80: 173-177.

– (1999a): *Paradelphomyia* (Diptera, Limoniidae) from the Baltic amber. – Acta Zool. Lituanica 9: 107-110.

– (1999b): New species of fossil mosquitoes (Diptera, Culicidae) from the Baltic amber. – Acta Zool. Lituanica 9: 111-114.

– (1999c): *Ormosia rondani* crane flies (Diptera, Limoniidae) from the Baltic amber. – Acta Zool. Lituanica 9: 183-200.

– (1999d): New *Cheilotrichia* crane flies (Diptera, Limoniidae) from Baltic amber. – Mitt. Geol.-Paläont. Inst. Univ. Hamburg.

POHL, H., KINZELBACH, R. (1995): Neufunde von Fächerflüglern aus dem Baltischen und Dominikanischen Bernstein (Strepsitera: Bohartillidae & Myrmecolacidae). – Mitt. Geol.-Paläolont. Inst. Univ. Hamburg, 78: 197-209.

POINAR, G. O. (1984): Fossil evidence of Nematode parasitism. – Rev. Nématologie, 7: 201-203; Paris.

– (1992): Life in amber. – Stanford Univ. Press, Stanford.

– (1993): Insects in Amber. – Ann. Rev. Entomol. 46: 145-159.

POINAR, G. O., GRIMALDI, D. A. (1990): Fossil And Extant Macrochelid Mites (Acari: Macrochelidae) Phoretic on Drosophilid Flies (Diptera: Drosphilidae). – J. New York Entomol. Soc. 98: 88-92.

POINAR, G. O., HESS, R. (1982): Ultrastructure of 40-million-year-old insect tissue. – Science 215: 1241-1242.

POINAR, G. O., POINAR, R. (1994): The Quest for Life in Amber. – Addison-Wesley, Reading.

– (1999): The Amber Forest – A Reconstruction of a Vanished World. – Princeton Univ. Press.

PONOMARENKO, A. G. (1975): A new Hymenopteran species from Baltic amber. – Paleontol. J. 9: 124-127; Washington.

– (1976): Corydalidae (Megaloptera) from the Cretaceous of northern Asia. – Entomol. Rev. 55: 114-122; Washington.

– (1977): Palaeozoic members of Megaloptera (Insecta). – Palaeont. J. 11: 73-81, Washington.

– (1995): The geological history of beetles. – in: PAKALUK, J., SLIPINSKI, S. A. (eds.): Biology, Phylogeny, and Classification of Coleoptera, Muz. Inst. Zool. PAN, Warszawa.

POPOV, Y. A. (1978): New species of Aradidae (Hemiptera) from the Baltic amber. – Prace Muzeum Ziemi 29: 137-140; Warszawa.

– (1996): Water measures from the Baltic amber (Heteroptera: Gerromorpha, Hydrometridae). – Mitt. Geol.-Paläolont. Inst. Univ. Hamburg, 79: 211-221.

– (2001): Fossil piesmatids from Baltic amber (Heteroptera, Pentatomomorpha, Piesmatidae). – Mitt. Geol.-Paläont. Inst. Univ. Hamburg, 85: 211-220.

POPOV, Y. A., HERCZEK, A. (1992): The first Isometopinae from Baltic Amber (Insecta: Heteroptera, Miridae). – Mitt. Geol.-Paläolont. Inst. Univ. Hamburg, 73: 241-258.

– (1993a): *Metoisops punctatus* sp.n., the second representative of the fossil genus *Metoisops* from Baltic amber (Heteroptera: Miridae: Isometopinae). – Ann. Upper Silesian Mus., Entomology, Suppl. 1: 51-56.

– (1993b): New data on Heteroptera in amber resins. – Ann. Upper Silesian Mus., Entomology, Suppl. 1: 7-12.

POPOV, Y. A. & PUTSHKOV P. V. (1998): *Redubinotus liedtkei* n.gen. n.sp. – The second *Centrocnemina* from the Baltic Amber (Heteroptera, Reduviidae, Cetrocnemidinae). – Ann. Upper Silesian Mus.(Entomol.), 8-9: 205-210.

POTONIÉ, R. (1925): Bernstein-Einschlüsse, ihre Freilegung und Untersuchung. – Der Naturforscher 1: 565-567.

PRESL, J. S. (1822): Additamenta ad faunam protogaeam, sistens descriptiones aliquot animalum in succino inclusorum. – Del. Prag, Hist. Nat. Spectr., 191pp.

PRIESNER, H. (1924): Bernstein-Thysanopteren. – Ent. Mitt. 13: 130-151.

– (1929): Bernstein-Thysanopteren II. – Bernstein-Forsch. 1: 111-138; Berlin.

PUTSHKOV, P. V., POPOV, Y. A. (1993): A remarkable nymph of a centrocneminous bug from the Baltic Amber (Insecta: Heteroptera, Reduviidae). – Mitt. Geol.-Paläolont. Inst. Univ. Hamburg, 75: 211-229.

– (1995): *Collarhamphus mixtus* n.gen. n.sp. – The first Collartidina (Heteroptera: Reduviidae, Emesinae) from the Baltic Amber. – Mitt. Geol.-Paläolont. Inst. Univ. Hamburg, 78: 179-187.

QUIEL, G. (1909): Zwei neue Coleopteren aus dem baltischen Bernstein. (Eocän bzw. Unteres Oligocän). – Berliner ent. Z., 54: 49-52.

– (1911): Bemerkungen über Coleopteren aus dem baltischen Bernstein. – Berliner ent. Z., 55: 181-192.

RASNITSYN, A. P. (1977): A new family of sawflies (Hymenoptera, Tenthredinoidea, Electrotomidae) from the Baltic amber. – Zool. Zhurn. 56: 1304-1308.

RASNITSYN, A. P., KULICKA, R. (1990): Hymenopteran Insects in Baltic amber with respects to the overall history of the order. – Prace Museum Ziemi 41: 53-64. Wroclaw.

REBEL, H. (1935): Bernstein-Lepidopteren. – Dtsch ent. Z. "Iris", 49 (1935-36): 162-186; Dresden.
– (1936): Mikrolepidopteren aus dem baltischen Bernstein. – Naturw. 24: 519-520.
– (1937): Zur Systematik der Bernstein-Lepidopteren. – Z. österr. Ent. Ver. 22: 1-3.
REINICKE, R. (1996): Bernstein im vorpommerischen Küstengebiet – in: GANZELEWSKI, M., SLOTTA, R. (Hrsg.): Bernstein – Tränen der Götter: 139-146.
RICE, P. C. (1980): Amber. The golden gem of the ages. – New York, London, 289 pp.
RICKER, W. E. (1935): New Canadian Perlids. – Can. Ent. Ottawa 67: 256-264.
RIEK, E. F. (1976): New Upper Permian Insects from Natal, South Africa. – Ann. Natal Mus. 22: 755-789, Pietermaritzburg.
RITZKOWSKI, S. (1990): Die Inklusen der ehemaligen Königsberger Bernsteinsammlung in Göttingen. – Prace Muzeum Ziemi 41: 149-153; Wroclaw.
– (1996): Die Geschichte der Bernsteinsammlung der Albertus-Universität zu Königsberg i.Pr. – in: GANZELEWSKI, M., SLOTTA, R. (Hrsg.): Bernstein – Tränen der Götter: 293-298.
– (1997): K-Ar-Altersbestimmung der bernsteinführenden Sedimente des Samlandes (Paläogen, Bezirk Kaliningrad) – Metalla (Sonderheft) 66: 19-23.
RÖDER, G. (1980): Eine neue *Cheilosia*-Art aus Baltischem Bernstein mit REM-Untersuchungen rezenter Arten. – Stuttg. Beitr. Naturk. B, 64: 1-18.
ROESLER, R. (1943): Über einige Copeognathengenera. – Stettiner ent. Z., 104: 1-14.
– (1944): Die Gattungen der Copeognathen. – Stettiner ent. Z., 105: 117-166.
ROEWER, C. F. (1939): Opilioniden im Bernstein. – Palaeobiologica 7: 1-5; Wien.
RÖSCHMANN, F. (1999): Revision of the evidence of *Tetracha carolina* (Coleoptera, Cicindelidae) in Baltic amber (Eocene-Oligocene). – Est. Mus. Cienc. Nat. de Alava 14: 205-209.
– (1995): Die Sciariden des Baltischen und des Sächsischen Bernsteins (Insecta, Diptera, Sciaridae). – Diss. Univ. Greifswald.
– (1997): Ökofaunistischer Vergleich von Nematocerenfaunen (Inscecta: Diptera: Sciaridae und Ceratopogonidae) des Baltischen und Sächsischen Bernsteins (Tertiär, Oligozän-Miozän). – Paläontol. Z. 71: 79-88; Berlin.
ROSEN, K. (1913): Die fossilen Termiten: eine kurze Zusammenfassung der bis jetzt bekannten Funde. – Trans. II. Ent. Congr. 1912: 318-335.
ROSS, A. (1998): Amber. – The Natural History Museum, London.
ROSS, E. S. (1956): A new genus of Embioptera from Baltic amber. – Mitt. geol. Staatsinst. Hamburg 25: 76-81.
– (1970): Biosystematics of the Embioptera. – Ann. Rev. Entomol. 15: 157-172.
ROSS, H. H. (1951): Phylogeny and biogeography of the caddisflies of the genera *Agapetus* and *Electragapetus* (Trichoptera: Rhyacophilidae). – J. Washington Acad. Sciences 41: 347-356.
RÜFFLE, L., HELMS, J. (1970): Waldsteppe und Insektenwelt im Bernstein, Beispiele aus der Bernsteinsammlung des Paläontologischen Museums. – Wiss. Z. Humboldt-Univ. Berlin, math.-naturwiss. Reihe, 19: 243-249.
RUNGE, W. (1868): Der Bernstein in Ostpreußen . – Berlin.

SALT, G. (1931): Three bees from Baltic amber. – Bernstein-Forsch. 2: 136-147.
SANDBERGER, V. F. (1887): Bemerkungen über einige Heliceen im Bernstein der preussischen Küste. – Schr. Naturf. Ges. Danzig 6: 137-141.
SAVCENCO, E. M. (1967): New fossil Limoniid-fly (Diptera, Limoniidae) from the Baltic amber. – Dopodivi Akademii Nauk Ukrainskoj RSR Ser. B, Nr. 5: 469-473; Kiev.
SAVKEVICH, S. (1969): K voprosu o vjilanii osobennostej fossilizacii na svojsta iskopaemych smaol (na primere baltiskogo jantarja). – Geochim. Sb., Leningrad 271: 312-323, Leningrad.
SAVKEVICH, S. (1996): Über die Möglichkeit der Herkunftsbestimmung des Bernstein und anderer bernsteinähnlicher fossiler Harze mittels verschiedener analytischer Methoden. – Int. Meeting. Fachhochschule Coburg.
SCHAUFUSS, C. (1891): Preussens Bernstein-Käfer. Neue Formen aus der Helm'schen Sammlung im Danziger Provinzialmuseum. – Berliner ent. Z., 36: 53-64.
– (1896): Preussens Bernsteinkäfer. II. Neue Formen aus der Helm'schen Sammlung im Danziger Provinzialmuseum. – Berliner ent. Z., 41: 51-54.
SCHAUFUSS, L. W. (1888): Einige Käfer aus dem baltischen Bernstein. – Berliner ent. Z., 32: 266-270.
– (1890a): Eine Staphylinengattung aus dem baltischen Bernsteine. – Ent. Nachr., 16: 69-70, Berlin.
– (1890b): Die Scydmaeniden des baltischen Bernsteines. – Nunquam otiosus, 3: 561-586, Dresden.
SCHAWALLER, W. (1978): Neue Pseudoscorpione aus dem baltischen Bernstein der Stuttgarter Bernsteinsammlung (Arachnida: Pseudoscorpionidea). – Stutt. Beitr. Naturk. B 42: 1-22.
– (1981a): Übersicht der Spinnen-Familien im Dominikanischen Bernmstein und anderen tertiären Harzen. – Stuttgarter Beitr. Naturk. B 77: 1-10.
– (1981b): Die Spinnenfamilie Hersiliidae in Dominikanischem Bernstein. – Stuttgarter Beitr. Naturk. B 79: 1-10.
– (1982): Spinnen der Familien Tetragnathidae, Uloboridae und Dipluridae in Dominkanischem Bernstein und allgemeine Gesichtspunkte (Arachnidae, Araneae). – Stuttgarter Beitr. Naturk. B 89: 1-19.
– (1984): Spinnentiere (Arachnida) im Dominikanischen Bernstein. – Stuttgarter Beitr. Naturk. C 18: 72-78.
SCHEDL, K. E. (1947): Die Borkenkäfer des baltischen Bernsteins. 58. Beitrag zur Morphologie und Systematik der Scolytidae und Platypodidae. – Zblatt Gesamtgebiet Entomologie, 2: 12-45; Klagenfurt.
– (1967): Bernsteinborkenkäfer aus dem Zoologischen Museum der Universität Kopenhagen. 247. Beitrag zur Systematik und Morphologie der Scolytoidea. – Entomol. Meddel., 35: 85-87; Kobenhavn.

– (1970): Bernsteinborkenkäfer aus dem Zoologischen Museum der Universität Kopenhagen II. 261. Beitrag zur Morphologie und Systematik der Scolytoidea (Coleoptera). – Entomol. Meddel., 38: 68-70; Kobenhavn.

SCHELLER, U., WUNDERLICH, J. (2001): First description of a fossil pauropod, *Eopauropus balticus* n.gen. n.sp. (Pauropoda, Pauropodidae), in Baltic amber. – Mitt. Geol.-Paläont. Inst. Univ. Hamburg, 85.

SCHLEE, D. (1969): Der Flügel von *Sphaeraspis* (Coccina), prinzipiell identisch mit Aphidina-Flügeln. – Stuttg. Beitr. Naturk. 211: 1-11.

– (1970): Insektenfossilien aus der unteren Kreide I. Verwandtschaftsforschung an fossilen und rezenten Aleyrodina (Insecta, Hemiptera). – Stuttgart. Beitr. Naturk. 213: 1-72.

– (1973): Harzkonservierte Vogelfedern aus der unteren Kreide. – J. Ornithologie 114: 207-219, Berlin.

– (1980): Bernstein-Raritäten – Farben, Strukturen, Fossilien, Handwerk. – Staatl. Museum Naturkunde Stuttgart. 88 S.

– (1984): Notizen über einige Bernsteine und Kopale aus aller Welt. – Stuttg. Beitr. Naturk., C, 18: 29-37.

– (1990): Das Bernstein-Kabinett. – Begleitheft zur Bernsteinausstellung im Museum am Löwentor, Stuttgart. Stuttg. Beitr. Naturk. C, 28: 1-100.

SCHLEE, D., GLÖCKNER, W. (1978): Bernstein – Bernsteine und Bernstein-Fossilien. – Stuttg. Beitr. Naturk. C, 8: 1-72.

SCHLIEPHAKE, G. (1990): Beiträge zur Kenntnis fossiler Franzenflügler (Thysanoptera, Insecta) aus dem Bernstein des Tertiärs. 1. Beitrag: Stenurothripidae. – Zoology (J. of Pure and Applied Zoology, Dehli) 2: 163-184.

SCHLIEPHAKE, G. (1993): Beiträge zur Kenntnis fossiler Franzenflügler (Thysanoptera, Insecta) aus dem Bernstein des Tertiärs. 2. Beitrag: Aeolothripidae (Melanthripinae) and Thripidae (Dendrothripinae and Thripinae). – Zool. Jb. Syst. 120: 215-251.

– (1997): Beitrag zur Kenntnis fossiler Franzenflügler (Thysanoptera) aus dem Bernstein des Tertiärs des Bitterfelder Raumes (3. Beitrag: Thripidae, Panchaetothripinae). – Entomol. Nachr. Ber. 4: 66-67, Dresden.

– (1999a): Fossil thrips (Thysanoptera, Insecta) of the Baltic (North- and Baltic Sea) and Saxonian (Bitterfeldian) Tertiary amber of Hoffeins collection. – Dtsch. Entomol. Z. 46: 83-100, Berlin.

– (1999b): On problems of fossil amber Thysanoptera. – Proc. 6th. Int. Symp. on Thysanoptera, Antalya, 1998: 129-137.

– (2000): Neue Franzenflügler aus dem Baltischen und Bitterfelder Bernstein (Insecta: Thysanoptera). – Mitt. Geol.-Paläont. Inst. Univ. Hamburg 84: 219-230.

SCHLÜTER, T. (1976): Die Fossilfalle Harz – der gegenwärtige Erforschungsstand. – Naturw. Rundschau 29: 350-354, Stuttgart.

– (1986): The Fossil Planipennia – a Review. – in: GEPP, J., ASPÖCK, H., HÖLZEL, H.: Recent Research in Neuropterology, Graz: 103-111.

SCHLÜTER, T., KÜHNE, W. G. (1975): Die einseitige Trübung von Harzinklusen – ein Indiz gleicher Bildungsumstände. – Entomol. germanica 2: 308-315, Stuttgart.

SCHUBERT, K. (1939): Mikroskopische Untersuchung pflanzlicher Einschlüsse des Bernsteins. I. – Bernstein-Forsch. 4: 23-44; Berlin.

– (1953): Mikroskopische Untersuchung pflanzlicher Einschlüsse des Bernsteins. 2. Teil: Rinden und Borken. – Palaeontogr. 93: 103-119; Stuttgart.

– (1961): Neue Untersuchungen über Bau und Leben der Bernsteinkiefern [*Pinus succinifera* (CONW.) emend.]. – Geol. Jb., Beih 45: 1-149; Hannover.

– (1964): Chemisch-physikalische Prozesse im Innern des Baltischen Bernsteins. – Natur und Museum 94: 259-264, Frankfurt a.M.

– (1965): Chemisch-physikalische Prozesse im Innern des Baltischen Bernsteins. 2. Die "Sonnen"- Flinten, Beugungs- und Brechungserscheinungen des Lichts an kapillaren Spaltflächen des Steininnern. – Natur und Museum 95: 261-270, Frankfurt a. M.

SCHÜLKE, M. (2000): Eine neue Art der Gattung *Bolitobius* LEACH in SAMOUELLE aus dem Baltischen Bernstein (Col., Staphylinidae, Tachyporinae). – Ent. Nachr. Ber. 44: 101-103.

SCHUMANN, E. (1885): Schnecken im Bernstein. – Malakozool. Blätter, N.F. 7: 100-101; Kassel.

SCHUMANN, H. (1984): Erstnachweis einer Raubfliege aus dem Sächsischen Bernstein. – Dtsch. ent. Z., N.F. 31: 217-223; Berlin.

SCHUMANN, H., WENDT, H. (1989): Zur Kenntnis der tierischen Inklusen des Sächsischen Bernsteins. – Dtsch. ent. Z., 36: 33-44; Berlin.

SCHUMANN, J. (1863): Preussische Diatomen. – Schr. Physik.-Ökonom. Ges. Königsberg 3: 166-192.

SCHUSTER, J. (1931): Über das Verhältnis der systematischen Gliederung, der geographischen Verbreitung und der paläontologischen Entwicklung der Cycadaceen. – Bot. Jb. Systematik, Pflanzengeschichte u. Pflanzengeographie 64: 165-260, Taf. IV-XI.

SELLNICK, M. (1919): Die Oribatiden der Bernsteinsammlung der Universität Königsberg i. Pr. – Schr. phys.-ökon. Ges. Königsberg (1918) 59: 21-42.

– (1927): Rezente und fossile Oribatiden (Acar. Oribat.). – Schr. phys.-ökon. Ges. Königsberg 65: 114-116.

– (1931): Milben im Bernstein. – Bernstein-Forsch. 2: 148-180.

SEREBRICKIJ, A.I. (1979): Der Bernstein der Nordukraine. – Vestnik. Leningrad. unta: 12 S.

SEREDSZUS, F., WICHARD, W. (2002): Buchonomyiinae in Baltic amber (Diptera, Chironomidae). – Studio dipterologica 9.

SHELLFORD, R. (1910): On a collection of Blattidae preserved in amber, from Prussia. – J. Linn. Soc. London 30: 336-355.

– (1911): The British Museum collection of Blattidae preserved in amber, from Prussia. – J. Linn. Soc. London 32: 59-69.

SHERCHIN, W. W. (1971): Über Rüsselkäfer (Insecta, Coleoptera) des baltischen Bernsteins. – Trans. Palaeont. Inst. Acad. Sci. USSR Moskau 130: 197-209.

SHULTZ, J. W. (1990): Evolutionary morphology and phylogeny of Arachnida. – Cladistics 6: 1-38.
SILVESTRI (1912): Die Thysanuren des baltischen Bernsteins. – Schr. phys.-ökon. Ges. Königsberg 53: 42-66.
SIVKOV, V., KRYLOV, A. (1997): Paleogene evolution of the ocean and some hypotheses with respect to the origin of the Baltic amber deposits – Metalla (Sonderheft); 66: 51-53.
SKALSKI, A. W. (1973): Studies on the Lepidoptera from fossil resins. Part II. *Epiborkhausenites obscurotrimaculatus* gen. et spec. nov. (Oecophoridae) and a Tineid Moth discovered in the Baltic amber. – Acta Palaeont. Pol. 18: 153-160.
– (1974): Studien an Lepidopteren aus fossilen Harzen. Teil V. Zwei neue Gattungen und Arten der Familie Tineidae aus dem baltischen Bernstein. – Beitr. Ent. 24: 97-104.
– (1977): Studies on the Lepidoptera from fossil resins. Part I. General remarks and descriptions of new genera and species in the families Tineidae and Oecophoridae from the Baltic amber. – Prace Muz. Ziemi 26: 3-24.
– (1985): Motyle (Lepidoptera) w bursztynie baltyckim. – Wiadomosci Entomologiczne, 6: 207-210; Warszawa.
– (1985): Stan badarn nad inkluzjami zwierzecymi w bursztynie baltyckim. – Wiadomosci Entomologiczne, 6: 167-178; Warszawa.
SLOTTA, R. (1996): Die Bernsteingewinnung im Samland (Ostpreußen) bis 1945 – in: GANZELEWSKI, M., SLOTTA, R. (Hrsg.): Bernstein – Tränen der Götter: 169-214.
SLOTTA, R., GANZELEWSKI, M. (1996): Die heutige Bernsteingewinnung und -verarbeitung in Jantarnyi – in: GANZELEWSKI, M., SLOTTA, R. (Hrsg.): Bernstein – Tränen der Götter: 249-268.
SNYDER, T. E. (1928): A new Reticulitermes from Baltic Sea amber. – Proc. Washington Ac. Sci. 18: 515-517.
SORG, M. (1986): Grabwespen der Gattung *Passaloecus* aus fossilen Harzen (Hymenoptera, Sphecoidea, Pemphredoninae). – Paläont. Z. 60: 277-284.
– (1988): Zur Phylogenie und Systematik der Bethylidae (Insecta: Hymenoptera, Chrysidoidea). – Geol. Inst. der Univ. Koeln (Sonderveröffentl.), 63: 1-146.
SPAHR, U. (1981a): Bibliographie der Bernstein- und Kopal-Käfer (Coleoptera). – Stuttg. Beitr. Naturk. B 72: 1-21.
– (1981b): Systematischer Katalog der Bernstein- und Kopal-Käfer (Coleoptera). – Stuttg. Beitr. Naturk. B 80: 1-107.
– (1985): Ergänzungen und Berichtigungen zu R. Keilbachs Bibliographie und Liste der Bernsteinfossilien – Ordnung Diptera. – Stuttg. Beitr. Naturk. B 111: 1-146.
– (1987): Ergänzungen und Berichtigungen zu R. Keilbachs Bibliographie und Liste der Bernsteinfossilien – Ordnung Hymenoptera. – Stuttg. Beitr. Naturk. B 127: 1-121.
– (1988): Ergänzungen und Berichtigungen zur Keilbachs Bibliographie und Liste der Bernsteinfossilien – Überordnung Hemipteroidea. – Stuttg. Beitr. Naturk. B 144: 1-60.
– (1989): Ergänzungen und Berichtigungen zu R. Keilbachs Bibliographie und Liste der Bernsteinfossilien – Überordnung Mecopteroidea. – Stuttg. Beitr. Naturk. B 157: 1-87.
– (1990): Ergänzungen und Berichtigungen zu R. Keilbachs Bibliographie und Liste der Bernsteinfossilien – "Apterygota". – Stuttg. Beitr. Naturk. B 166: 1-23.
– (1992): Ergänzungen und Berichtigungen zu R. Keilbachs Bibliographie und Liste der Bernsteinfossilien – Klasse Insecta. – Stuttg. Beitr. Naturk. B 182: 1-102.
– (1993a): Ergänzungen und Berichtigungen zu R. Keilbachs Bibliographie und Liste der Bernsteinfossilien – verschiedene Tiergruppen, ausgenommen Insecta und Araneae. – Stuttg. Beitr. Naturk. B 194, 1-77.
– (1993b): Systematischer Katalog und Bibliographie der Bernstein- und Kopal-Flora. – Stuttg. Beitr. Naturk. B 195: 1-99, Stuttgart.
STACH, J. (1922): Eine neue *Sminthurus*art aus der Bernsteinfauna. – Bull. intern. Acad. Pol: 53-61.
– (1972): Apterygota (Insecta) from the Baltic amber. – Przeglad zool. 16: 416-420.
STANICZEK, A. H., BECHLY, G. (2001): First fossil record of the mayfly family Baetiscidae from Baltic amber (Insecta: Ephemeroptera). – Stuttg. Beitr. Ser. B.
STANNARD, L. J. (1956): Two new fossil Thrips from Baltic amber (Thysanoptera; Terebrantia). – Fieldiana, Zoology, 34: 453-460; Chicago.
STARÝ, P. (1973): A revision of the fossil Aphidiidae (Hymenoptera). – Ann. Zool. Bot., 87: 1-22; Bratislava.
STEIN, J. P. E. (1877): Drei merkwürdige Bernstein-Insekten. – Mitt. München ent. Ver. 1: 28-30.
STEYSKAL, G. C. (1972): A catalogue of species and key to the genera of the family Diopsidae (Diptera: Acalyptratae). – Stuttg. Beitr. Naturk. 234: 1-20.
STRÜMPEL, H. (1983): Homoptera (Pflanzensauger). – Handbuch der Zoologie, 4 (2) 19: 1-222, Berlin.
STUCKENBERG, B. R. (1974): A new genus and two new species of Athericidae (Diptera) in Baltic amber. – Ann. Natal. Mus. 22: 275-288.
SZABÓ, J (1975). Einige phylogenetische Folgerungen aufgrund der fossilen Psychodiden (Diptera, Nematocera) von Europa und Amerika. – Acta Biologica Debrecina 12: 65-68.
SZADZIEWSKI, R. (1988): Biting midges (Diptera, Ceratopogonidae) from Baltic amber. – Polskie Pismo entomol. 58: 3-283.
– (1990): Biting midges (Diptera, Ceratopogonidae) from Sakhalin amber. – Prace Museum Ziemi 41: 77-81, Warschau.
– (1993): Biting midges (Diptera, Ceratopogonidae) from Miocene Saxonian amber. – Acta zool. cracov. 35: 603-656, Krakow.
– (1995): The oldest fossil Corethrellidae (Diptera) from Lower Cretaceous Lebanese amber. – Acta zool. cracov. 38: 177-181, Krakow.
– (1996): Biting midges from Lower Cretaceous amber of Libanon and Upper Cretaceous Siberian amber of Taimyr (Diptera, Ceratopogonidae). – Studia dipterologica 3: 23-86.
– (1998): New mosquitoes from Baltic amber (Diptera: Culicidae). – Polskie Pismo entomol. 67: 233-244.

– (1998): A new species of the predaceous midge genus *Metahelea* from Baltic amber (Diptera: Ceratopogonidae). – Polskie Pismo entomol. 67: 245-253.

– (2000): Biting midges (Diptera: Ceratopogonidae) from the Lower Cretaceaous amber of Jordan. – Polskie Pismo entomol. 69: 251-256.

Szadziewski, R., Grogan, W. L. (1994): Biting midges from Dominican amber. I. A new fossil species of *Baeodasymyia* (Diptera, Ceratopogonidae). – Proc. Ent. Soc. Washington 96: 219-229.

Szadziewski, R., Krzeminski, W., Kutscher, M. (1994): A new species of Corethrella (Diptera, Corethrellidae) from Miocene Saxonian amber. – Acta zool. cracov. 37: 87-90, Krakow.

Szadziewski, R., Schlüter, T. (1992): Biting midges (Diptera, Ceratopogonidae) from upper Cretaceous (Cenomanian) amber of France. – Ann. Soc. Entomol. France 28: 73-81.

Szadziewski, R., Szadziewski, M. M. (1985): *Culex erikae* sp. n. (Diptera, Culicidae) from the Baltic amber. – Polskie Pismo Entomologiczne, 55: 513-518; Warszawa.

Szwedo, J., Gebicki, C. (1998): *Ambericarda skalskii* gen. et sp.n. from Baltic amber (Homoptera: Cicadellidae). – Polskie Pismo Entomol. 67: 179-184.

– (1999): Representatives of Petalocephalini (Homoptera, Cicadellidae: Ledrinae) in Baltic amber (Eocene –Oligocene). – Est. Mus. Cienc.Nat. de Alava, **14** (2): 199-204.

Szwedo, J., Kulicka, R. (1999a): Auchenorrhyncha (Insecta, Homoptera) in Baltic amber from the collection of the Museum of the Earth, Warsaw. – Est. Mus. Cienc.Nat. de Alava, **14**: 173-176.

– (1999b): Inclusions of Auchenorrhyncha in Baltic amber (Insecta Homoptera). – Est. Mus. Cienc.Nat. de Alava, **14**: 1977-197.

Thomas, B. R. (1969). Kauri-resins – modern and fossil. – Org.Geochem (1969): 599-618.

Thomas, K. (1858): Einige Bemerkungen über eine den Bernstein und seine Begleitung betreffende Sammlung ostpreußischer Mineralien. – Neue Preuß. Prov.-Blätter, 3. Folge, 1: 218-232; Königsberg.

Thuróczy, C. (1983): *Stethynium townesi* sp. n. from the Baltic Oligocene. (Hymenoptera, Mymaridae). – Reichenbachia 21: 123-125; Dresden.

Tilgner, E. H. (2001): The fossil record of Phasmida (Insecta: Neoptera). – Insect Syst. Evol., 31: 473-480.

Tornquist, A. (1910): Die Geologie von Ostpreußen. – Berlin.

Trjapitzin, V. A. (1963): A new hymenopteron genus from Baltic amber. – Pal. Zhurn. Moskva 3: 89-95.

Trjapitzin, V. A., Manukyan, A. R. (1995): *Electrocampe sugonjaevi* gen. et sp. n., the first record of a fossil Tertiary tetracampid (Hymenoptera, Tetracampidae, Mongolocampinae) – Amber & Fossils 1: 17-22.

Turkin, N. I. (1997): Preliminary results of mikroscopic research of tangential wood imprints in Baltic amber – Metalla (Sonderheft) 66: 55-56.

Tutkovski, P. A. (1911): Der Bernstein des Gouvernements Wolhynien. – Shitomir.

Tutskij, W. (1997): Geologie und Entstehung der Bernsteinvorkommen im Nordwesten der Ukraine – Metalla (Sonderheft) 66: 57-62.

Uhnann, E. (1939): Hispinen aus baltischem Bernstein. 81. Beitrag zur Kenntnis der Hispinen. (Col. Chrys.). – Bernstein-Forsch. 4: 18-22; Berlin (de Gruyter).

Ulke, T. (1947): A new genus and species of Curculionidae (Coleoptera) in Baltic amber. – Notulae naturae 193; 1-7, Philadelphlia.

Ulmer, G. (1910): Über Bernsteintrichopteren. – Zool. Anz. 36: 449-453.

– (1912): Die Trichopteren des Baltischen Bernsteins. – Beitr. Naturk. Preussens 10: 1-380, Königsberg.

Ulrich, H., Schmelz, R. M. (2001): Enchytraeidae as prey of Dolichopodidae, recent and in Baltic amber (Clitellata: “Oligochaeta”; Insecta: Diptera). – Bonn. zool. Beitr. 49.

Ulrich, W. (1927): Über das bisher einzige Strepsipteron aus dem baltischen Bernstein und über eine Theorie der Mengeinenbiologie. – Z. Morphol. Ökol. 8: 45-62.

Usinger, R. L. (1939): *Protepiptera*, a new genus of Achilidae from Baltic amber (Hemiptera, Fulgoroidea). – Psyche 46: 65-67.

– (1941): Two new species of Aradidae from Baltic amber (Hemiptera). – Psyche, 48: 95-100.

– (1942): An annectent genus of Cimicoidea from Baltic amber (Hemiptera). – Psyche 49: 41-46.

Vachon, M. (1949): Ordre des Pseudoscorpions. – Traité de Zoologie 6: 437-481.

Vassilishin, I. S. & Pantschenko, V. I. (1996): Bernstein in der Ukraine. – in: Ganzelewski, M. & Slotta, R. (Hrsg.): Bernstein – Tränen der Götter: 77-82.

Vavra, N. (1982): Bernstein und andere fossile Harze. – Z. Dtsch. Gemmol. Ges. 31: 213-254; Idar-Oberstein.

– (1984): Bernstein – die Tier- und Pflanzenwelt fossiler Harze. – Schr.Verein z. Verbr. naturwiss. Kenntn. Wien 122/123: 67-96, Wien.

Vinken, R. (1987): The Northwest European Tertiary Basin. – Geol. Jb., A, 100, 1-267, Hannover.

Voigt, E. (1935): Die Erhaltung von Epithelzellen mit Zellkernen von Chromatophoren und Corium in fossiler Froschhaut aus der mitteleozänen Braunkohle des Geiseltales. – Nova Acta Leopoldina, N.F. 3, 14, 339-360, Halle.

– (1937): Paläohistologische Untersuchungen an Bernsteineinschlüssen. – Paläontol. Z. 19: 35-46; Berlin.

– (1938): Eine neue Methode zur mikroskopischen Untersuchung von Bernsteineinschlüssen. – Forsch. Fortschr. 14: 55-56.

– (1952a): Ein Haareinschluß mit Phthirapteren Eiern im Bernstein. – Mitt. geol. Staatsinstitut Hamburg, 21: 59-74.

– (1952b): Ein seltener Fund: Nissen im Bernstein. – Die Umschau, 52: 432; Frankfurt.

Voss, E. (1953): Einige Rhynchophoren der Bernsteinfauna (Col.). – Mitt. geol. Staatsinstitut Hamburg, 22: 119-140.

– (1957): *Archimetrioxena electrica* Voss und ihre Beziehungen zu rezenten Formenkreisen, (Col., Curc.). – Dtsch ent. Z, N.F. 4: 95-102; Berlin.
– (1972): Einige Rüsselkäfer der Tertiärzeit aus baltischem Bernstein. (Coleoptera, Curculionidea). – Steenstrupia, 2: 167-181; Kobenhavn.

WAGNER, H. (1924): Ein neues Apion aus dem baltischen Bernstein. (Col. Curcul.). – Dtsch ent. Z. 1924: 134-136; Berlin.
WAGNER, R., HOFFEINS, C., HOFFEINS, H. W. (2000a): Beschreibung einer weiteren Art der Gattung *Heleodromia* (Diptera, Empididae, Hemerodromiinae) aus dem Bitterfelder Bernstein. – Studia dipterologica 7: 19-23.
– (2000b): A fossil nymphomyiid (Diptera) from the Baltic and Bitterfelder amber. – Systematic Entomology 25: 115-120.
WASMANN, E. (1926a): Die *Arthropterus*-Formen des baltischen Bernsteins. (267. Beitrag zur Kenntnis der Myrmecophilen). – Zool. Anz. 68: 225-232; Leipzig.
– (1926b): Die Paussidengattungen des baltischen Bernsteins. (265. Beitrag zur Kenntnis der Myrmecophilen). – Zool. Anz. 68: 25-30; Leipzig.
– (1927): Die Paussiden des baltischen Bernsteins und die Stammesgeschichte der Paussiden. – Tijdschrift voor Entomologie 70: 63-69.
– (1929): Die Paussiden des baltischen Bernsteins und die Stammesgeschichte der Paussiden. 270. – Bernstein-Forsch. 1: 1-110, Berlin.
– (1933): Eine ameisenmordende Gastwanze (*Proptilocerus dolosus* n.g.n.sp.) im Baltischen Bernstein. – Bernstein-Forschungen 3: 1-3, Berlin.
WEGIEREK, P. (1990): Aphid species from the Collection of the Baltic amber in the Museum of the Earth, Polish Academy of Sciences in Warsaw (Part one). – Prace Museum Ziemi, 41: 83-90.
– (1996a): Aphid species (Homoptera: Aphidinea) from the collection of the Baltic amber in the Museum of the Earth, Polish Academy of Sciences, Warsaw (Part two). – Prace Museum Ziemi, 44: 25-39.
– (1996b): Aphid species (Homoptera: Aphidinea) from the collection of the Baltic amber in the Museum of the Earth, Polish Academy of Sciences, Warsaw (Part three). – Prace Muzeum Ziemi 44: 41-44.
WEIDNER, H. (1952): Insektenleben im Bernsteinwald. (Ein Bericht über die Bernsteinsammlung des Geologischen Staatsinstitutes Hamburg.) – Entomologische Zeitschrift 62: 62-72 u. 88; Stuttgart.
WEIDNER, H. (1955): Die Bernstein-Termiten der Sammlung des Geologischen Staatsinstitutes Hamburg. – Mitt. Geol. Staatsinst. Hamburg 24: 55-74.
– (1956): Die Bernstein-Heuschrecken (Saltatoptera) der Sammlung des Geologischen Staatsinstitutes Hamburg (Orthopteroidea). – Mitt. Geol. Staatsinst. Hamburg 25: 82-103.
– (1956): Kotballen von Termiten im Bernstein. – Veröfft. Überseemus. Bremen A 2: 363-364.
– (1958): Einige interessante Insektenlarven aus der Bernsteininklusen-Sammlung des Geologischen Staatsinstituts Hamburg (Odonata, Coleoptera, Megaloptera, Planipennia). – Mitt. Geol. Staatsinst. Hamburg 27: 50-68.
– (1964): Eine Zecke, *Ixodus succineus* sp. n. im baltischen Bernstein. – Veröff. Überseemus. Bremen A 3: 143-151.
– (1970): Isoptera (Termiten). – In: Handbuch der Zoologie, 4 (2) 14: 1-89, Berlin.
– (1972): Copeognatha (Staubläuse). – In: Handbuch der Zoologie, 4 (2) 16: 1-94: Berlin, New York (De Gruyter).
WEITSCHAT, W. (1987): Bernstein der Insel Sylt. – In: Fossilien von Sylt (Hrsg: HACHT, U.): 109-121; Hamburg.
– (1996): Bernstein in der Deutschen Bucht und in Jütland auf 3., 4., 5. oder 6. Lagerstätte – in: GANZELEWSKI, M., SLOTTA, R. (Hrsg.): Bernstein – Tränen der Götter: 77-82.
– (1997): Bitterfelder Bernstein – ein eozäner Bernstein auf miozäner Lagerstätte – Metalla (Sonderheft) 66: 71-84.
WEITSCHAT, W., WICHARD, W. (1992): Farbeffekte bei Komplexaugen von Köcherfliegen (Insecta: Trichoptera) im Bernstein. – Mitt. Geol.-Paläont. Inst. Univ. Hamburg 73: 223-233.
WERMUTH, H. (1966): Der Status von *Platydactylus minutus* GIEBEL 1862 (Reptilia, Sauria, Gekkonidae). – Stuttg. Beitr. Naturk. 163: 1-6.
WETTSTEIN, R. VON (1891): Der Bernstein und die Bernsteinbäume. – Schr. Vereines Verbr. Naturw. Kenntnisse in Wien 31: 363-386.
WETZEL, W. (1939): Miozäner Bernstein im Westbaltikum. – Z. dtsch. Geol. Ges. 91: 815-822, Stuttgart.
– (1953): Mikropaläontologische Untersuchung des schleswig-holsteinischen Bernsteins. – Neues Jb. Geol. Palaeontol., Monatshefte, 1953: 311-321; Stuttgart.
WEYGOLDT, P. (1966): Moos- und Bücherskorpione. – Die Neue Brehm Bücherei 365, Ziemsen Verlag, Wittenberg.
– (1969): The Biology of Pseudoscorpions. – Harvard Univ. Press, Cambridge.
WHALLEY, P. (1978): New taxa of fossil and recent Micropterigidae with a discussion of their evolution and a comment on the evolution of Lepidoptera (Insecta). – Ann. Transvaal Mus. 31: 71-86; Pretoria.
WHEELER, W. M. (1910): Fossil ants. – In: Ants. Their structure, development and behaviour: 166-175; Columbia Univ. Biol. Series 9; New York.
– (1914): The ants of the Baltic Amber. – Schrift. phys. ökonom. Ges. Königsberg 55: 1-142.
– (1915): The ants of the Baltic amber. – Schr. Phys.-ökon. Ges. Königsberg, 55: 1-142.
WHITING, M. F., WHEELER, W. C. (1994): Insect homeotic transformation. – Nature 368: 696.
WHITING, M. F., CARPENTER, J. C., WHEELER, Q. D., WHEELER, W. C. (1997): The Strepisptera problem: phylogeny of the holometabolous insect orders inferred from 18S and 28S ribosomal DNA sequences and morphology. – Syst. Biol. 46: 1-68.
WICHARD, W. (1984): Fossil caddisflies in fossil resins. – Series Entomologica 30: 441-444; The Hague (Junk).
– (1986): Köcherfliegen des Baltischen Bernsteins. 1. *Marilia altrocki* sp.n. (Trichoptera, Odontoceridae) der Bernsteinsammlung Bachofen-Echt. – Mitt. Bayer. Staatsslg. Paläont. hist. Geol. 26: 33-40, München.

– (1988): Die Köcherfliegen – Trichoptera. – Die Neue Brehm-Bücherei 512 (2. Aufl.), Ziemsen Verlag, Wittenberg.
– (1997): Schlammfliegen aus Baltischem Bernstein (Megaloptera, Sialidae). – Mitt. Geol.-Paläont. Inst. Univ. Hamburg, 80: 197-211.
– (2000): Die Larve einer Köcherfliege im Baltischen Bernstein (Insecta, Trichoptera). – Mitt. Geol.-Paläont. Inst. Univ. Hamburg 84: 261-266.
– (2001): The Caddisflies (Trichoptera) of Baltic Amber – Fossil Larvae. – in MEY, (ed.): 10th Int. Symp. On Trichoptera.
WICHARD, W., ARENS, W., EISENBEIS, G. (1995): Atlas zur Biologie der Wasserinsekten. – G. Fischer, Stuttgart.
WICHARD, W., CASPERS, N. (1991): Caddis flies of Baltic Amber. 2. Fossil species of the genus *Rhyacophila*. – Proc. 6th Int. Symp. Trichoptera, Lodz-Zakopane 1989: 447-45, Univ Press, Poznan.
WICHARD, W., SUKATCHEVA, I. (1992): Köcherfliegen des Baltischen Bernsteins. 3. *Lithax herrlingi* n.sp., eine fossile Art der Goeridae (Insecta: Trichoptera). – Mitt. Geol.-Paläont. Inst. Univ. Hamburg 73: 217-222.
WICHARD, W., WEITSCHAT, W. (1996): Wasserinsekten im Bernstein – eine paläobiologische Studie. – Entomol. Mitt. Löbbecke-Museum+Aquazoo, Beih., 4, 1-122, Düsseldorf.
WILLEMSTEIN, S. C. (1978): Preliminary results of a pollen analysis of insects enclosed in Baltic amber. – Abstracts Ann. Meeting Palaeontol. Ges. Paleontol. Ass., Maastricht.
WILLMANN, R. (1989): Evolution und phylogenetisches System der Mecoptera (Insecta: Holometabola). – Abh. Senckenberg. Naturforsch. Ges., 544: 1-153; Frankfurt.
WITZTUM, A., PARTHASARATHY, M. V. (1985): Observation with the scanning electron microscope of a fly captured in amber. – Proc. Ent. Soc. Washington, 87: 676-677.
WOOD, D. M., BORKENT, A. (1989): Phylogeny an Classification of the Nematocera. – in: MCALPINE, J. F. (Editor): Manual of Nearctic Diptera. Vol. 3: 1333-1370, Hull, Quebec.
WOODLEY, N. E. (1989): Phylogeny an Classification of the "Orthorrhaphous" Brachycera. – in: MCALPINE, J. F. (Editor): Manual of Nearctic Diptera. Vol. 3: 1371-1396, Hull, Quebec.
WUNDERLICH, J. (1983): Zur Konservierung von Bernstein-Einschlüssen und über den "Bitterfelder Bernstein". – Neue Ent. Nachr. 4: 11-13; Keltern.
– (1986): Fossile Spinnen in Bernstein und ihre heute lebenden Verwandten. – Spinnenfauna gestern und heute, 1: 283 S.; Wiesbaden (E. Bauer bei Quelle & Meyer).
– (1988): Die fossilen Spinnen im Dominikanischen Bernstein. – Beiträge zur Araneologie, 2: 378 S.; Straubenhardt (Verlag J. Wunderlich).
– (1993a): Die ersten fossilen Becherspinnen (Fam. Cyatholipidae) in Baltischem und Bitterfelder Bernstein (Arachnida: Araneae). – Mitt. Geol.-Paläolont. Inst. Univ. Hamburg, 75: 231-241.
– (1993b): Die ersten fossilen Speispinnen (Fam. Scytodidae) im Baltischen Bernstein (Arachnida: Araneae). – Mitt. Geol.-Paläolont. Inst. Univ. Hamburg, 75: 243-247.
– (1996): Nachträge zum Buch von Bachofen-Echt: Übersehene und neue Funde aus großen Gruppen (I) und eine aktuelle Übersicht (II). – als Anhang zu: BACHOFEN-ECHT, A.: Der Bernstein und seine Einschlüsse. Neudruck J. Wunderlich Verlag, Straubenhardt.

YABLOKOV-KHNZORIAN, S. M. (1960): Neue Käfer aus baltischem Bernstein. – Pal. Zhurn. 1960: 90-101.
– (1961): Helodidae (Coleoptera) from Baltic amber. – Pal. Zhurn. 1961: 107-116.
– (1961): New Elateridae from the Baltic amber. Pal. Zhurn. 1961: 84-97.
– (1962): Some Sternoxia (Coleoptera) from Baltic amber. – Pal. Zhurn. 1962: 81-89.
YEATES, D. K. (1992): Towards a monophyletic Bombyliidae (Diptera): The removal of the Proratinae (Diptera: Scenopinidae). – Amer. Mus. Novitates 3051: 1-30.
– (1994): The cladistics and classification of the Bombyliidae (Diptera: Asiloidea). – Bull. Amer. Mus. Nat. Hist. 219: 1-191.

ZACHOS, J. C., LOHMANN, K. C, WALKER, J. C., WISE, S. W. (1993): Abrupt climate change and transient climates during the Paleogene: A marine perspective. – J. Geology 101: 191-213.
ZACHOS, J. K., STOTT, L. D., LOHMANN, K. C. (1994): Evolution of the early Cenozoic marine temperatures. – Palaeooceanology 9: 353-387, London.
ZADDACH, E. G. (1864): Ein Amphipode im Bernstein. – Schr. phys.-ökon. Ges., Königsberg 5: 1-12.
– (1868): Amber, its origin and history, as illustrated by the geology of Samland. – Quaterly J. Science 5: 167-185; London.
ZADZIEWSKI, R. (1985): Biting midges of the genus *Eohelea* PETRUNKEVITCH (Insecta, Diptera, Ceratopogonidae) from Baltic amber (in the collection of the Museum of the Earth). – Prace Muzeum Ziemi 37: 123-129; Warszawa.
ZAITZEV, V. F. (1981): Vergleich der tertiären (Oligozän und Miozän) und rezenten Bombyliidenfauna Europas (Diptera, Bombyliidae). – Acta ento. jugoslavica 17: 103-106; Zagreb.
ZANG, R. (1905): Coleoptera Longicornia aus der Berendt'schen Bernsteinsammlung. – Sb. der Ges. naturf. Fr. Berlin, 1905: 232-245.
– (1905): Über Coleoptera Lamellicornia aus dem baltischen Bernstein. – Sb. Ges. naturf. Fr. Berlin, 1905: 197-205.
ZANON, D. V. (1929): Le Diatomee dell'ambra. – Studi Etruschi 3: 427-450.
ZERCHE, L. (1999): Eine neue Art der Gattung *Adinopsis* CAMERON aus dem Baltischen Bernstein. – Beitr. Ent. Berlin 49: 97-105.
ZERICHIN, V. V. (1968): O novych nachodkach Curculionoidea iz baltijskogo jantarja. – Bjulleten Moskowskogo Obscestva ispytatelej Prirody, Otd. geol., 43: 156; Moskva.

– (1971): The zoogeographical relationships of Baltic amber Rhynchophora (excluding Scolytidae). – In: Proc. 13th Int. Congr. Entomology, Moscow 1968, 1: 322-323, Leningrad.

ZEUNER, F. E. (1939): Fossil Orthoptera Ensifera. – Brit. Mus. nat. Hist. London 321 S.

ZEUNER, F. E., MANNING, F. J. (1976): A monograph on fossil bees (Hymenoptera: Apoidea). – Bull. British Museum (Nat. Hist.), Geol. 27: 149-268, London.

ZOMPRO, O. (2001): The Phasmatodea and *Raptophasma* n. gen., Orthoptera incertae sedis, in Baltic amber (Insecta Orthoptera). – Mitt. Geol. Paläont. Inst. Univ. Hamburg, **85**.

ZWICK, P. (1967): Beschreibung der aquatischen Larve von *Neurorthus fallax* (RAMBUR) und Errichtung der neuen Planipennierfamilie Neurorthidae fam. nov. – Gewässer und Abwässer 44/45: 65-86, Düsseldorf.

4 Index of Scientific Names

D

E

F

G

H

N

O

Q

R

S

W

X

Y

Z